John K. McCarthy • Jonathan Benjamin
Trevor Winton • Wendy van Duivenvoorde
Editors

3D Recording and Interpretation for Maritime Archaeology

Editors
John K. McCarthy
Department of Archaeology
Flinders University
Adelaide, SA, Australia

Jonathan Benjamin
Department of Archaeology
Flinders University
Adelaide, SA, Australia

Trevor Winton
Department of Archaeology
Flinders University
Adelaide, SA, Australia

Wendy van Duivenvoorde
Department of Archaeology
Flinders University
Adelaide, SA, Australia

https://doi.org/10.1007/978-3-030-03635-5

This volume has been produced with the generous support of

HONOR FROST FOUNDATION

Additional support provided by

Flinders
UNIVERSITY

wessexarchaeology

Under the patronage of

UNESCO

United Nations
Educational, Scientific and
Cultural Organization

Unitwin Network
for Underwater
Archaeology

Foreword

Archaeology is a discipline that works natively in four dimensions. Whether it is excavation, survey or lab-based analysis, our drive is to untangle and reveal the nature of relationships across time and space. Since the birth of the modern discipline, we have sought ways to capture this data in a precise and accurate manner, from the use of plane tables and survey chains to photogrammetry, laser scanners and geophysics. Over the last 15 years, we have seen a remarkable shift in capability, and nowhere is this more apparent than in maritime archaeology. Here, research interests regularly straddle the terrestrial–marine boundary, requiring practitioners to adapt to different environmental constraints whilst delivering products of comparable standards. Where in the past those working on sites underwater had to rely on tape measures alone, photogrammetric survey has now become ubiquitous, generating rich 3D datasets. This cheap, flexible and potentially highly accurate method has helped to remove differences in data quality above and below water. When matched to the reduction in cost for regional swath bathymetric surveys underwater, and Digital Elevation Models derived from satellite and airborne sensors on land, the context of archaeological work has undergone a revolution, fully transitioning into three, and at times four, digital dimensions.

This volume is thus timely, charting the point where we move from novelty to utility and, with that, a loss of innocence. Thus, it helps to move the discipline forward, not only thinking about how we draw on these techniques to generate data but also how we use this data to engage others. It is increasingly clear that generating 3D data is no longer the main challenge, and archaeologists can focus on what we require of that data and how best we can make it serve our purpose. At the same time, we need to remain flexible enough to recognise the potential for new ways of doing things of new aesthetics of representation and new modes of communication. There is something inherent in the dynamism of archaeology that ensures that scholars always feel they are lucky to be working in the era that they are, the white heat of 'science' during the birth of processualism, the intellectual challenges of 'post-processualism' and now the rich, unpredictable and democratic nature of 3D digital data. These truly are exciting times.

Southampton, UK
23 August 2018

Fraser Sturt

Acknowledgements

The editors wish to thank Flinders University College of Humanities, Arts and Social Sciences and the UNESCO UNITWIN Network for Underwater Archaeology, which hosted the workshop *3D Modelling and Interpretation for Underwater Archaeology*, held on 24–26 November 2016. Open access for this volume has been made possible by the generous support of the Honor Frost Foundation. Additional support for the production of this volume has been provided by Flinders University and Wessex Archaeology. Thanks are due to the contributors and the numerous peer reviewers who have helped to ensure high-quality of this edited volume. Finally, the editors would like to thank their families for their support and patience during the preparation of this book.

Contents

The Rise of 3D in Maritime Archaeology

1

John McCarthy, Jonathan Benjamin, Trevor Winton, and Wendy van Duivenvoorde

Abstract

This chapter provides an overview of the rise of 3D technologies in the practice of maritime archaeology and sets the scene for the following chapters in this volume. Evidence is presented for a paradigm shift in the discipline from 2D to 3D recording and interpretation techniques which becomes particularly evident in publications from 2009. This is due to the emergence or improvement of a suite of sonar, laser, optical and other sensor-based technologies capable of capturing terrestrial, intertidal, seabed and sub-seabed sediments in 3D and in high resolution. The general increase in available computing power and convergence between technologies such as Geographic Information Systems and 3D modelling software have catalysed this process. As a result, a wide variety of new analytical approaches have begun to develop within maritime archaeology. These approaches, rather than the sensor technologies themselves, are of most interest to the maritime archaeologist and provide the core content for this volume. We conclude our discussion with a brief consideration of key issues such as survey standards, digital archiving and future directions.

Keywords

3D applications · 3D reconstruction · 3D mapping · Shipwrecks · Submerged landscapes · Marine survey

J. McCarthy (✉) · J. Benjamin · T. Winton · W. van Duivenvoorde
Maritime Archaeology Program, Flinders University, Adelaide, SA, Australia
e-mail: john.mccarthy@flinders.edu.au; jonathan.benjamin@flinders.edu.au; wint0062@flinders.edu.au; wendy.vanduivenvoorde@flinders.edu.au

1.1 Background

The need for a volume focused on the use of 3D technologies in maritime archaeology has become increasingly apparent to practitioners in the field. This is due to an exponential increase in the application of several distinct 3D recording, analysis and interpretation techniques which have emerged and become part of the maritime archaeologist's toolbox in recent years. In November of 2016, a workshop on this theme was hosted by the UNESCO UNITWIN Network for Underwater Archaeology and Flinders University, Maritime Archaeology Program, in Adelaide, South Australia. The UNITWIN Network (2018) is a UNESCO twinning network of universities involved in education and research of maritime and underwater archaeology. The criteria for full membership requires that each university must offer a dedicated degree in maritime or underwater archaeology. Membership (full and associate members) of the Network currently stands at 30 universities worldwide and the network continues to grow as more universities with existing courses are added. Flinders University chaired the Network as its elected Coordinator (2015–2018), which was passed on to Southampton University at the end of 2018. The workshop in Adelaide and this publication have been undertaken in line with the objectives of the UNITWIN Network which include promotion of 'an integrated system of research, training, information and documentation activities in the field of archaeology related to underwater cultural heritage and related disciplines.' A major element of the workshop was group discussion and many participants in the workshop noted an urgent need for stronger communication and collaboration between maritime archaeologists working in the areas of 3D applications. This volume was inspired by the group discussions held at the workshop and is the first collection of studies devoted exclusively to discussion of 3D technologies for maritime archaeology. As such it is hoped that it will make an important contribution towards fulfilling the aims of the Network.

J. K. McCarthy et al. (eds.), *3D Recording and Interpretation for Maritime Archaeology*, Coastal Research Library 31,
https://doi.org/10.1007/978-3-030-03635-5_1

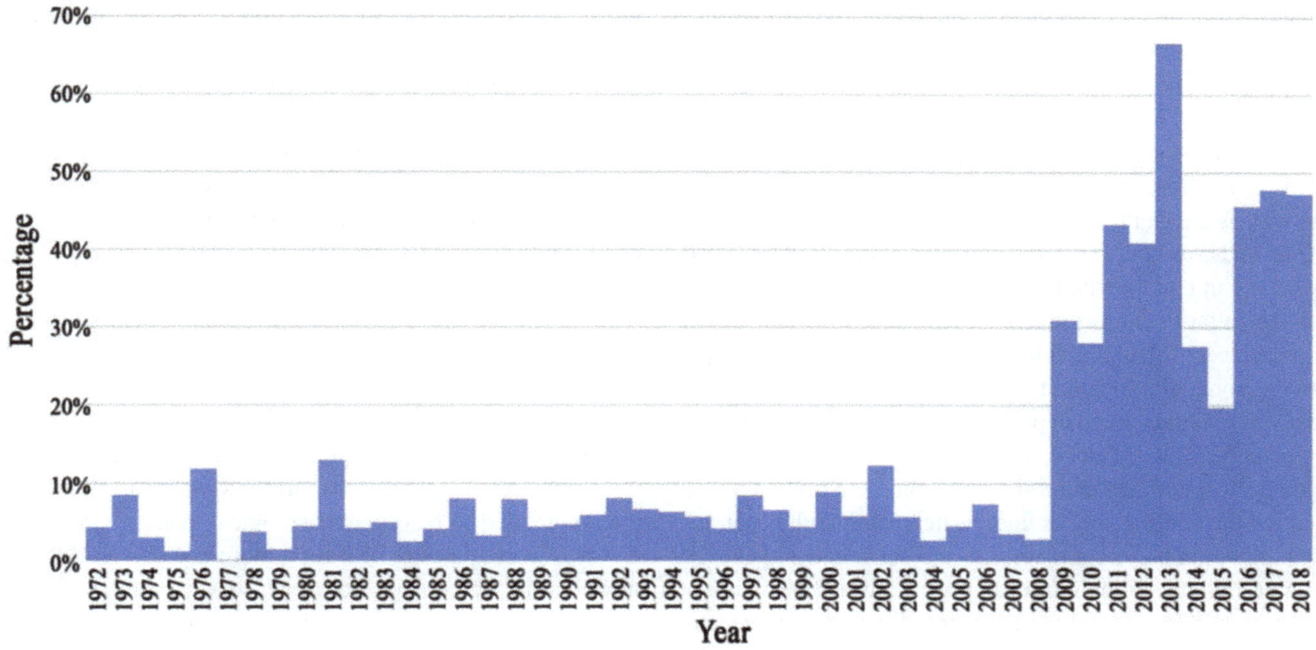

Fig. 1.1 The percentage of *International Journal of Nautical Archaeology* (IJNA) articles by year which mention the phrase 3D or related variations, from 1972 to mid-2018

The recent and rapid adoption of 3D techniques is well known by practitioners of maritime archaeology but can be illustrated for those outside the discipline by tracing use of the term 3D and related variants in papers published in the *International Journal of Nautical Archaeology* (IJNA). As the longest running periodical focused on maritime archaeology (founded in 1972) a review of the IJNA serves as a useful indicator of activity in the field. A search was undertaken of all IJNA articles (including references) using the citation analysis software *Publish or Perish* (Harzing 2007), which draws on the Google Scholar database. The search covered the period 1972 to mid-2018 and returned 466 published articles that include the term 3D (or similar variants) from a total of 3400 articles. A breakdown by year demonstrates clearly that use of the term in the journal was consistently low from the first edition up to 2009 when values jumped from roughly 6% to over 20%, up to a maximum of 65%. While some of these articles may only mention 3D applications in passing, this nevertheless illustrates a noteworthy step change within the discipline (Fig. 1.1).

1.2 The Importance of 3D for Maritime Archaeology

The general shift towards greater use of 3D sensors and workspaces is not exclusive to archaeology and can be seen in many other disciplines. Although archaeology encompasses many different perspectives and approaches, it is, by definition, grounded in the physical remains of the past. A standardized 2D record has been the accepted standard for recording sites during the twentieth century. This includes the production of scaled plans in which the third dimension was indicated using symbolic conventions, such as spot heights and hachure lines. Such outputs remain in use but as 3D surveys have become more popular there is increased recognition that flattening of an archaeological feature creates more abstraction (Campana 2014; Morgan and Wright 2018). This leads to some interesting debates on the tensions between capturing the most accurate and objective surveys possible and the archaeologist's ultimate goal of cultural interpretation. So successful has been the research on high-resolution 3D sensors for maritime archaeology in the last decade that Drap et al. (Chap. 9) can now state that 'In a way, building a 3D facsimile of an archaeological site is not itself a matter of archaeological research even in an underwater context.' Menna et al. (2018) have provided an overview of the main passive and active sensors generating 3D data for maritime archaeology at present, categorized with respect to their useable scale, depth and applicable environment, with a list of key associated publications for each. There will always be a need for research into *technical* improvements in 3D survey techniques but research into new *analytical* techniques founded upon these 3D survey datasets is just beginning. The chapters in this volume demonstrate this in a wide variety of innovative and exciting ways.

Broadly, maritime archaeology is the study of the human past, through material culture and physical remains, that specifically relate the interaction between people and bodies of water and there are numerous factors that make data capture and analysis in 3D particularly important to the maritime archaeologist. There is a greater reliance upon recording techniques that capture data quickly in maritime archaeology (Flatman 2007, 78–79), especially in subaquatic environments where maritime archaeology fieldwork often occurs. This is mainly because of the cost of vessels and equipment, as well as the fact that divers can spend only short periods of time under water. Until recently, maritime archaeologists working in complex underwater surveys or excavations had to rely almost entirely upon difficult and time-consuming manual techniques. A single measurement required a diver to swim around the site taking several tape measurements from datums to obtain a single position (Rule 1989). This manual approach still has a place; however, since 2006, high resolution 3D capture has increasingly become the first choice of survey method for wrecks underwater, using both sonar and photogrammetric techniques. Of the sonar techniques, the use of high resolution multibeam has allowed 3D capture of vast areas of the seabed in 3D at resolutions of up to a metre and of individual exposed wrecks at much higher resolutions. Demonstration of the value of high resolution multibeam for wrecks was perhaps first clearly demonstrated by the RASSE (Bates et al. 2011) and ScapaMap projects (Calder et al. 2007), described as 'the most influential in illustrating the potential for multibeam in archaeology and the most pertinent to multibeam use for deepwater shipwreck studies' (Warren et al. 2010, 2455). Multibeam data are increasingly gathered on a national scale by governmental agencies and often made available to maritime archaeologists to underpin their site-specific studies. Work on the Scapa wrecks continues with demonstration of extremely high-quality survey and visualization for large metal wrecks (Rowland and Hyttinen 2017).

Representing another step change in 3D recording, underwater photogrammetry is now capable of highly detailed surveys of large wreck standing well above the seabed. Good examples include the *Mars* Project—involving comprehensive 3D survey of an incredibly well-preserved shipwreck in the Swedish Baltic (Eriksson and Rönnby 2017) and the Black Sea Project—where deep-sea ROVs are being used to 3D survey some of the oldest intact shipwrecks ever discovered (Pacheco-Ruiz et al. 2018). Photogrammetry even facilitates 3D survey of the spaces inside large vessels, as demonstrated by the early results of the *Thistlegorm* Project (2018)—a comprehensive survey in 3D of one of the most well-known and dived wrecks in the world. Other important 3D sensing techniques for the marine environment also emerged around the same time, including lidar bathymetry (Doneus et al. 2013, 2015), 3D sub-bottom profilers (Gutowski et al. 2015; Missiaen et al. 2018; Plets et al. 2008; Vardy et al. 2008) and Electrical Resistivity Tomography (ERT) (Simyrdanis et al. 2016; Passaro et al. 2009; Ranieri et al. 2010; Simyrdanis et al. 2015, 2018). These are enormously important due to their ability to non-invasively recover 3D data from shallow water (lidar bathymetry) sites and from below the seabed (sub-bottom profilers and ERT), but due to cost and availability are not nearly as widely used as multibeam and photogrammetry. On a final note regarding terminology, Agisoft rebranded Photoscan as Metashape with the release of Version 1.5 at the end of 2018. In order to avoid confusion, the term 'Photoscan/Metashape' is used throughout this volume for all versions.

1.3 Photogrammetry

One of the most rapidly adopted and widely used techniques photogrammetry, or Structure from Motion, is now frequently applied to record archaeological material underwater—it is worth pausing here for a more detailed look at the impact of the technique. Underwater survey of complex features is a frequent task for maritime archaeologists, who aim to achieve a standard of recording equal to terrestrial site investigations. Excavations at Cape Gelidonya (Bass et al. 1967) are often described as the first attempt to apply this standard. For some detailed wreck excavations, achieving this standard has required an investment of time and money that far exceeds terrestrial excavation, particularly for deep wreck sites. The excavation and survey dives required for the wreck at Uluburun reached a total of 22,413 dives to depths of between 44 m and 61 m (Lin 2003, 9), with all the attendant cost and risk that goes with such high figures. Since it is possible to carry out high quality photography under water, it is understandable that the potential to recover measurements from photographs should have been of interest from the earliest underwater surveys. Despite some successes in the earliest experiments by underwater archaeologists (Bass 1966), the use of photogrammetry failed to generate significant levels of interest for the first 30 years of the discipline as it remained technical and time consuming (Green 2004, 194–202). Outside of archaeology, major developments in algorithms and mathematical models were slowly accruing in the field of photogrammetry (Micheletti et al. 2015, 2–3), eventually leading to the advent of automated software packages that removed much of the overhead for technical knowledge. These software packages were created for use in terrestrial contexts, but scientific divers quickly realized that they could be applied underwater with some simple adaptations (McCarthy and Benjamin 2014). There has been a flurry of publication in maritime archaeology (Menna et al. 2018, 11–14), much of which has focussed exclusively on the technical challenges of achieving higher quality and accuracy. Photogrammetry has also been extremely effective for archaeological survey when used with multi-rotor aerial drones, which first began to make an impact in archaeological publication circa 2005

(Campana 2017, 288). Paired with software such as Photoscan/Metashape and Pix4D from 2011, drones have become effective tools for coastal, intertidal and even shallow water survey for maritime archaeology (see Benjamin et al. Chap. 14 for a more detailed discussion).

A brief note on terminology for photogrammetry is necessary as it is a broad term. Defined by the *Oxford English Dictionary* (2018) as 'the use of photography in surveying and mapping to ascertain measurements between objects' photogrammetry has been in use as a mapping technique since the mid-nineteenth century, primarily from airborne cameras. The modern convergence of different technologies and workflows from various disciplines utilising photogrammetry at close range has created confusion in terminology within maritime archaeological publications (McCarthy and Benjamin 2014, 96). The rise of highly automated and integrated software packages such as Visual SfM, Photoscan/Metashape, Reality Capture, PhotoModeler, Pix4D and Autodesk's ReCap software, although built on the same principles as 'traditional photogrammetry' are far more automated and produce a high-resolution 3D model with little or no operator intervention. As a result, they have a much greater impact on the discipline of archaeology and related sciences. It is necessary to differentiate these types of workflow from previous techniques, but several competing terms have been used in parallel, even by the same researchers. The term 'automated photogrammetry' (Mahiddine et al. 2012) has been used by some, in recognition of the much higher degree of automation in these workflows. Unfortunately, this can be confusing as there have been many incremental steps toward automation of photogrammetry prior to the appearance of these software packages. One of the most widely used terms at present is 'computer vision' (Van Damme 2015a; Yamafune 2016), the most detailed defence of which in the field of maritime archaeology is provided by Van Damme (2015b, 4–13). Computer vision and photogrammetry are converging technologies—the subtle difference, however, is that photogrammetry has a greater emphasis on the geometric integrity of the 3D model. Others have used 'multi-image photogrammetry' (Balletti et al. 2015; McCarthy 2014; McCarthy and Benjamin 2014; Yamafune et al. 2016) as earlier applications of photogrammetry have been mainly based on use of stereo pairs. Another popular term appearing with increasing frequency in the literature is Structure from Motion (SfM). Remondino et al. (2017, 594) define SfM as a two-step process 'a preliminary phase where 2D features are automatically detected and matched among images and then a bundle adjustment (BA) procedure to iteratively estimate all camera parameters and 3D coordinates of 2D features.' While this definition covers the core of the process used within these software packages and has a strong analogy to traditional photogrammetry, SfM does not necessarily cover the process of meshing or texturing commonly applied at the end of the workflow.

In practice, the umbrella term 'photogrammetry' appears to have become the most popular term in archaeology to refer to this specific approach, because other types of photogrammetry are now far less commonly used by practicioners. Due to a lack of consensus at present, the editors of this volume have deliberately not attempted to standardize use of the term across the chapters. In contrast, the use of the form '3D' has been adopted over alternatives such as '3-D' or 'three dimensional' throughout, following the argument by Woods (2013).

1.4 Beyond Survey

Contributors to this volume have demonstrated meaningful results using both simple approaches, from use of 3D scan data to undertake volumetric calculations, through to complex approaches such as use of machine learning. In addition to enhanced levels of prospection and survey, there are an increasing variety of new possibilities opening up as a result of advances in 3D analysis for ship and aviation wrecks. In part, this is driven by general rise in available computing power and an ongoing convergence between technologies such as Geographic Information Systems and 3D modelling software. This has encouraged use of 3D software in a general way. Tanner (2012) provides a good example of this through the use of 3D scans to calculate hydrostatic performance of vessels.

Some authors have demonstrated simple and effective analytical applications for 3D data. Semaan et al. (Chap. 5) demonstrate use of photogrammetric surveys of stone anchors to make more accurate assessments of their volume, offering insights into vessel size. A particularly interesting application of photogrammetry for maritime archaeology is the use of legacy photogrammetry data; using old photographs to generate 3D data. While there has been at least one example of this in terrestrial archaeology (Discamps et al. 2016), suitable photographic datasets are hard to find in the archives as there are rarely enough photographs of archaeological subjects with sufficient coverage and overlap to process in this way. Maritime archaeologists, however, have relied heavily on orthomosaic photography since the first surveys of underwater wrecks in the 1960s. Even as a manually overlapped patchwork of separate prints, photos provided important additional details once the archaeologist was back on dry land. As a result, there are likely to be many opportunities to revisit these datasets. Green has done just this (Chap. 3), reprocessing vertical photos from two shipwreck excavations undertaken in 1969 and 1970. The quality of the results suggests an enormous future potential for similar work and for new insights based on this recovered 3D data. Hunter et al., in their contribution (Chap. 6), consider whether a similar approach might be useable for single images. In their chapter, several historical photos of a ship taken throughout the course of its lifetime are used to generate a 3D model of the changing ship through a semi-automated process that provides new insights into the life of a historical shipwreck.

Public dissemination represents a major opportunity for 3D technologies to enhance maritime archaeology. The sharing of 3D survey data and of reconstructions in 3D has a particular appeal for maritime archaeology, as the majority of the public are not divers. Many sections of society cannot experience shipwrecks in person, for many reasons including opportunity, physical capacity and financial factors. The potential of 'virtual museums' for maritime archaeology was first discussed by Kenderdine (1998) but the first substantial projects did not begin until around 2004 (Adams 2013, 93–94) and interest continues to grow (Alvik et al. 2014; Chapman et al. 2010; Drap et al. 2007; Haydar et al. 2008; Sanders 2011). The iMareCulture (2018) project is amongst the most substantial current developments; the EU-funded collaboration between 11 partners in 8 countries, integrates archaeological data into virtual reality and further advances the practice by gamifying the experience (Bruno et al. 2016, 2017; Liarokapis et al. 2017; Philbin-Briscoe et al. 2017; Skarlatos et al. 2016). Woods et al. provide an excellent example of virtual reality for maritime archaeology in this volume (Chap. 13), with one of the most comprehensively captured maritime landscapes yet released. Crucially, this project demonstrates impact via its wide dissemination to the public through a variety of interactive and virtual reality platforms. Another emerging 3D dissemination strategy is the use of online 3D model sharing platforms (Galeazzi et al. 2016). Both Europeana (2018), the EU digital platform for cultural heritage, and the popular Sketchfab (2018) website, began their 3D model hosting services in 2012. While Sketchfab does not conform to archaeological digital archiving standards, it has proven popular and hosts 3D models of hundreds of professional and avocational maritime archaeological sites and objects. Firth et al. (Chap. 12) volume demonstrate the potential power of simple tools like Sketchfab have when combined with professional archaeological input, in this case combining scans of a builder's scale model with high resolution multibeam survey of the wreckage of the same First World War ship—an outlet that has so far achieved over 20,000 views.

Given the widespread use of superficially realistic pseudo-historical animations and simulations in popular culture, particularly in film and television (Gately and Benjamin 2018), it is critical that genuinely researched outputs, based on archaeological data and created for educational purposes, have transparent and scientifically grounded authenticity. The chapter by Suarez et al. (Chap. 8) on procedural modelling for nautical archaeology offers one potential solution in this regard for, as noted by Frankland and Earl (2012, 66), 'the interpretive process an archaeologist undergoes whilst creating a reconstruction using procedural modelling is recorded and made explicit.' In other words, every interpretation and assumption made by the archaeologist is codified as a rule in the procedure used to generate the final model, and may in theory, be deconstructed or modified in light of new evidence. Suarez et al.'s chapter is one of the most developed attempts to apply procedural modelling in the field of archaeology to date and demonstrates the enormous potential for this approach to change the way we approach historical ship reconstruction (Chap. 8).

For submerged landscape applications, working in 3D offers major benefits. 'To create a useful maritime archaeological landscape formation model, archaeological space and time must be analysed in three dimensions, including the surface and water column in addition to the sea floor' (Caporaso 2017, 17). After all, the study of submerged prehistory is reliant on landscape change over time, sea-level change, geomorphology and sediment modelling. There, it is necessary to understand site formation processes when prospecting for submerged archaeological sites. This has been demonstrated in the Southern North Sea (Gaffney et al. 2007) where 3D deep seismic survey gathered by the oil industry was used to model a vast submerged landscape which would have been occupied by Mesolithic Europeans. In researching a submerged archaeological site, the modern sea level imposes a division of the landscape that can interfere with the archaeologist's interpretation of that site. Through an integrated suite of 3D technologies, this division can effectively be erased. This has been amply demonstrated by the work undertaken at the submerged Greek settlement of Pavlopetri (Henderson et al. 2013; Johnson-Roberson et al. 2017; Mahon et al. 2011) where detailed reconstructions of the city have been extrapolated from wide-area 3D photogrammetric survey. In this volume, the chapter by Benjamin et al. (Chap. 14) also demonstrates this through a series of case studies, culminating in a submerged Mesolithic landscape captured in 3D across and beyond the intertidal zone. This chapter addresses the critical issue of theory in the discipline and asks how these new tools are influencing the way we engage with Maritime Cultural Landscapes, providing a much-needed balance to a volume that is necessarily centred on technology.

As well as facilitating long-term accurate monitoring of maritime archaeological sites over time, 3D geophysical techniques offer far more detailed non-destructive surveys of shallow-buried archaeological material. Article 2 of the 2001 UNESCO Convention on the Protection of the Underwater Cultural Heritage prioritizes in situ preservation. Although still not widely available, there have been a few projects that have demonstrated sub-seabed surveys in estuarine and coastal locations in high resolution 3D without the need for excavation. Perhaps the earliest example is by Quinn et al.

(1997) who published the geophysical evidence for paleo-scour marks at the *Mary Rose* site. Subsequent technical evolution of 3D sub-bottom profiling systems, include a 3D Chirp reconstruction of the wreck of *Grace Dieu* (Plets et al. 2008, 2009) and Missiaen et al's (2018) parametric 3D imaging of submerged complex peat exploitation patterns. Two further ground-breaking studies on this subject appear in this volume. Winton's chapter on *James Matthews* also uses a parametric sub-bottom profiler to build up a detailed model of a previously excavated and reburied wreck, allowing a quantitative assessment of data quality (Chap. 10). In a similar way, the chapter by Simyrdanis et al. (Chap. 11) demonstrates a new technology using Electrical Resistivity Tomography to recover the shape of a buried vessel in a riverine context. These chapters clearly demonstrate the future importance of this approach. It is also telling that both chapters have been able to incorporate use of 3D reconstructions of their vessels.

1.5 Future Directions

A comprehensive review of all 3D technologies likely to become part of maritime archaeology is beyond the scope of this chapter, though some of the techniques with significant potential are highlighted. In the concluding section of the *Oxford Handbook of Maritime Archaeology,* Martin (2011, 1094) considers the trajectory of maritime archaeology and asked whether the role of the diver was threatened by advances in remote sensing. In another chapter of the handbook, Sanders (2011) speculated that we might soon be wearing 'location-aware wearable computers linked to a 3D-based semantic Internet with the capability of projection-holographic imagery of distant, hard-to-access, or lost maritime sites.' Since those words were written they have already come partly true through the rise of the internet-linked GPS-enabled smartphones and portable virtual and augmented reality headsets. Indeed, augmented reality has enormous potential for maritime archaeology through the use of augmented displays for scientific divers (Morales et al. 2009).

It is easy to see the potential benefits of overlaying sonar and photogrammetric models of underwater archaeological sites on the diver's vision, particularly in low visibility. Augmented and virtual reality systems may also help to give the non-diving public an immersive experience of exploring underwater sites, perhaps even while in a swimming pool (Yamashita et al. 2016). Management of maritime archaeological sites will certainly be facilitated by these new technologies. Effective in situ management requires a priori 3D information to identify lateral extent, height and/or depth of burial of archaeological material on the site, their material type and state of deterioration. In terms of more accurately understanding site formation processes, Quinn and Boland (2010) demonstrated how multiple fine-scale 3D bathymetric models can be used in time lapse sequence and Quinn and Smyth (2017) showed how 3D ship models can be incorporated into sediment scour analyses. The cost of high quality 3D survey is now at the point that it is likely that states will begin to develop 3D versions of their national inventories of maritime archaeological sites. Radić Rossi et al. in this volume present ground-breaking work on a sixteenth-century wreck in Croatia (Chap. 4), where 3D survey has been used to generate 2D plans, site condition has been monitored in 3D over multiple field seasons and the archaeological remains have been fitted to a 3D reconstruction of the vessel.

In his consideration of the future of photogrammetry for underwater archaeology, Drap et al. (2013) highlighted a number of future applications of the technology, including the merging of data from optic and acoustic sensors and has stated that once the technical challenge of high resolution and accurate survey was overcome, the 'main problem now is to add semantic to this survey and offering dynamic link between geometry and knowledge' and at that stage suggested that pattern recognition and the development of ontologies would be key steps (Drap et al. 2013, 389). In a wide-ranging contribution to this volume, Drap et al. develop these ideas further, including use of virtual reality, the application of machine learning to the recognition of archaeological objects visible in the 3D survey data and experimentation with 3D reconstruction from single images.

1.6 Standards

The wave of technological innovation has occurred in such a short space of time that knowledge sharing through publication has often proved inadequate, with many practitioners developing workflows in relative isolation from their peers. While this has led to a flowering of experimentation and innovation and is part of the natural process of technological change, it has also caused duplication, wasted effort and a general sense of a discipline working in unconnected silos. A greater problem is that the adoption of these new workflows risks seducing the discipline away from the rigorous standards using traditional recording techniques, which have developed over many decades.

To some extent the approach toward standardization will vary by technique and will depend on whether maritime archaeologists work with technical specialists or whether an attempt is made to make a technique part of their own workflow. This echoes the early debate on whether archaeologists should train as divers or vice versa (Muckelroy 1978, 30–32). Some techniques such as bathymetric lidar survey are likely to remain within the hands of highly specialized technicians, while the simple nature and low cost of photogrammetry means that many archaeologists have taken it entirely into their own hands. This technique, however, has many hidden complexities and Huggett (2017) has highlighted the potential danger of blind reliance on technologies that processes and transform data in ways not generally understood by the user.

As Remondino et al. (2017, 591) state 'nowadays many conferences are filled with screenshots of photogrammetric models and cameras floating over a dense point cloud. Nonetheless object distortions and deformations, scaling problems and non-metric products are very commonly presented but not understood or investigated.' A small number of guidance documents have begun to appear for photogrammetry. Perhaps the most detailed in the English language for capture using current techniques is that by Historic England, which includes case studies for maritime archaeology (Historic England 2017, 102–106). This guidance includes important sections on the use and configuration of control networks, calculation of accuracy as well as formats and standards for archiving of digital data.

Austin et al. (2009) have written guidance for marine remote sensing and photogrammetry, focused mainly on data management and archiving, although this is already quite dated after less than a decade. At the time of writing, there is no detailed formal guidance focused on underwater photogrammetry. While most of the important information is available in journal publications, such sources tend to present case studies with specific workflows which are still experimental in many ways. Shortis, who has been heavily involved in the development of photogrammetry for scientific recording, has provided a chapter for this volume that discusses these issues (Chap. 2). Numerous authors have also highlighted the risks of disruption of archiving standards in this period of rapid transition to digital technologies (Austin et al. 2009; Jeffrey 2012). One possible solution to this challenge is the publication of supplementary digital data alongside academic papers (Castro and Drap 2017, 46) and the *International Journal of Nautical Archaeology* has taken the first step in this direction by publishing an online 3D model alongside an article (Cooper et al. 2018). A similar facility has also been offered to the authors of the current volume. While not a complete solution equivalent to a national infrastructure for comprehensive digital archiving, this approach does provide an improved record of digital archaeological investigations compared to a 2D publication and this will facilitate further reuse and reinterpretation of data.

1.7 Conclusions

The timing of the great leap in interest in 3D seen in IJNA articles from 2009 onwards can be correlated with the introduction or maturation of several different 3D survey techniques and 3D dissemination tools. Some of these had a long history, such as photogrammetry, but had evolved from niche technical forms into accessible tools with wide appeal. After several decades of relatively incremental refinement of manual and low-resolution survey methods, and highly abstracted and symbolized 2D modes of analysis and dissemination, a watershed has been reached in the last decade whereby maritime archaeology has rapidly added 3D digital practices to its core toolbox. The need for enhancements of these survey techniques (as well as research into new technologies) continues, however, high-resolution data capture in 3D is now possible across submerged, terrestrial and coastal, marine and freshwater environments both shallow and deep. Practitioners are developing a fluency in 3D working practices to deal with these datasets and this has led to a flowering of different analytical approaches that were not possible in the past. The review of changes in the past decades suggests that it would be foolhardy to predict the future direction of technologies but it is clear that changes will continue. If anything, advances are likely to accelerate. It is more important than ever that practitioners defend the discipline's scientific status, through the maintenance of standards as they relate to recording, analysis, interpretation, dissemination and archiving of archaeology in 3D.

References

Adams JR (2013) Experiencing shipwrecks and the primacy of vision. In: Adams JR, Ronnby J (eds) Interpreting shipwrecks: maritime archaeological approaches. The Highfield Press, Southampton, pp 85–96

Alvik R, Hautsalo V, Klemelä U, Leinonen A, Matikka H, Tikkanen S, Vakkari E (2014) The *Vrouw Maria* underwater project 2009–2012 final report. National Board of Antiquities, vol 2. National Board of Antiquities, Helsinki

Austin BT, Bateman J, Jeffrey S, Mitcham J, Niven K (2009) Marine remote sensing and photogrammetry: a guide to good practice. VENUS Virtual Exploration of Underwater Site. Information Society Technologies http://guides.archaeologydataservice.ac.uk/g2gp.pdf?page=VENUS_Toc&xsl=test.xsl&ext=.pdf. Accessed 5 Aug 2018

Balletti C, Beltrame C, Costa E, Guerra F, Vernier P (2015) Photogrammetry in maritime and underwater archaeology: two marble wrecks from Sicily. In: Pezzati L, Targowski P (eds) Proceedings SPIE 9527: optics for arts, architecture, and archaeology V, 95270M (30 June 2015). SPIE Optical Metrology, Munich, pp 1–12. https://doi.org/10.1117/12.2184802

Bass GF (1966) Archaeology under water. Praeger, New York

Bass GF, Throckmorton P, Taylor JP, Hennessy JB, Shulman AR, Buchholz H-G (1967) Cape Gelidonya: a Bronze Age shipwreck. Trans Am Philos Soc 57(8):1–177. https://doi.org/10.2307/1005978

Bates CR, Lawrence M, Dean M, Robertson P (2011) Geophysical methods for wreck-site monitoring: the Rapid Archaeological Site Surveying and Evaluation (RASSE) programme. Int J Naut Archaeol 40(2):404–416. https://doi.org/10.1111/j.1095-9270.2010.00298.x

Bruno F, Lagudi A, Barbieri L, Muzzupappa M, Ritacco G, Cozza A, Cozza M, Peluso R, Lupia M, Cario G (2016) Virtual and augmented reality tools to improve the exploitation of underwater archaeological sites by diver and non-diver tourists. In: Ioannides M, Fink E, Moropoulou A, Hagedorn-Saupe M, Fresa A, Liestøl G, Rajcic V,

Grussenmeyer P (eds) Digital heritage. Progress in cultural heritage: documentation, preservation, and protection, EuroMed 2016. Lecture notes in computer science, vol 10058. Springer, Cham, pp 269–280. https://doi.org/10.1007/978-3-319-48496-9_22

Bruno F, Lagudi A, Ritacco G, Philpin-Briscoe O, Poullis C, Mudur S, Simon B (2017) Development and integration of digital technologies addressed to raise awareness and access to European underwater cultural heritage: an overview of the H2020 i-MARECULTURE project. Oceans. 2017 – Aberdeen. Aberdeen: 1–10. https://doi.org/10.1109/OCEANSE.2017.8084984?src=document

Calder BR, Forbes B, Mallace D (2007) Marine heritage monitoring with high-resolution survey tools: Scapa flow 2001–2006. In: US hydrographic conference, pp 1–23

Campana S (2014) 3D modeling in archaeology and cultural heritage—theory and best practice. In: Campana S, Remondino F (eds) 3D recording and modelling in archaeology and cultural heritage. BAR International Series 2598. British Archaeological Reports, pp 7–12

Campana S (2017) Drones in archaeology: state-of-the-art and future perspectives. Archaeol Prospect 24(4):275–296. https://doi.org/10.1002/arp.1569

Caporaso A (2017) A dynamic processual maritime archaeological landscape formation model. In: Caporaso A (ed) Formation processes of maritime archaeological landscapes, When the land meets the sea book series (ACUA). Springer, Cham, pp 7–31. https://doi.org/10.1007/978-3-319-48787-8

Castro F, Drap P (2017) A arqueologia marítima e o futuro—maritime archaeology and the future. Revista Latino-Americana de Arqueologia. For Hist 11:40–55

Chapman P, Bale K, Drap P (2010) We all live in a virtual submarine. IEEE Comput Graph Appl 30(1):85–89. https://doi.org/10.1109/MCG.2010.20

Cooper JP, Wetherelt A, Eyre M (2018) From boatyard to museum: 3D laser scanning and digital modelling of the Qatar Museums watercraft collection. Int J Naut Archaeol 47(2):1–24. https://doi.org/10.1111/1095-9270.12298

Discamps E, Muth X, Gravina B, Lacrampe-Cuyaubère F, Chadelle JP, Faivre JP, Maureille B (2016) Photogrammetry as a tool for integrating archival data in archaeological fieldwork: examples from the Middle Palaeolithic sites of Combe-Grenal, Le Moustier, and Regourdou. J Archaeol Sci 8:268–276. https://doi.org/10.1016/j.jasrep.2016.06.004

Doneus M, Doneus N, Briese C, Pregesbauer M, Mandlburger G, Verhoeven GJ (2013) Airborne laser bathymetry—detecting and recording submerged archaeological sites from the air. J Archaeol Sci 40(4):2136–2151. https://doi.org/10.1016/j.jas.2012.12.021

Doneus M, Miholjek I, Mandlburger G, Doneus N, Verhoeven GJ, Briese C, Pregesbauer M (2015) Airborne laser bathymetry for documentation of submerged archaeological sites in shallow water. Int Arch Photogramm, Remote Sens Spat Inf Sci Arch 40(5W5):99–107. https://doi.org/10.5194/isprsarchives-XL-5-W5-99-2015

Drap P, Seinturier J, Scaradozzi D, Gambogi P, Long L, Gauch F (2007) Photogrammetry for virtual exploration of underwater archaeological sites. Proceedings of the 21st International Symposium of CIPA, Athens, Greece, 1–6 October 2007, 6 pp

Drap P, Merad DD, Mahiddine A, Seinturier J, Peloso D, Boï J-M, Chemisky B, Long L (2013) Underwater photogrammetry for archaeology: what will be the next step? Int J Herit Digit Era 2(3):375–394. https://doi.org/10.1260/2047-4970.2.3.375

Eriksson N, Rönnby J (2017) Mars (1564): the initial archaeological investigations of a great 16th-century Swedish warship. Int J Naut Archaeol 46(1):92–107. https://doi.org/10.1111/1095-9270.12210

Europeana (2018) https://www.europeana.eu. Accessed 23 July 2018

Flatman J (2007) The origins and ethics of maritime archaeology part I. Public Archaeol 6(2):77–97. https://doi.org/10.1179/175355307X230739

Frankland T, Earl G (2012) Authority and authenticity in future archaeological visualisation. In: Hohl M (ed) Making visible the invisible: art, design and science in data visualisation. University of Huddersfield, Huddersfield, pp 62–68

Gaffney VL, Thomson K, Fitch S (eds) (2007) Mapping Doggerland: the Mesolithic landscapes of the Southern North Sea. Archaeopress, Oxford

Galeazzi F, Callieri M, Dellepiane M, Charno M, Richards J, Scopigno R (2016) Web-based visualization for 3D data in archaeology: the ADS 3D viewer. J Archaeol Sci 9:1–11. https://doi.org/10.1016/j.jasrep.2016.06.045

Gately I, Benjamin J (2018) Archaeology hijacked: addressing the historical misappropriations of maritime and underwater archaeology. J Marit Archaeol 13(1):15–35. https://doi.org/10.1007/s11457-017-9177-8

Green J (2004) Maritime archaeology: a technical handbook, 2nd edn. Elsevier, London

Gutowski M, Malgorn J, Vardy M (2015) 3D sub-bottom profiling—high resolution 3D imaging of shallow subsurface structures and buried objects. In: Proceedings of MTS/IEEE OCEANS 2015. Genova, Discovering Sustainable Ocean Energy for a New World, pp 1–7

Harzing AW (2007) Publish or Perish. https://harzing.com/resources/publish-or-perish. Accessed 7 Aug 2018

Haydar M, Maidi M, Roussel D, Drap P, Bale K, Chapman P (2008) 'Virtual exploration of underwater archaeological sites: visualization and interaction in mixed reality environments. In: Ashley M, Hermon S, Proenca A, Rodriguez-Echavarria K (eds) Proceedings of VAST: international symposium on virtual reality, archaeology and intelligent cultural heritage. The Eurographics Association, pp 141–148. https://doi.org/10.2312/VAST/VAST08/141-148

Henderson JC, Pizarro O, Johnson-Roberson M, Mahon I (2013) Mapping submerged archaeological sites using stereo-vision photogrammetry. Int J Naut Archaeol 42(2):243–256. https://doi.org/10.1111/1095-9270.12016

Historic England (2017) Photogrammetric applications for cultural heritage: guidance for good practice. Swindon

Huggett J (2017) The apparatus of digital archaeology. Internet Archaeol (44). https://doi.org/10.11141/ia.44.7

iMareCulture (2018) http://imareculture.weebly.com. Accessed 1 Aug 2018

Jeffrey S (2012) A new digital dark age? collaborative web tools, social media and long-term preservation. World Archaeol 44(4):553–570. https://doi.org/10.1080/00438243.2012.737579

Johnson-Roberson M, Bryson M, Friedman A, Pizarro O, Troni G, Ozog P, Henderson JC (2017) High-resolution underwater robotic vision-based mapping and three-dimensional reconstruction for archaeology. J Field Robot 34:625–643. https://doi.org/10.1002/rob.21658

Kenderdine S (1998) Sailing on the silicon sea: the design of a virtual maritime museum. Arch Mus Inform 12(1):17–38

Liarokapis F, Kouřil P, Agrafiotis P, Demesticha S, Chmelík J, Skarlatos D (2017) 3D Modelling and mapping for virtual exploration of underwater archaeology assets. Int Arch Photogramm, Remote Sens Spat Inf Sci Arch XLII-2/W3(March):425–431. https://doi.org/10.5194/isprs-archives-XLII-2-W3-425-2017

Lin SS (2003) Lading of the Late Bronze Age ship at Uluburun. MA thesis, Texas A&M University

Mahiddine A, Seinturier J, Peloso D, Boï J-M, Drap P, Merad D D, Long L (2012) Underwater image preprocessing for automated photogrammetry in high turbidity water: an application on the Arles-Rhone XIII Roman wreck in the Rhodano river, France. In: Proceedings of the 18th international conference on virtual systems and multimedia (VSMM 2012), Milan, Italy, 2–5 September 2012. Virtual Systems in the Information Society, pp 189–194. https://doi.org/10.1109/VSMM.2012.6365924

Mahon I, Pizarro O, Johnson-Roberson M, Friedman A, Williams SB, Henderson JC (2011) Reconstructing Pavlopetri: mapping the world's oldest submerged town using stereo-vision. In: Proceedings of the IEEE international conference on robotics and automation, 9–13 May 2011. IEEE, Shanghai, pp 2315–2321. https://doi.org/10.1109/ICRA.2011.5980536

Martin P (2011) Conclusion: future directions. In: Catsambis A, Ford B, Hamilton DL (eds) The Oxford handbook of maritime archaeology. Oxford University Press, Oxford, pp 1085–1101. https://doi.org/10.1093/oxfordhb/9780199336005.001.0001

McCarthy J (2014) Multi-image photogrammetry as a practical tool for cultural heritage survey and community engagement. J Archaeol Sci 43:175–185. https://doi.org/10.1016/j.jas.2014.01.010

McCarthy JK, Benjamin J (2014) Multi-image photogrammetry for underwater archaeological site recording: an accessible, diver-based approach. J Marit Archaeol 9(1):95–114. https://doi.org/10.1007/s11457-014-9127-7

Menna F, Agrafiotis P, Georgopoulos A (2018) State of the art and applications in archaeological underwater 3D recording and mapping. J Cult Herit 2017:1–18. https://doi.org/10.1016/j.culher.2018.02.017

Micheletti N, Chandler JH, Lane SN (2015) Structure from motion (SfM) photogrammetry. British Society for Geomorphology. Geomorphol Tech 2(2):12

Missiaen T, Evangelinos D, Claerhout C, De Clercq M, Pieters M, Demerre I (2018) Archaeological prospection of the nearshore and intertidal area using ultra-high resolution marine acoustic techniques: results from a test study on the Belgian coast at Ostend-Raversijde. Geoarchaeology 33(3):386–400. https://doi.org/10.1002/gea.21656

Morales R, Keitler P, Maier P, Klinker G (2009) An underwater augmented reality system for commercial diving operations. Oceans 2009(1–3):794–801

Morgan C, Wright H (2018) Pencils and pixels: drawing and digital media in archaeological field recording. J Field Archaeol 43(2):1–16. https://doi.org/10.1080/00934690.2018.1428488

Muckelroy K (1978) Maritime archaeology. Cambridge University Press, Cambridge

Oxford English Dictionary (2018) Oxford Dictionaries. https://www.oxforddictionaries.com/. Accessed 23 July 2018

Pacheco-Ruiz R, Adams J, Pedrotti F (2018) 4D modelling of low visibility underwater archaeological excavations using multi-source photogrammetry in the Bulgarian Black Sea. J Archaeol Sci 100:120–129. https://doi.org/10.1016/j.jas.2018.10.005

Passaro S, Budillon F, Ruggieri S, Bilotti G, Cipriani M, Di Maio R, D'Isanto C, Giordano F, Leggieri C, Marsella E, Soldovieri MG (2009) Integrated geophysical investigation applied to the definition of buried and outcropping targets of archaeological relevance in very shallow water. Il Quaternario (Ital J Quat Sci) 22(1):33–38

Philbin-Briscoe O, Simon B, Mudur S, Poullis C, Rizvic S, Boskovic D, Katsouri I, Demesticha S, Skarlatos D (2017) A serious game for understanding ancient seafaring in the Mediterranean Sea. In: Proceedings of the 2017 9th international conference on virtual worlds and games for serious applications, Athens, 6–8 September 2017, pp 1–5. https://doi.org/10.1109/VS-GAMES.2017.8055804

Plets RMK, Dix JK, Best AI (2008) Mapping of the buried Yarmouth roads wreck, Isle of Wight, UK, using a chirp sub-bottom profiler. Int J Naut Archaeol 37(2):360–373. https://doi.org/10.1111/j.1095-9270.2007.00176.x

Plets RMK, Dix JK, Adams JR, Bull JM, Henstock TJ, Gutowski M, Best AI (2009) The use of a high-resolution 3D Chirp sub-bottom profiler for the reconstruction of the shallow water archaeological site of the *Grace Dieu* (1439), River Hamble, UK. J Archaeol Sci 36(2):408–418. https://doi.org/10.1016/j.jas.2008.09.026

Quinn R, Boland D (2010) The role of time-lapse bathymetric surveys in assessing morphological change at shipwreck sites. J Archaeol Sci 37(11):2938–2946

Quinn R, Smyth TAG (2017) Processes and patterns of flow, erosion, and deposition at shipwreck sites: a computational fluid dynamic simulation. Archaeol Anthropol Sci 2017:11. https://doi.org/10.1007/s12520-017-0468-7

Quinn R, Bull JM, Dix JK, Adams JR (1997) The *Mary Rose* site—geophysical evidence for palaeo-scour marks. Int J Naut Archaeol 26(1):3–16. https://doi.org/10.1111/j.1095-9270.1997.tb01309.x

Ranieri G, Loddo F, Godio A, Stocco S, Cosentino PL, Capizzi P, Messina P, Savini A, Bruno V, Cau MA, Orfila M (2010) Reconstruction of archaeological features in a Mediterranean coastal environment using non-invasive techniques. In: Making history interactive: computer applications and quantitative methods in archaeology: proceedings of the 37th international conference, Williamsburg, Virginia, USA, 22–26 March. CAA2009. BAR International Series S2079. 19(2000), pp 330–337

Remondino F, Nocerino E, Toschi I, Menna F (2017) A critical review of automated photogrammetric processing of large datasets. Int Arch Photogramm, Remote Sens Spat Inf Sci XLII-2/W5:591–599. https://doi.org/10.5194/isprs-archives-XLII-2-W5-591-2017

Rowland C, Hyttinen K (2017) Photogrammetry in depth: revealing HMS Hampshire. In: Bowen JP, Diprose G, Lambert N (eds) Proceedings of electronic visualisation and the arts (EVA 2017), London, UK, 11–13 July 2017. EVA, London, pp 358–364. https://doi.org/10.14236/ewic/EVA2017.72

Rule N (1989) The Direct Survey Method (DSM) of underwater survey, and its application underwater. Int J Naut Archaeol 18(2):157–162. https://doi.org/10.1111/j.1095-9270.1989.tb00187.x

Sanders DH (2011) Virtual reconstruction of maritime sites and artifacts. In: Catsambis A, Ford B, Hamilton DL (eds) The Oxford handbook of maritime archaeology. Oxford University Press, Oxford, pp 305–326. https://doi.org/10.1093/oxfordhb/9780199336005.001.0001

Simyrdanis K, Papadopoulos N, Kim J-H, Tsourlos P, Moffat I (2015) Archaeological investigations in the shallow seawater environment with electrical resistivity tomography. Near Surf Geophys 13(6):601–611. https://doi.org/10.3997/1873-0604.2015045

Simyrdanis K, Papadopoulos N, Cantoro G (2016) Shallow off-shore archaeological prospection with 3-D electrical resistivity tomography: the case of Olous (Modern Elounda), Greece. Remote Sens 8(11):897. https://doi.org/10.3390/rs8110897

Simyrdanis K, Moffat I, Papadopoulos N, Kowlessar J, Bailey M (2018) 3D mapping of the submerged *Crowie* barge using electrical resistivity tomography. Int J Geophys:1–11. https://doi.org/10.1155/2018/6480565

Skarlatos D, Agrafiotis P, Balogh T, Bruno F, Castro F, Petriaggi BD, Demesticha S, Doulamis A, Drap P, Georgopoulos A, Kikillos F, Kyriakidis P, Liarokapis F, Poullis C, Rizvic S (2016) Project iMARECULTURE: advanced VR, iMmersive serious games and augmented REality as tools to raise awareness and access to European underwater CULTURal heritagE. In: Ioannides M et al (eds) Digital heritage: progress in cultural heritage: documentation, preservation, and protection, EuroMed 2016. Lecture notes in computer science, vol 10058. Springer, Cham, pp 805–813. https://doi.org/10.1007/978-3-319-48496-9_64

Sketchfab (2018) https://sketchfab.com. Accessed 23 July 2018

Tanner P (2012) The application of 3D laser scanning for recording vessels in the field. Naut Archaeol Soc Newsl: 1–4

The Thistlegorm Project (2018) The Thistlegorm project: an archaeological survey of the SS *Thistlegorm*, Red Sea, Egypt. http://thethistlegormproject.com. Accessed 1 Aug 2018

The UNITWIN Network (2018) http://www.underwaterarchaeology.net. Accessed 1 Aug 2018

Van Damme T (2015a) Computer vision photogrammetry for underwater archaeological site recording: a critical assessment. MA thesis, University of Southern Denmark

Van Damme T (2015b) Computer vision photogrammetry for underwater archaeological site recording in a low-visibility environment.

Int Arch Photogramm Remote Sens Spat Inf Sci XL-5/W5:231–238. https://doi.org/10.5194/isprsarchives-XL-5-W5-231-2015
Vardy ME, Dix JK, Henstock TJ, Bull JM, Gutowski M (2008) Decimeter-resolution 3D seismic volume in shallow water: a case study in small-object detection. Geophysics 73(2):B33–B40
Warren D, Wu C-W, Church RA, Westrick R (2010) Utilization of multibeam bathymetry and backscatter for documenting and planning detailed investigations of deepwater archaeological sites. In: Proceedings of Offshore Technology Conference, 3–6 May, Houston, pp 1–8. https://doi.org/10.4043/20853-MS
Woods A (2013) 3D or 3-D: a study of terminology, usage and style. Eur Sci Editing 39(3):59–62
Yamafune K (2016) Using computer vision photogrammetry (Agisoft Photoscan) to record and analyze underwater shipwreck sites. PhD dissertation, Texas A&M University
Yamafune K, Torres R, Castro F (2016) Multi-image photogrammetry to record and reconstruct underwater shipwreck sites. J Archaeol Method Theory 24(3):703–725. https://doi.org/10.1007/s10816-016-9283-1
Yamashita S, Zhang X, Rekimoto J (2016) AquaCAVE: augmented swimming environment with immersive surround-screen virtual reality. In: Proceedings of the 29th annual symposium on user interface software and technology, pp 183–184. https://doi.org/10.1145/2984751.2984760

Camera Calibration Techniques for Accurate Measurement Underwater

2

Mark Shortis

Abstract

Calibration of a camera system is essential to ensure that image measurements result in accurate estimates of locations and dimensions within the object space. In the underwater environment, the calibration must implicitly or explicitly model and compensate for the refractive effects of waterproof housings and the water medium. This chapter reviews the different approaches to the calibration of underwater camera systems in theoretical and practical terms. The accuracy, reliability, validation and stability of underwater camera system calibration are also discussed. Samples of results from published reports are provided to demonstrate the range of possible accuracies for the measurements produced by underwater camera systems.

Keywords

Underwater photography · Optics · Refraction · Accuracy · Validation

2.1 Introduction

2.1.1 Historical Context

Photography has been used to document the underwater environment since the invention of the camera. In 1856 the first underwater images were captured on glass plates from a camera enclosed in a box and lowered into the sea (Martínez 2014). The first photographs captured by a diver date to 1893 and in 1914 the first movie was shot on film from a spherical observation chamber (Williamson 1936). Various experiments with camera housings and photography from submersibles followed during the next decades, but it was only after the invention of effective water-tight housings in 1930s that still and movie film cameras were used extensively underwater. In the 1950s the use of SCUBA became more widespread; several underwater feature movies were released and the first documented uses of underwater television cameras to record the marine environment were conducted (Barnes 1952). A major milestone in 1957 was the invention of the first waterproof 35 mm camera that could be used both above and under water, later developed into the Nikonos series of cameras with interchangeable, water-tight lenses.

The first use of underwater images in conjunction with photogrammetry for heritage recording was the use of a stereo camera system in 1964 to map a late Roman shipwreck (Bass 1966). Other surveys of shipwrecks using pairs of Nikonos cameras controlled by divers (Hohle 1971), mounted on towed body systems (Pollio 1972) or mounted on submersibles (Bass and Rosencrantz 1977) soon followed. Subsequently a variety of underwater cameras have been deployed for traditional mapping techniques and cartographic representations, based on diver-controlled systems (Henderson et al. 2013) and ROVs (Drap et al. 2007). Digital images and modelling software have been used to create models of artefacts such as anchors and amphorae (Green et al. 2002). These analyses of the stereo pairs utilized the traditional techniques of mapping from stereo photographs, developed for topographic mapping from aerial photography. These first applications of photogrammetry to underwater archaeology were motivated by the well-documented advantages of the technique, especially the non-contact nature of the measurements, the impartiality and accuracy of the measurements, and the creation of a permanent record that could be reanalysed and repurposed later (Anderson 1982). Stereo photogrammetry has the disadvantage that the measurement capture and analysis is a complex task that requires specific techniques and expertise, however this

This is a revised version based on a paper originally published in the online access journal Sensors (Shortis 2015).

M. Shortis (✉)
School of Science, RMIT University, Melbourne, VIC, Australia
e-mail: mark.shortis@rmit.edu.au

J. K. McCarthy et al. (eds.), *3D Recording and Interpretation for Maritime Archaeology*, Coastal Research Library 31,
https://doi.org/10.1007/978-3-030-03635-5_2

complexity can be ameliorated by the documentation of operations at the site and in the office (Green 2016; Green et al. 1971).

2.1.2 Modern Systems and Applications

More recent advances in equipment and techniques have dramatically improved the efficacy of the measurement technique and the production of deliverables. There is an extensive range of underwater-capable, digital cameras with high-resolution sensors that can capture both still images and video sequences (Underwater Photography Guide 2017). Rather than highly constrained patterns of stereo photographs and traditional, manual photogrammetric solutions, many photographs from a single camera and the principle of Structure from Motion (SfM) (Pollefeys et al. 2000) can be used to automatically generate a detailed 3D model of the site, shipwreck or artefacts. SfM has been used effectively to map archaeological sites (McCarthy 2014; McCarthy and Benjamin 2014; Skarlatos et al. 2012; Van Damme 2015), compare sites before and after the removal of encrustations (Bruno et al. 2013) and create models for the artefacts from a shipwreck (Balletti et al. 2015; Fulton et al. 2016; Green et al. 2002; McCarthy and Benjamin 2014). Whilst there are some practical considerations that must be respected to obtain an effective and complete 3D virtual model (McCarthy and Benjamin 2014), the locations of the photographs are relatively unconstrained, which is a significant advantage in the underwater environment.

Based on citations in the literature (Mallet and Pelletier 2014; Shortis et al. 2009a), however, marine habitat conservation, biodiversity monitoring and fisheries stock assessment dominate the application of accurate measurement by underwater camera systems. The age and biomass of fish can be reliably estimated based on length measurement and a length-weight or length-age regression (Pienaar and Thomson 1969; Santos et al. 2002). When combined with spatial or temporal sampling in marine ecosystems, or counts of fish in an aquaculture cage or a trawl net, the distribution of lengths can be used to estimate distributions of or changes in biomass, and shifts in or impacts on population distributions. Underwater camera systems are now widely employed in preference to manual methods as a non-contact, non-invasive technique to capture accurate length information and thereby estimate biomass or population distributions (Shortis et al. 2009a). Underwater camera systems have the further advantages that the measurements are accurate and repeatable (Murphy and Jenkins 2010), sample areas can be very accurately estimated (Harvey et al. 2004) and the accuracy of the length measurements vastly improves the statistical power of the population estimates when sample counts are very low (Harvey et al. 2001).

Underwater stereo-video systems have been used in the assessment of wild fish stocks with a variety of cameras and modes of operation (Klimley and Brown 1983; Mallet and Pelletier 2014; McLaren et al. 2015; Santana-Garcon et al. 2014; Seiler et al. 2012; Watson et al. 2009), in pilot studies to monitor length frequencies of fish in aquaculture cages (Harvey et al. 2003; Petrell et al. 1997; Phillips et al. 2009) and in fish nets during capture (Rosen et al. 2013). Commercial systems such as the AKVAsmart, formerly VICASS (Shieh and Petrell 1998), and the AQ1 AM100 (Phillips et al. 2009) are widely used in aquaculture and fisheries.

There are many other applications of underwater photogrammetry. Stereo camera systems were used to conduct the first accurate seabed mapping applications (Hale and Cook 1962; Pollio 1971) and have been used to measure the growth of coral (Done 1981). Single and stereo cameras have been used for monitoring of submarine structures, most notably to support energy exploration and extraction in the North Sea (Baldwin 1984; Leatherdale and Turner 1983), mapping of seabed topography (Moore 1976; Pollio 1971), 3D models of sea grass meadows (Rende et al. 2015) and inshore sea floor mapping (Doucette et al. 2002; Newton 1989). A video camera has been used to measure the shape of fish pens (Schewe et al. 1996), a stereo camera has been used to map cave profiles (Capra 1992) and digital still cameras have been used underwater for the estimation of sponge volumes (Abdo et al. 2006). Seafloor monitoring has been carried out in deep water using continuously recorded stereo video cameras combined with a high resolution digital still camera (Shortis et al. 2009b). A network of digital still camera images has been used to accurately characterize the shape of a semi-submerged ship hull (Menna et al. 2013).

2.1.3 Calibration and Accuracy

The common factor for all these applications of underwater imagery is a designed or specified level of accuracy. Photogrammetric surveys for heritage recording, marine biomass or fish population distributions are directly dependent on the accuracy of the 3D measurements. Any inaccuracy will lead to significant errors in the measured dimensions of artefacts (Capra et al. 2015), under- or over-estimation of biomass (Boutros et al. 2015) or a systematic bias in the population distribution (Harvey et al. 2001). Other applications such as structural monitoring or seabed mapping must achieve a specified level of accuracy for the surface shape.

Calibration of any camera system is essential to achieve accurate and reliable measurements. Small errors in the perspective projection must be modelled and eliminated to prevent the introduction of systematic errors in the measurements. In the underwater environment, the calibration of the cameras is of even greater importance because the effects of refraction through the air, housing and water interfaces must be incorporated.

Compared to in-air calibration, camera calibration under water is subject to the additional uncertainty caused by attenuation of light through the housing port and water media, as well as the potential for small errors in the refracted light path due to modelling assumptions or non-uniformities in the media. Accordingly, the precision and accuracy of calibration under water is always expected to be degraded relative to an equivalent calibration in-air. Experience demonstrates that, because of these effects, underwater calibration is more likely to result in scale errors in the measurements.

2.2 Calibration Approaches

2.2.1 Physical Correction

In a limited range of circumstances calibration may be unnecessary. If a high level of accuracy is not required, and the object to be measured approximates a 2D planar surface, a straightforward solution is possible.

Correction lenses or dome ports such as those described in Ivanoff and Cherney (1960) and Moore (1976) can be used to provide a near-perfect central projection under water by eliminating the refraction effects. Any remaining, small errors or imperfections can either be corrected using a grid or graticule placed in the field of view, or simply accepted as a small deterioration in accuracy. The correction lens or dome port has the further advantage that there is little, if any, degradation of image quality near the edges of the port. Plane camera ports exhibit loss of contrast and intensity at the extremes of the field of view due to acute angles of incidence and greater apparent thickness of the port material.

This simplified approach has been used, either with correction lenses or with a pre-calibration of the camera system, to carry out two-dimensional mapping. A portable control frame with a fixed grid or target reference is imaged before deployment or placed against the object to measured, to provide both calibration corrections as well as position and orient the camera system relative to the object. Typical applications of this approach are shipwreck mapping (Hohle 1971), sea floor characterization surveys (Moore 1976), length measurements in aquaculture (Petrell et al. 1997) and monitoring of sea floor habitats (Chong and Stratford 2002).

If accuracy is a priority, however, and especially if the object to be measured is a 3D surface, then a comprehensive calibration is essential. The correction lens approach assumes that the camera is a perfect central projection and that the entrance pupil of the camera lens coincides exactly with the centre of curvature of the correction lens. Any simple correction approach, such as a graticule or control frame placed in the field of view, will be applicable only at the same distance. Any significant extrapolation outside of the plane of the control frame will inevitably introduce systematic errors.

2.2.2 Target Field Calibration

The alternative approach of a comprehensive calibration translates a reliable technique from in-air into the underwater environment. Close range calibration of cameras is a well-established technique that was pioneered by Brown (1971), extended to include self-calibration of the camera(s) by Kenefick et al. (1972) and subsequently adapted to the underwater environment (Fryer and Fraser 1986; Harvey and Shortis 1996). The mathematical basis of the technique is reviewed in Granshaw (1980).

The essence of this approach is to capture multiple, convergent images of a fixed calibration range or portable calibration fixture to determine the physical parameters of the camera calibration (Fig. 2.1). A typical calibration range or fixture is based on discrete targets to precisely identify measurement locations throughout the camera fields of view from the many photographs (Fig. 2.1). The targets may be circular dots or the corners of a checkerboard. Coded targets or checkerboard corners on the fixture can be automatically recognized using image analysis techniques (Shortis and Seager 2014; Zhang 2000) to substantially improve the efficiency of the measurements and network processing. The ideal geometry and a full set of images for a calibration fixture are shown in Figs. 2.2 and 2.3, respectively.

A fixed test range, such as the 'Manhattan' object shown in Fig. 2.1, has the advantage that accurately known target coordinates can be used in a pre-calibration approach. The disadvantage, however, is that the camera system must be transported to the range and then back to the deployment location. In comparison, accurate information for the positions of the targets on a portable calibration fixture is not required, as coordinates of the targets can be derived as part of a self-calibration approach. Hence, it is immaterial if the portable fixture distorts or is dis-assembled between calibrations, although the fixture must retain its dimensional integrity during the image capture.

Scale within the 3D measurement space is determined by introducing distances measured between pre-identified targets into the self-calibration network (El-Hakim and Faig 1981). The known distances between the targets must be reliable and accurate, so known lengths are specified between targets on the rigid arms of the fixture or between the corners of the checkerboard.

In practice, cameras are most often pre-calibrated using a self-calibration network and a portable calibration fixture in

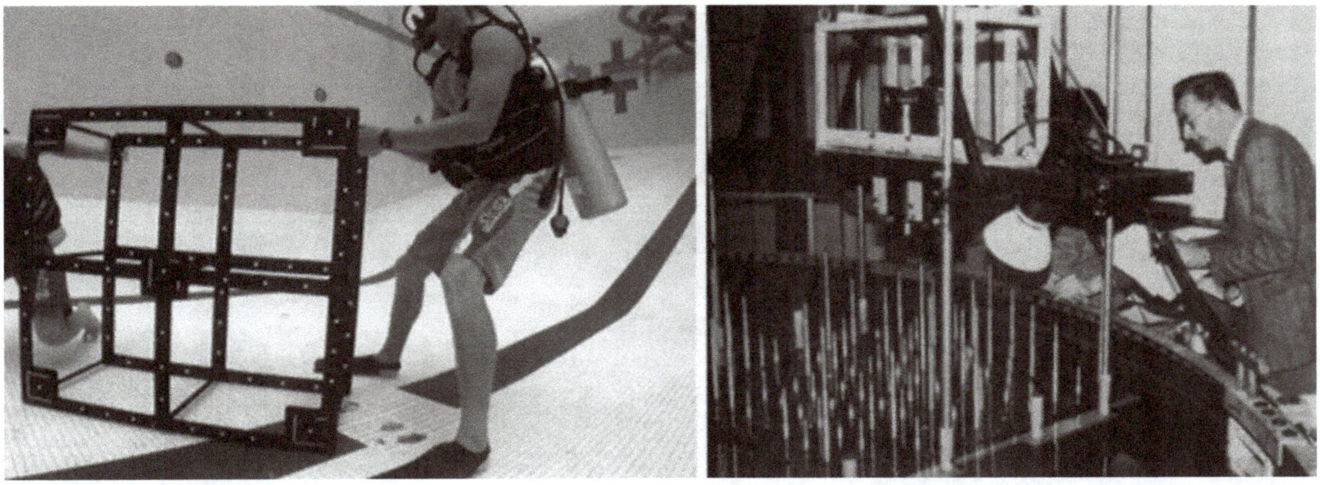

Fig. 2.1 Typical portable calibration fixture (left, courtesy of NOAA) and test range. (Right, from Leatherdale and Turner 1983)

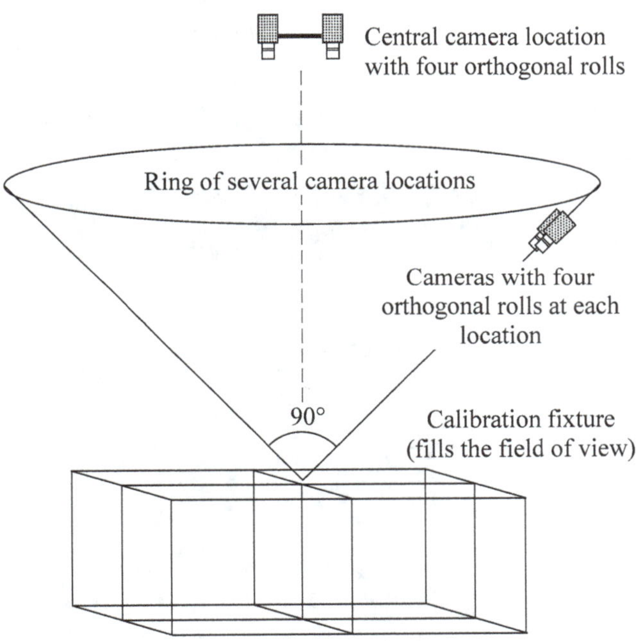

Fig. 2.2 The ideal geometry for a self-calibration network

a venue convenient to the deployment. The refractive index of water is insensitive to temperature, pressure or salinity (Newton 1989), so the conditions prevailing for the pre-calibration can be assumed to be valid for the actual deployment of the system to capture measurements. The assumption is also made that the camera configurations, such as focus and zoom, and the relative orientation for a multi camera system, are locked down and undisturbed. In practice this means that the camera lens focus and zoom adjustments must be held in place using tape or a lock screw, and the connection between multiple cameras, usually a base bar between stereo cameras, must be rigid. A close proximity between the locations of the calibration and the deployment minimizes the risk of a physical change to the camera system.

The process of self-calibration of underwater cameras is straightforward and quick. The calibration can take place in a swimming pool, in an on-board tank on the vessel or, conditions permitting, adjacent to, or beneath, the vessel. The calibration fixture can be held in place and the cameras manoeuvred around it, or the calibration fixture can be manipulated whilst the cameras are held in position, or a combination of both approaches can be used (Fig. 2.3). For example, a small 2D checkerboard may be manipulated in front of an ROV stereo-camera system held in a tank. A large, towed body system may be suspended in the water next to a wharf and a large 3D calibration fixture manipulated in front of the stereo video cameras. In the case of a diver-controlled stereo-camera system, a 3D calibration fixture may be tethered underneath the vessel and the cameras moved around the fixture to replicate the network geometry shown in Fig. 2.2.

There are very few examples of in situ self-calibrations of camera systems, because this type of approach is not readily adapted to the dynamic and uncontrolled underwater environment. Nevertheless, there are some examples of a single camera or stereo camera in situ self-calibration (Abdo et al. 2006; Green et al. 2002; Schewe et al. 1996). In most cases a pre- or post-calibration is conducted anyway to determine an estimate of the calibration of the camera system as a contingency.

2.3 Calibration Algorithms

2.3.1 Calibration Parameters

Calibration of a camera system is necessary for two reasons. First, the internal geometric characteristics of the cameras must be determined (Brown 1971). In photogrammetric practice, camera calibration is most often defined by physical

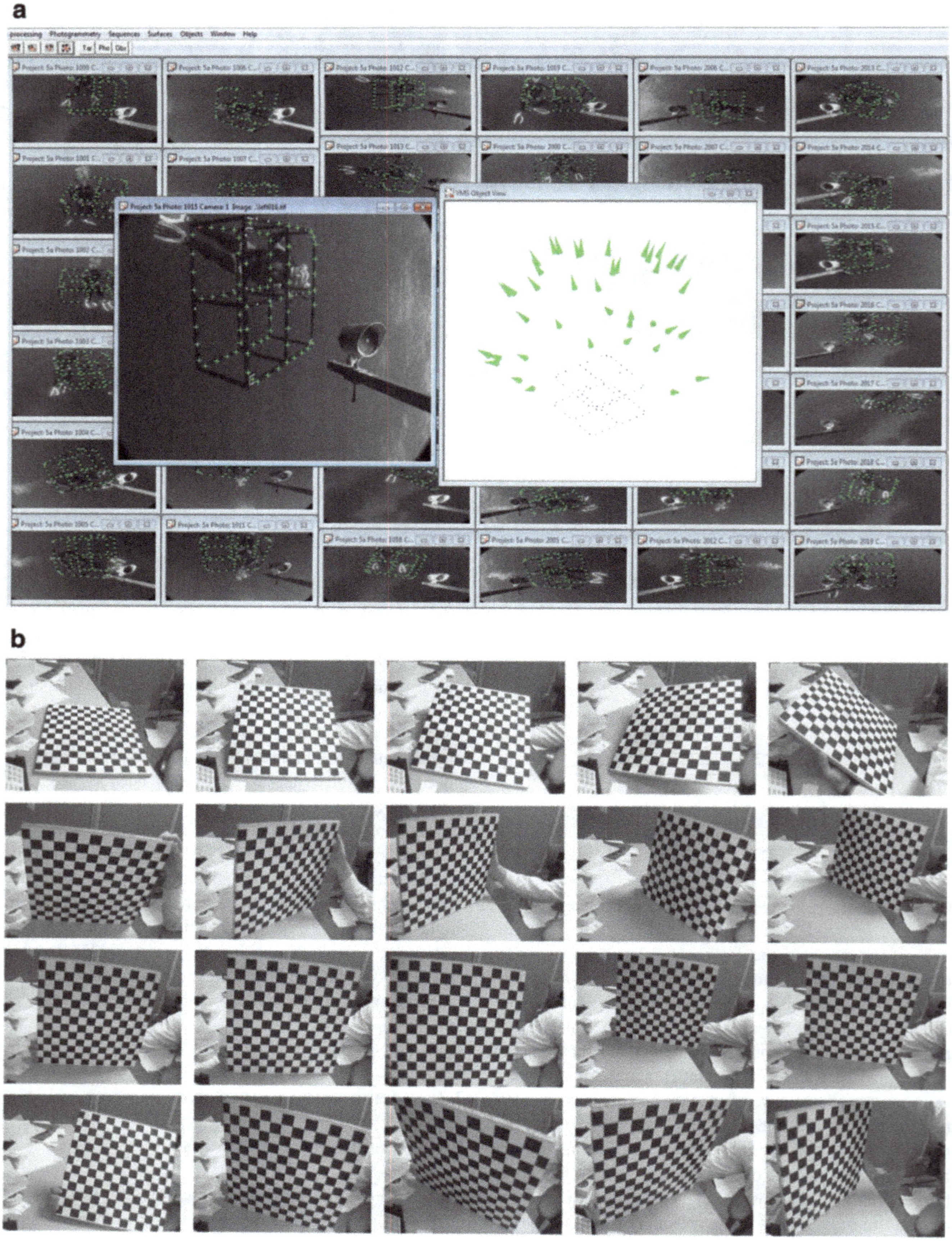

Fig. 2.3 Top: a set of calibration images from an underwater stereo-video system using a 3D calibration fixture. Both the cameras and the object have been rotated to acquire the convergent geometry of the network. Bottom: a set of calibration images of a 2D checkerboard for a single camera calibration, for which only the checkerboard has been rotated. (From Bouguet 2017)

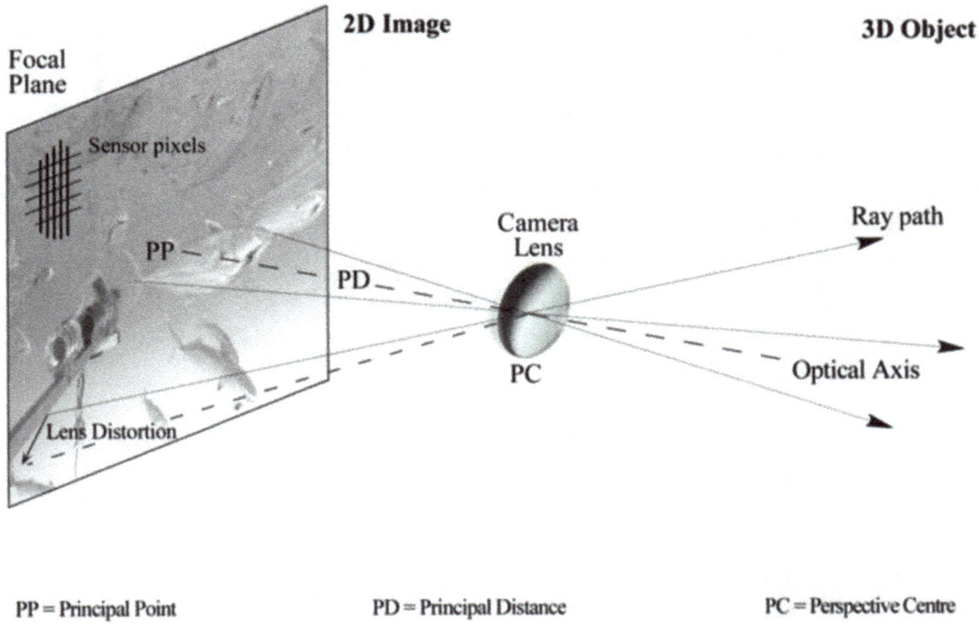

Fig. 2.4 The geometry of perspective projection based on physical calibration parameters

parameter set (Fig. 2.4) comprising principal distance, principal point location, radial (Ziemann and El-Hakim 1983) and decentring (Brown 1966) lens distortions, plus affinity and orthogonality terms to compensate for minor optical effects (Fraser et al. 1995; Shortis 2012). The principal distance is formally defined as the separation, along the camera optical axis, between the lens perspective centre and the image plane. The principal point is the intersection of the camera optical axis with the image plane.

Radial distortion is a by-product of the design criteria for camera lenses to produce very even lighting across the entire field of view and is defined by an odd-ordered polynomial (Ziemann and El-Hakim 1983). Three terms are generally sufficient to model the radial lens distortion of most cameras in-air or in-water. SfM applications such as *Agisoft Photoscan/Metashape* (Agisoft 2017) and *Reality Capture* (Capturing Reality 2017) offer up to five terms in the polynomial; however, these extra terms are redundant except for camera lenses with extreme distortion profiles.

Decentring distortion is described by up to four terms (Brown 1971), but in practice only the first two terms are significant. This distortion is caused by the mis-centring of lens components in a multi-element lens and the degree of mis-centring is closely associated with the quality of the manufacture of the lens. The magnitude of this distortion is much less than radial distortion (Figs. 2.6 and 2.7) and should always be small for simple lenses with few elements when calibrated in-air.

Second, the relative orientation of the cameras with respect to one another, or the exterior orientation with respect to an external reference, must be determined. Also known as pose estimation, both the location and orientation of the camera(s) must be determined. For the commonly used approach of stereo cameras, the relative orientation effectively defines the separation of the perspective centres of the two lenses, the pointing angles (omega and phi rotations) of the two optical axes of the cameras and the roll angles (kappa rotations) of the two focal plane sensors (Fig. 2.5).

2.3.2 Absorption of Refraction Effects

In the underwater environment the effects of refraction must be corrected or modelled to obtain an accurate calibration. The entire light path, including the camera lens, housing port and water medium, must be considered. By far the most common approach is to correct the refraction effects using absorption by the physical camera calibration parameters. Assuming that the camera optical axis is approximately perpendicular to a plane or dome camera port, the primary effect of refraction through the air-port and port-water interfaces will be radially symmetric around the principal point (Li et al. 1996). This primary effect can be absorbed by the radial lens distortion component of the calibration parameters. Figure 2.6 shows a comparison of radial lens distortion from calibrations in-air and in-water for the same camera, demonstrating the compensation effect for the radial distortion profile. There will also be some small, asymmetric effects caused by, for example, alignment errors between the optical axis and the housing port, and perhaps non-uniformities in the thickness or material of the housing. These secondary effects can be absorbed by calibration parameters such as the decentring lens distortion and the affinity term. Figure 2.7 shows a comparison of decentring lens distortion from calibrations in-air and in-water of the same camera. Similar

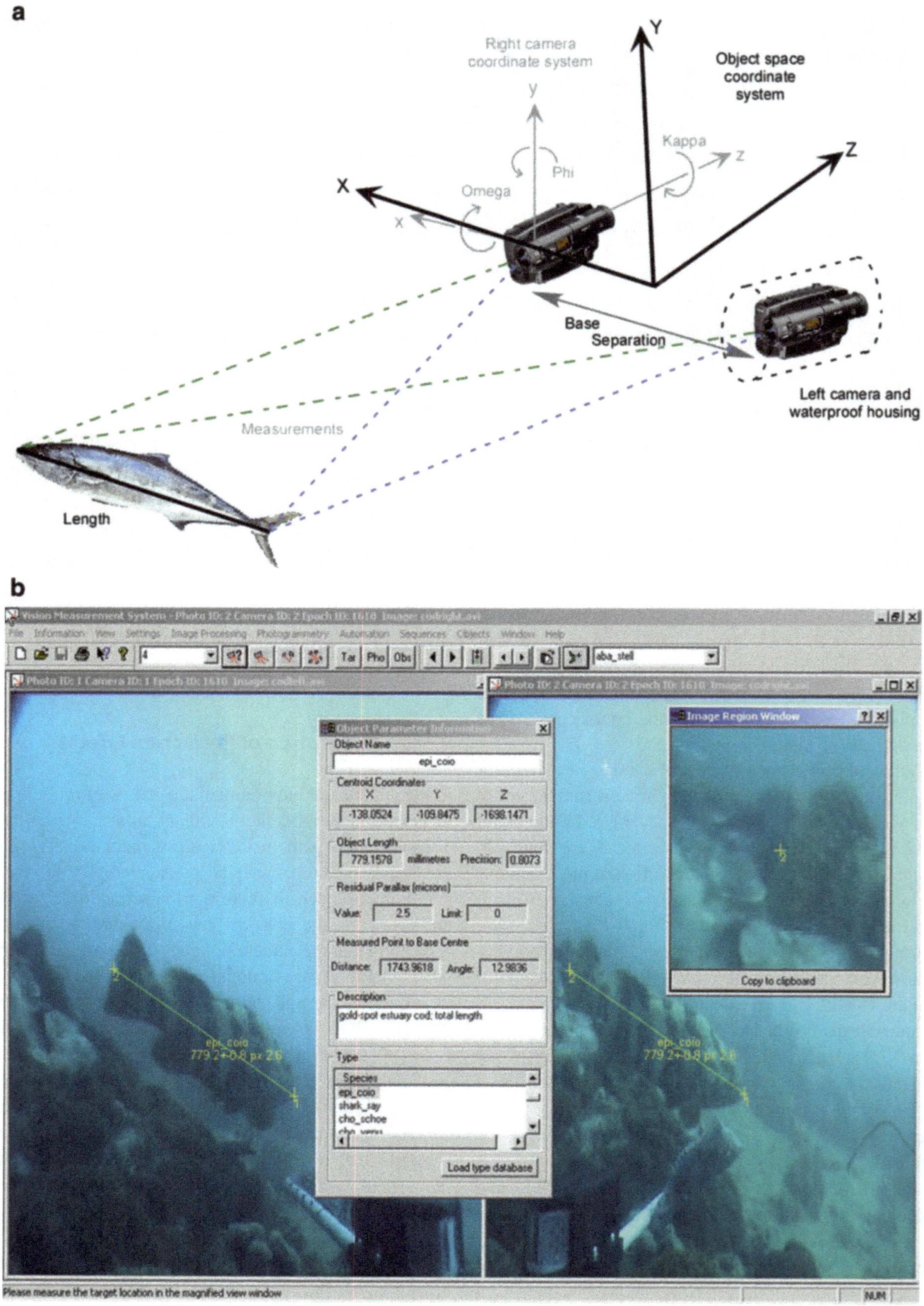

Fig. 2.5 Schematic view of a stereo-image measurement of a length from 3D coordinates (top) and view of a measurement interface (bottom). (Courtesy E.S. Harvey)

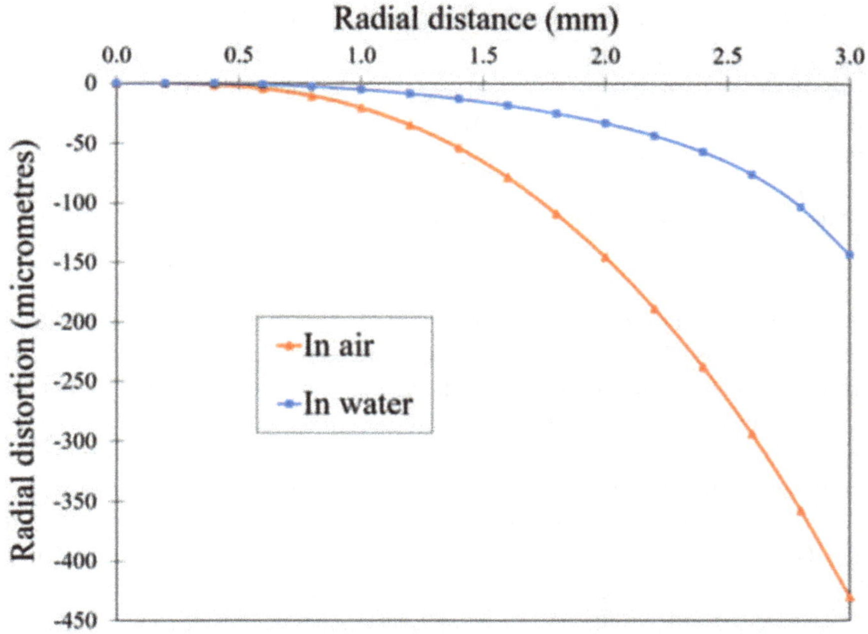

Fig. 2.6 Comparison of radial lens distortion from in-air and in-water calibrations of a GoPro Hero4 camera operated in HD video mode

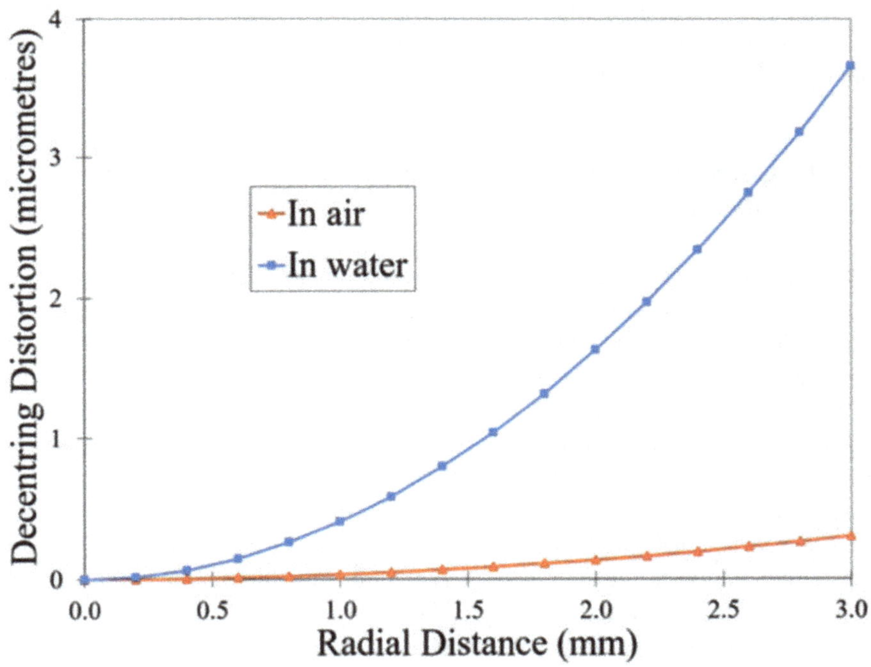

Fig. 2.7 Comparison of decentring lens distortion from in-air and in-water calibrations of a GoPro Hero4 camera operated in HD video mode. Note the much smaller range of distortion values (vertical axis) compared to Fig. 2.6

changes in the lens distortion profiles are demonstrated in Fryer and Fraser (1986) and Lavest et al. (2000).

Table 2.1 shows some of the calibration parameters for the in-air and in-water calibrations of two GoPro Hero4 cameras. The ratios of the magnitudes of the parameters indicate whether there is a contribution to the refractive effects. As could be expected, for a plane housing port, the principal distance is affected directly, whilst changes in parameters such as the principal point location and the affinity term may include the combined influences of secondary effects, correlations with other parameters and statistical fluctuation. These results are consistent for the two cameras, consistent with other cameras tested, and Lavest et al. (2000) presents similar outcomes from in-air versus in-water calibrations for flat ports. Very small percentage changes to all parameters, including the principal distance, are reported in Bruno et al. (2011) for housings with dome ports. This result is in accord with the expected physical model of the refraction.

Table 2.1 Comparison of parameters from in-air and in-water calibrations for two GoPro Hero4 camera used in HD video mode

Camera	GoPro Hero4 #1			GoPro Hero4 #2		
Parameter	In-air	In-water	Ratio	In-air	In-water	Ratio
PPx (mm)	0.080	0.071	0.88	−0.032	−0.059	1.82
PPy (mm)	−0.066	−0.085	1.27	−0.143	−0.171	1.20
PD (mm)	3.676	4.922	1.34	3.658	4.898	1.34
Affinity	−6.74E-03	−6.71E-03	1.00	−6.74E-03	−6.84E-03	1.01

The disadvantage of the absorption approach for the refractive effects is that there will always be some systematic errors which are not incorporated into the model. The effect of refraction invalidates the assumption of a single projection centre for the camera (Sedlazeck and Koch 2012), which is the basis for the physical parameter model. The errors are most often manifest as scale changes when measurements are taken outside of the range used for the calibration process. Experience over many years of operation demonstrates that, if the ranges for the calibration and the measurements are commensurate, then the level of systematic error is generally less than the precision with which measurements can be extracted. This masking effect is partly due to the elevated level of noise in the measurements, caused by the attenuation and loss of contrast in the water medium.

2.3.3 Geometric Correction of Refraction Effects

The alternative to the simple approach of absorption is the more complex process of geometric correction, effectively an application of ray tracing of the light paths through the refractive interfaces. A two-phase approach is developed in Li et al. (1997) for a stereo camera housing with concave lens covers. An in-air calibration is carried out first, followed by an in-water calibration that introduces 11 lens cover parameters such as the centre of curvature of the concave lens and, if not known from external measurements, refractive indices for the lens covers and water. A more general geometric correction solution is developed for plane port housings in Jordt-Sedlazeck and Koch (2012). Additional unknowns in the solution are the distance between the camera perspective centre and the housing, and the normal of the plane housing port, whilst the port thickness and refractive indices must be known. Using ray tracing, Kotowski (1988) develops a general solution to refractive surfaces that, in theory, can accommodate any shape of camera housing port. The shape of the refractive surface and the refractive indices must be known. Maas (2015), develops a modular solution to the effects of plane, parallel refraction surfaces, such as a plane camera port or the wall of a hydraulic testing facility, which can be readily included in standard photogrammetric tools.

A variation on the geometric correction is the perspective centre shift or virtual projection centre approach. A specific solution for a planar housing port is developed in Telem and Filin (2010). The parameters include the standard physical parameters, the refractive indices of glass and water, the distance between the perspective centre and the port, the tilt and direction of the optical axis with respect to the normal to the port, and the housing interface thickness. A modified approach neglects the direction of the optical axis and the thickness of thin ports, as these factors can be readily absorbed by the standard physical parameters. Again, a two-phase process is required: first a 'dry' calibration in-air and then a 'wet' calibration in-water (Telem and Filin 2010). A similar principle is used in Bräuer-Burchardt et al. (2015), also with a two-phase calibration approach.

The advantage of these techniques is that, without the approximations in the models, the correction of the refractive effects is exact. The disadvantages are the requirements for two phase calibrations and necessary data such as refractive indices. Further, in some cases the theoretical solution is specific to a housing type, whereas the absorption approach has the distinct advantage that it can be used with any type of underwater housing.

As well as the common approaches described above, some other investigations are worthy of note. The Direct Linear Transformation (DLT) algorithm (Abdel-Aziz and Karara 1971) is used with three different techniques in Kwon and Casebolt (2006). The first is essentially an absorption approach, but used in conjunction with a sectioning of the object space to minimize the remaining errors in the solution. A double plane correction grid is applied in the second approach. In the last technique a formal refraction correction model is included with the requirements that the camera-to-interface distance and the refractive index must be known. A review of refraction correction methods for underwater imaging is given in Sedlazeck and Koch (2012). The perspective camera model, ray-based models and physical models are analysed, including an error analysis based on synthetic data. The analysis demonstrates that perspective camera models incur increasing errors with increasing distance and tilt of the refractive surfaces, and only the physical model of refraction correction permits a complete theoretical compensation.

2.3.4 Relative Orientation

Once the camera calibration is established, single camera systems can be used to acquire measurements when used in conjunction with reference frames (Moore 1976) or sea floor reference marks (Green et al. 2002). For multi-camera systems the relative orientation is required as well as the camera calibration. The relative orientation can be included in the self-calibration solution as a constraint (King 1995) or can be computed as a post-process based on the camera positions and orientations for each set of synchronized exposures (Harvey and Shortis 1996). In either case it is important to detect and eliminate outliers, usually caused by lack of synchronization, which would otherwise unduly influence the calibration solution or the relative orientation computation. Outliers caused by synchronization effects are more common for systems based on camcorders or video cameras in separate housings, which typically use an external device such as a flashing LED light to synchronize the images to within one video frame (Harvey and Shortis 1996).

In the case of post-processing, the exterior orientations for the sets of synchronized exposures are initially in the frame of reference of the calibration fixture, so each set must be transformed into a local frame of reference with respect to a specific baseline between the cameras. In the case of stereo cameras, the local frame of reference is adopted as the centre of the baseline between the camera perspective centres, with the axes aligned with the baseline direction and the mean optical axis pointing direction (Fig. 2.5). The final parameters for the precise relative orientation are adopted as the mean values for all sets in the calibration network, after any outliers have been detected and eliminated.

2.4 Calibration Reliability and Stability

2.4.1 Reliability Factors

The reliability and accuracy of the calibration of underwater camera systems is dependent on a number of factors. Chief amongst the factors are the geometry and redundancy for the calibration network. A high level of redundant information—provided by many target image observations on many exposures—produces high reliability so that outliers in the image observations can be detected and eliminated. An optimum 3D geometry is essential to minimize correlations between the parameters and ensure that the camera calibration is an accurate representation of the physical model (Kenefick et al. 1972). It should be noted, however, that it is not possible to eliminate all correlations between the calibration parameters. Correlations are always present between the three radial distortion terms and between the principal point and two decentring terms.

The accuracy of the calibration parameters is enhanced if the network of camera and target locations meets the following criteria:

1. The camera and target arrays are 3D in nature. 2D arrays are a source of weak network geometry. 3D arrays minimize correlations between the internal camera calibration parameters and the external camera location and orientation parameters.
2. The many, convergent camera views approach a 90° intersection at the centre of the target array. A narrowly grouped array of camera views will produce shallow intersections, weakening the network and thereby decreasing the confidence with which the calibration parameters are determined.
3. The calibration fixture or range fills the field of view of the camera(s) to ensure that image measurements are captured across the entire format. If the fixture or range is small and centred in the field of view, then the radial and decentring lens distortion profiles will be defined very poorly because measurements are captured only where the distortion signal is small in magnitude.
4. The camera(s) are rolled around the optical axis for different exposures so that 0°, 90°, 180° and 270° orthogonal rotations are spread throughout the calibration network. A variety of camera rolls in the network also minimizes correlations between the internal camera calibration parameters and the external camera location and orientation parameters.

If these four conditions are met, the self-calibration approach can be used to simultaneously and confidently determine the camera calibration parameters, camera exposure locations and orientations, and updated target coordinates (Kenefick et al. 1972).

In recent years there has been an increasing adoption of a calibration technique using a small 2D checkerboard and a freely available Matlab solution (Bouguet 2017). The main advantages of this approach are the simplicity of the calibration fixture and the rapid measurement and processing of the captured images, made possible by the automatic recognition of the checkerboard pattern (Zhang 2000). A practical guide to the use of this technique is provided in Wehkamp and Fischer (2014).

The small size and 2D nature of the checkerboard, however, limits the reliability and accuracy of measurements made using this technique (Boutros et al. 2015). The technique is equivalent to a fixed test range calibration rather than a self-calibration, because the coordinates of the checkerboard corners are not updated. Any inaccuracy in the coordinates, especially if the checkerboard has variations from a true 2D plane, will introduce systematic errors into the calibration. Nevertheless, the 2D fixture can produce a calibration suitable for measurements at short ranges and with modest accuracy requirements. AUV and diver-operated stereo camera systems pre-calibrated with this technique have been used to capture fish length measurements (Seiler et al. 2012; Wehkamp and Fischer 2014) and tested for the 3D re-construction of artefacts (Bruno et al. 2011).

2.4.2 Stability Factors

The stability of the calibration for underwater camera systems has been well documented in published reports (Harvey and Shortis 1998; Shortis et al. 2000). As noted previously, the basic camera settings such as focus and zoom must be consistent between the calibration and deployments—usually ensured through the use of tape or a locking screw to prevent the settings from being inadvertently altered. For cameras used in-air, other factors are related to the handling of the camera—especially when the camera is rolled about the optical axis or a zoom lens is employed—and the quality of the lens mount. Any distortion of the camera body or movement of the lens or optical elements will result in variation of the relationship between the perspective centre and the CMOS or CCD imager at the focal plane, which will disturb the calibration (Shortis and Beyer 1997). Fixed focal length lenses are preferred over zoom lenses to minimise the instabilities.

The most significant sensitivity for the calibration stability of underwater camera systems, however, is the relationship between the camera lens and housing port. Rigid mounting of the camera in the housing is critical to ensure that the total optical path from the image sensor to the water medium is consistent (Harvey and Shortis 1998). Testing and validation have shown that calibration is only reliable if the camera in the housing is mounted on a rigid connection to the camera port (Shortis et al. 2000). This applies to both a single deployment and multiple, separate deployments of the camera system. Unlike correction lenses and dome ports, a specific position and alignment within the housing is unnecessary, but the distance and orientation of the camera lens relative to the housing port must be consistent. The most reliable option is a direct, mechanical linkage between the camera lens and the housing port that can consistently re-create the physical relationship. The consistency of distance and orientation is especially important for portable camcorders because they must be regularly removed from the housings to retrieve storage media and replenish batteries.

Finally, for multi-camera systems—in-air or in-water—their housings must have a rigid mechanical connection to a base bar to ensure that the separation and relative orientation of the cameras is also consistent. Perturbation of the separation or relative orientation often results in apparent scale errors, which can be readily confused with refractive effects. Figure 2.8 shows some results of repeated calibrations of a GoPro Hero 2 stereo-video system. The variation in the parameters between consecutive calibrations demonstrates a comparatively stable relative orientation but a more unstable camera calibration, in this case caused by a non-rigid mounting of the camera in the housing.

2.5 Calibration and Validation Results

2.5.1 Quality Indicators

The first evaluation of a calibration is generally the internal consistency of the network solution that is used to compute the calibration parameters, camera locations and orientations, and if applicable, updated target coordinates. The 'internal' indicator is the Root Mean Square (RMS) error of image measurement, a metric for the internal 'fit' of the least squares estimation solution (Granshaw 1980). Note that in general the measurements are based on an intensity weighted centroid to locate the centre of each circular target in the image (Shortis et al. 1995).

To allow comparison of different cameras with different spacing of the light sensitive elements in the CMOS or CCD imager, the RMS error is expressed in fractions of a pixel. In ideal conditions in-air, the RMS image error is typically in the range of 0.03–0.1 pixels (Shortis et al. 1995). In the underwater environment, the attenuation of light and loss of contrast, along with small non-uniformities in the media, degrades the RMS error into the range of 0.1–0.3 pixels (Table 2.2). This degradation is a combination of a larger statistical signature for the image measurements and the influence of small, uncompensated systematic errors. In conditions of poor lighting or poor visibility the RMS error deteriorates rapidly (Wehkamp and Fischer 2014).

The second indicator that is commonly used to compare the calibration, especially for in-air operations, is the proportional error, expressed as the ratio of the RMS error in the 3D coordinates of the targets to the largest dimension of the object. This 'external' indicator provides a standardized, relative measure of precision in the object space. In the circumstance of a camera calibration, the largest dimension is the diagonal span of the test range volume, or the diagonal span of the volume envelope of all imaged locations of the calibration fixture. Whilst the RMS image error may be favourable, the proportional error may be relatively poor if the object is contained within a small volume or the geometry of the calibration network is poor. Table 2.2 presents a sample of some results for the precision of calibrations. It is evident that the proportional error can vary substantially, however an average figure is approximately 1:5000.

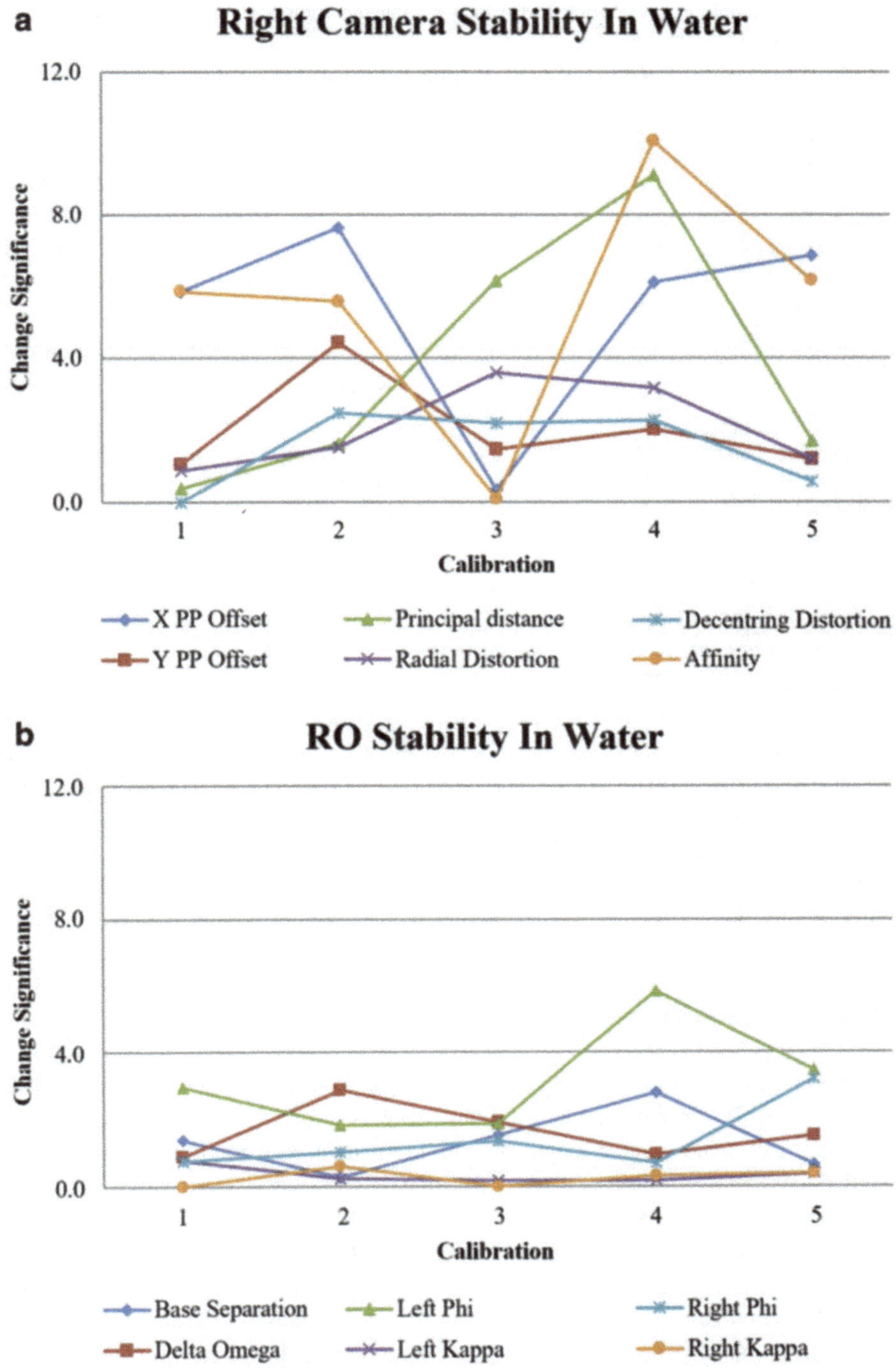

Fig. 2.8 Stability of the right camera calibration parameters (top) and the relative orientation parameters (bottom) for a GoPro Hero 2 stereo-video system. The vertical axis is the change significance of individual parameters between consecutive calibrations (Harvey and Shortis 1998)

2.5.2 Validation Techniques

As a consequence of the potential misrepresentation by proportional error, independent testing of the accuracy of underwater camera systems is essential to ensure the validity of 3D locations, length, area or volume measurements. For stereo and multi-camera systems, the primary interest is length measurements that are subsequently used to estimate the size of artefacts or the biomass of fish. One validation technique is to use known distances on the rigid components of the calibration fixture (Harvey et al. 2003), however this has some limitations.

As already noted, the circular, discrete targets are dissimilar to the natural feature points of a fish snout or an anchor tip, and they are measured by different techniques. The variation in size and angle of the distance on the calibration fixture may not correlate well with the size and orientation of the measurement. In particular, measurements of objects of interest are often taken at greater ranges than that of the calibration fixture, partly due to expediency in surveys and partly because the calibration fixture must be close enough to the cameras to fill a reasonable portion of the field of view. Given the approximations in the refraction models, it is important that accuracy validations are carried out at ranges greater

Table 2.2 A sample of some published results for the precision of underwater camera calibrations

Technique	RMS image error (pixels)	RMS XYZ error (mm)	Proportional error
Absorption (Harvey and Shortis 1996)	0.1–0.3	0.1–0.5	1:3000–1:15000
Absorption (Schewe et al. 1996)	0.3	40–200	1:500
Geometric correction (Li et al. 1997)	1.0	10	1:210
Perspective shift (Telem and Filin 2010)	0.3	2.0	1:1000
Absorption (Menna et al. 2015)	0.2–0.25	1.9	1:32000

Note that Schewe et al. (1996) used observations of a mobile fish pen and the measurements used by Li et al. (1997) were made to the nearest whole pixel

than the average range to the calibration fixture. Further, it has been demonstrated that the accuracy of length measurements is dependent on the separation of the cameras in a multi-camera system (Boutros et al. 2015) and significantly affected by the orientation of the artefact relative to the cameras (Harvey and Shortis 1996; Harvey et al. 2002). Accordingly, validation of underwater video measurement systems is typically carried out by introducing a known length, such as a rod or a fish silhouette, which is measured manually at a variety of ranges and orientations within the field of view (Fig. 2.9).

2.5.3 Validation Results

In the best-case scenario of clear visibility and high contrast targets, the RMS error of validation measurements is typically less than 1 mm over a length of 1 m, equivalent to a length accuracy of 0.1%. In realistic, operational conditions using fish silhouettes or validated measurements of live fish, length measurements have an accuracy of 0.2–0.7% (Boutros et al. 2015; Harvey et al. 2002, 2003, 2004; Telem and Filin 2010). The accuracy is somewhat degraded if a simple correction grid is used (Petrell et al. 1997) or a simplified calibration approach is adopted (Wehkamp and Fischer 2014). A sample of published results of validations based on known lengths or geometric objects is given in Table 2.3.

McCarthy and Benjamin (2014) presents some validation results from direct comparisons between a 3D virtual model generated by photogrammetry and taped measurements taken by divers. The artefacts in this case were cannons lying on the sea floor and the 3D information was derived from a self-calibration, SfM solution. An accurate scale for the mesh was provided by a 1 m length bar placed within the site. The average difference for long measurements was found to be 3% and, for the longest distances, differences were typically less than 1%. Shorter distances tended to exhibit much larger errors, however the comparisons are detrimentally influenced by the inability to choose exactly corresponding points of reference for the virtual model and the tape measurements.

Two different types of underwater cameras are evaluated in a preliminary study of accuracy for the monitoring of coral reefs (Guo et al. 2016). In-air and underwater calibrations were undertaken, validated by an accurately known target fixture and 3D point cloud models of cinder blocks. The targets on the calibration frame were divided into 12 control points and 33 check points for the calibration networks. Based on the approximate 1 m span of the fixture, the proportional errors underwater range from 1:2500 to 1:7000. Validation based on comparisons of in-air and underwater SfM 3D models of the cinder blocks indicated RMS errors of the order of 1–2 mm, corresponding to an accuracy in the range of 0.1–0.2%.

Validations of biomass estimates of Southern Bluefin Tuna measured in aquaculture pens (Harvey et al. 2003) and sponges measured in the field (Abdo et al. 2006) have shown that volumes can be estimated with an accuracy of the order of a few percent. The Southern Bluefin Tuna validation was based on distances such as body length and span, made by a stereo-video system and compared to a length board and caliper system of manual measurement. Each Southern Bluefin Tuna in a sample of 40 fish was also individually weighed. The stereo-video system produced an estimate of better than 1% for the total biomass (Harvey et al. 2003). Triangulation meshes on the surface of simulated and live specimens were used to estimate the volume of sponges. The resulting errors were 3–5%, and no worse than 10%, for individual sponges (Abdo et al. 2006). Greater variability is to be expected for the estimates of the sponge volumes, because of the uncertainty associated with the assumed shape of the unseen substrate surface beneath each sponge.

By the nature of conversion from length to weight, errors can be amplified significantly. Typical regression functions are power series with a near cubic term (Harvey et al. 2003; Pienaar and Thomson 1969; Santos et al. 2002). Accordingly, inaccuracies in the calibration and the precision of the measurement may combine to produce unacceptable results. A simulation is employed by Boutros et al. (2015) to demonstrate clearly that the predicted error in the biomass of a fish, based on the error in the length, deteriorates rapidly with range from the cameras, especially with a small 2D calibration fixture and a narrow separation between the stereo cameras. Errors in the weight in excess of 10% are possible, reinforcing the need for validation testing throughout the expected range of measurements. Validation at the most distant ranges, where errors in biomass can approach 40%, is critical to ensure that an acceptable level of accuracy is maintained.

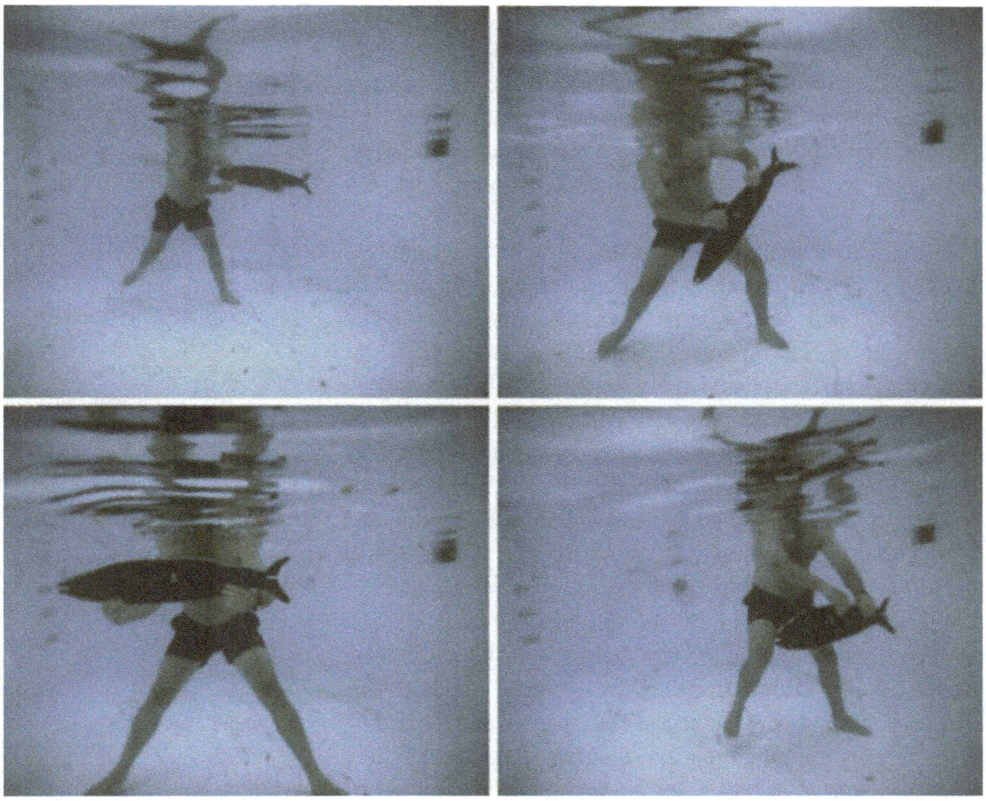

Fig. 2.9 Example of a fish silhouette validation in a swimming pool. (Courtesy of E.S. Harvey)

Table 2.3 A sample of some published results for the validation of underwater camera calibrations

Technique	Validation	Percentage error (%)
Absorption (Harvey and Shortis 1996)	Length measurement of silhouettes or rods throughout the volume	0.2–0.7
Lens distortion grid (Petrell et al. 1997)	Calliper measurements of Chinook Salmon	1.5
Absorption (Harvey et al. 2003)	Calliper measurements of Southern Bluefin Tuna	0.2
Perspective shift (Telem and Filin 2010)	Flat reference plate and straight-line reconstruction	0.4
Absorption (Menna et al. 2015)	Similarity transformation between above and below water networks	0.3
Radial lens distortion correction (Wehkamp and Fischer 2014)	Distances on checkerboard	0.9–1.5
Absorption (Boutros et al. 2015)	Length measurements of a rod throughout the volume	0.5
Perspective shift (Bräuer-Burchardt et al. 2015)	Flat reference plate and distance between spheres	0.4–0.7

2.6 Conclusions

This chapter has presented a review of different calibration techniques that incorporate the effects of refraction from the camera housing and the water medium. Calibration of underwater camera systems is essential to ensure the accuracy and reliability of measurements of marine fauna, flora or artefacts. Calibration is a key process to ensure that the analysis of biomass, population distribution or dimensions is free of systematic errors.

Irrespective of whether an implicit absorption or an explicit refractive model is used in the calibration of underwater camera systems, it is clear from the sample of validation results that an accuracy of the order of 0.5% of the measured dimensions can be achieved. Less favourable results are likely when approximate methods, such as 2D planar correction grids, are used. The configuration of the underwater camera system is a significant factor that has a primary influence on the accuracy achieved. The advantage of photogrammetric systems, however, is that the configuration can be readily adapted to suit the desired or specified accuracy.

Understanding all the complexities of calibration and applying an appropriate technique may be daunting for anyone entering this field of endeavour for the first time. The first consideration should always be the accuracy requirements or expectations for the underwater measurement or modelling task. There is a clear correlation between the level of accuracy achieved and the complexity of the calibration. If accuracy is not a priority then calibration can be ignored completely, with the understanding that there is a significant risk of systematic errors in any measurements or models. The use of 2D calibration objects is a compromise between accuracy requirements and the complexity of the calibration approach, but has gained popularity despite the potential for systematic errors in the measurements. At the other end of the scale, for the most stringent accuracy requirements, in-situ self-calibration of a high quality, high stability underwater camera system using a 3D object and an optimal network geometry is critical.

Lack of understanding of the interplay between calibration and systematic errors in the measurements can be exacerbated by 'black box' systems that incorporate an automatic assignment of calibration parameters. Systems such as *Agisoft Photoscan/Metashape* (2017) and *Pix4D* (2017) incorporate 'adaptive' calibration that selects the parameters based on the geometry of the network, without requiring any intervention by the operator of the software. Whilst the motivation for this functionality is clearly to aid the operator, and the operator can intervene if they wish, the risk here is that the software may tend to nominate too many parameters to minimize errors and achieve the 'best' possible result. The additional, normally redundant, terms for the radial and decentring distortion parameters will only exaggerate this effect in most circumstances. The over-parameterization leads to over-fitting by the least squares estimation solution, produces overly optimistic estimates of errors and precisions, and generates systematic distortions in the derived model.

Irrespective of the approach to calibration, however, validation of measurements is the ultimate test of accuracy. The very straightforward task of introducing a known object into the field of view of the camera(s) and measuring lengths at a variety of locations and ranges produces an independent assessment of accuracy. This is a highly recommended, rapid test that can evaluate the actual accuracy against the specified or expected level based on the chosen approach. The system configuration and choice of calibration technique can be modified accordingly for subsequent measurement or modelling tasks until an optimum outcome is achieved.

Essential further reading for anyone entering this field are a guide to underwater cameras such as the Underwater Photography Guide (2017) and practical advice on heritage recording underwater such as Green (2016, Chap. 6), and McCarthy (2014). A practical guide to the procedure for the calibration technique based on the 2D checkerboard given by Bouguet (2017) is provided by Wehkamp and Fischer (2014). For more information on the use of 3D calibration objects, see Fryer and Fraser (1986), Harvey and Shortis (1996), Shortis et al. (2000), and Boutros et al. (2015).

Acknowledgements This is a revised version based on a paper originally published in the online access journal *Sensors* (Shortis 2015). The author acknowledges the Creative Commons licence and notes that this is an updated version of the original paper.

References

Abdel-Aziz YI, Karara HM (1971) Direct linear transformation from comparator coordinates into object space coordinates in close-range photogrammetry. In: Proceedings of the symposium on close-range photogrammetry. American Society of Photogrammetry, Falls Church, VA, pp 1–18

Abdo DA, Seager JW, Harvey ES, McDonald JI, Kendrick GA, Shortis MR (2006) Efficiently measuring complex sessile epibenthic organisms using a novel photogrammetric technique. J Exp Mar Biol Ecol 339(1):120–133

Agisoft (2017) Agisoft Photoscan. http://www.agisoft.com/. Accessed 27 Oct 2017

Anderson RC (1982) Photogrammetry: the pros and cons for archaeology. World Archaeol 14(2):200–205

Baldwin RA (1984) An underwater photogrammetric measurement system for structural inspection. Int Arch Photogramm 25(A5):9–18

Balletti C, Beltrame C, Costa E, Guerra F, Vernier P (2015) Underwater photogrammetry and 3d reconstruction of marble cargos shipwreck. Int Arch Photogramm Remote Sens Spat Inf Sci XL-5/W5:7–13. https://doi.org/10.5194/isprsarchives-XL-5-W5-7-2015

Barnes H (1952) Underwater television and marine biology. Nature 169:477–479

Bass GF (1966) Archaeology under water. Thames and Hudson, Bristol

Bass GF, Rosencrantz DM (1977) The ASHREAH—a pioneer in search of the past. In: Geyer RA (ed) Submersibles and their use in oceanography and ocean engineering. Elsevier, Amsterdam, pp 335–350

Bouguet J (2017) Camera calibration toolbox for MATLAB. California Institute of Technology. http://www.vision.caltech.edu/bouguetj/calib_doc/index.html. Accessed 27 Oct 2017

Boutros N, Harvey ES, Shortis MR (2015) Calibration and configuration of underwater stereo-video systems for applications in marine ecology. Limnol Oceanogr Methods 13(5):224–236

Bräuer-Burchardt C, Kühmstedt P, Notni G (2015) Combination of air- and water-calibration for a fringe projection based underwater 3d-scanner. In: Azzopardi G, Petkov N (eds) Computer analysis of images and patterns. 16th international conference, CAIP 2015, Valletta, Malta, 2–4 September 2015 Proceedings. Part I: Image processing, computer vision, pattern recognition, and graphics, vol 9257. Springer, Basel, pp 49–60

Brown DC (1966) Decentring distortion of lenses. Photogramm Eng 22:444–462

Brown DC (1971) Close range camera calibration. Photogramm Eng 37(8):855–866

Bruno F, Bianco G, Muzzupappa M, Barone S, Razionale AV (2011) Experimentation of structured light and stereo vision for underwater 3D reconstruction. ISPRS J Photogramm Remote Sens 66(4):508–518

Bruno F, Gallo A, De Filippo F, Muzzupappa M, Petriaggi BD, Caputo P (2013) 3D documentation and monitoring of the experimental cleaning operations in the underwater archaeological site of Baia (Italy). Proceedings of Digital Heritage International Congress (DigitalHeritage), IEEE, vol 1. Institute of Electrical and Electronics Engineer, Piscataway, NJ, pp 105–112

Capra A (1992) Non-conventional system in underwater photogrammetry. Int Arch Photogramm Remote Sens 29(B5):234–240

Capra A, Dubbini M, Bertacchini E, Castagnetti C, Mancini F (2015) 3D reconstruction of an underwater archaeological site: comparison between low cost cameras. Int Arch Photogramm Remote Sens Spat Inf Sci XL-5/W5:67–72. https://doi.org/10.5194/isprsarchives-XL-5-W5-67-2015

Capturing Reality (2017) Reality capture. https://www.capturingreality.com/. Accessed 27 Oct 2017

Chong AK, Stratford P (2002) Underwater digital stereo-observation technique for red hydrocoral study. Photogramm Eng Remote Sens 68(7):745–751

Done TJ (1981) Photogrammetry in coral ecology: a technique for the study of change in coral communities. In: Gomez ED (ed) 4th international coral reef symposium. Marine Sciences Center, University of the Philippines, Manila, vol 2, pp 315–320

Doucette JS, Harvey ES, Shortis MR (2002) Stereo-video observation of nearshore bedforms on a low energy beach. Mar Geol 189(3–4):289–305

Drap P, Seinturier J, Scaradozzi D, Gambogi P, Long L, Gauch F (2007) Photogrammetry for virtual exploration of underwater archaeological sites. In: Proceedings of the 21st international symposium of CIPA, Athens, 1–6 October 2007, 6 pp

El-Hakim SF, Faig W (1981) A combined adjustment of geodetic and photogrammetric observations. Photogramm Eng Remote Sens 47(1):93–99

Fraser CS, Shortis MR, Ganci G (1995) Multi-sensor system self-calibration. In: El-Hakim SF (ed) Videometrics IV, vol 2598. SPIE, pp 2–18

Fryer JG, Fraser CS (1986) On the calibration of underwater cameras. Photogramm Rec 12(67):73–85

Fulton C, Viduka A, Hutchison A, Hollick J, Woods A, Sewell D, Manning S (2016) Use of photogrammetry for non-disturbance underwater survey—an analysis of in situ stone anchors. Adv Archaeol Pract 4(1):17–30

Granshaw SI (1980) Bundle adjustment methods in engineering photogrammetry. Photogramm Rec 10(56):181–207

Green J (2016) Maritime archaeology: a technical handbook. Routledge, Oxford

Green JN, Baker PE, Richards B, Squire DM (1971) Simple underwater photogrammetric techniques. Archaeometry 13(2):221–232

Green J, Matthews S, Turanli T (2002) Underwater archaeological surveying using PhotoModeler, VirtualMapper: different applications for different problems. Int J Naut Archaeol 31(2):283–292. https://doi.org/10.1006/ijna.2002.1041

Guo T, Capra A, Troyer M, Gruen A, Brooks AJ, Hench JL, Schmitt RJ, Holbrook SJ, Dubbini M (2016) Accuracy assessment of underwater photogrammetric three dimensional modelling for coral reefs. Int Arch Photogramm Remote Sens Spat Inf Sci XLI-B5:821–828. https://doi.org/10.5194/isprs-archives-XLI-B5-821-2016

Hale WB, Cook CE (1962) Underwater microcontouring. Photogramm Eng 28(1):96–98

Harvey ES, Shortis MR (1996) A system for stereo-video measurement of sub-tidal organisms. Mar Technol Soc J 29(4):10–22

Harvey ES, Shortis MR (1998) Calibration stability of an underwater stereo-video system: implications for measurement accuracy and precision. Mar Technol Soc J 32(2):3–17

Harvey ES, Fletcher D, Shortis MR (2001) Improving the statistical power of visual length estimates of reef fish: comparison of divers and stereo-video. Fish Bull 99(1):63–71

Harvey ES, Shortis MR, Stadler M, Cappo M (2002) A comparison of the accuracy of measurements from single and stereo-video systems. Mar Technol Soc J 36(2):38–49

Harvey ES, Cappo M, Shortis MR, Robson S, Buchanan J, Speare P (2003) The accuracy and precision of underwater measurements of length and maximum body depth of Southern Bluefin Tuna (*Thunnus maccoyii*) with a stereo-video camera system. Fish Res 63:315–326

Harvey ES, Fletcher D, Shortis MR, Kendrick G (2004) A comparison of underwater visual distance estimates made by SCUBA divers and a stereo-video system: implications for underwater visual census of reef fish abundance. Mar Freshw Res 55(6):573–580

Henderson J, Pizarro O, Johnson-Roberson M, Mahon I (2013) Mapping submerged archaeological sites using stereo-vision photogrammetry. Int J Naut Archaeol 42(2):243–256. https://doi.org/10.1111/1095-9270.12016

Hohle J (1971) Reconstruction of an underwater object. Photogramm Eng 37(9):948–954

Ivanoff A, Cherney P (1960) Correcting lenses for underwater use. J Soc Motion Picture Telev Eng 69(4):264–266

Jordt-Sedlazeck A, Koch R (2012) Refractive calibration of underwater cameras. In: Fitzgibbon A, Lazebnik S, Perona P, Sato Y, Schmid C (eds) Computer vision—ECCV 2012: 12th European conference on Computer Vision, Florence, Italy, 7–13 October 2012. Proceedings. Part I: Image processing, computer vision, pattern recognition and graphics, vol 7572. Springer, Berlin, Heidelberg, pp 846–859

Kenefick JF, Gyer MS, Harp BF (1972) Analytical self-calibration. Photogramm Eng Remote Sens 38(11):1117–1126

King BR (1995) Bundle adjustment of constrained stereo pairs—mathematical models. Geomatics Res Australas 63:67–92

Klimley AP, Brown ST (1983) Stereophotography for the field biologist: measurement of lengths and three-dimensional positions of free-swimming sharks. Mar Biol 74:175–185

Kotowski R (1988) Phototriangulation in multi-media photogrammetry. Int Arch Photogramm Remote Sens 27(B5):324–334

Kwon YH, Casebolt JB (2006) Effects of light refraction on the accuracy of camera calibration and reconstruction in underwater motion analysis. Sports Biomech 5(1):95–120

Lavest JM, Rives G, Lapresté JT (2000) Underwater camera calibration. In: Vernon D (ed) Computer vision—ECCV 2000: 6th European conference on Computer Vision Dublin, Ireland, 26 June–2 July 2000, Proceedings. Part I. Lecture notes in computer science, vol 1842. Springer, Berlin, pp 654–668

Leatherdale JD, Turner DJ (1983) Underwater photogrammetry in the North Sea. Photogramm Rec 11(62):151–167

Li R, Tao C, Zou W, Smith RG, Curran TA (1996) An underwater digital photogrammetric system for fishery geomatics. Int Arch Photogramm Remote Sens 31(B5):319–323

Li R, Li H, Zou W, Smith RG, Curran TA (1997) Quantitative photogrammetric analysis of digital underwater video imagery. IEEE J Ocean Eng 22(2):364–375

Maas H-G (2015) A modular geometric model for underwater photogrammetry. Int Arch Photogramm Remote Sens Spat Inf Sci XL-5/W5:139–141. https://doi.org/10.5194/isprsarchives-XL-5-W5-139-2015 2015

Mallet D, Pelletier D (2014) Underwater video techniques for observing coastal marine biodiversity: a review of sixty years of publications (1952–2012). Fish Res 154:44–62

Martínez A (2014) A souvenir of undersea landscapes: underwater photography and the limits of photographic visibility, 1890–1910. História, Ciências, Saúde-Manguinhos 21:1029–1047

McCarthy J (2014) Multi-image photogrammetry as a practical tool for cultural heritage survey and community engagement. J Archaeol Sci 43:175–185. https://doi.org/10.1016/j.jas.2014.01.010

McCarthy J, Benjamin J (2014) Multi-image photogrammetry for underwater archaeological site recording: an accessible, diver-based approach. J Marit Archaeol 9(1):95–114. https://doi.org/10.1007/s11457-014-9127-7

McLaren BW, Langlois TJ, Harvey ES, Shortland-Jones H, Stevens R (2015) A small no-take marine sanctuary provides consistent protection for small-bodied by-catch species, but not for large-bodied, high-risk species. J Exp Mar Biol Ecol 471:153–163

Menna F, Nocerino E, Troisi S, Remondino F (2013) A photogrammetric approach to survey floating and semi-submerged objects. In: Remondino F, Shortis MR (eds) Videometrics, range imaging, and applications XII, vol 8791. SPIE, paper 87910H

Moore EJ (1976) Underwater photogrammetry. Photogramm Rec 8(48):748–763

Murphy HM, Jenkins GP (2010) Observational methods used in marine spatial monitoring of fishes and associated habitats: a review. Mar Freshw Res 61:236–252

Newton I (1989) Underwater photogrammetry. In: Karara HM (ed) Non-topographic photogrammetry. American Society for Photogrammetry and Remote Sensing, Bethesda, pp 147–176

Petrell RJ, Shi X, Ward RK, Naiberg A, Savage CR (1997) Determining fish size and swimming speed in cages and tanks using simple video techniques. Aquac Eng 16(1–2):63–84

Phillips K, Boero Rodriguez V, Harvey E, Ellis D, Seager J, Begg G, Hender J (2009) Assessing the operational feasibility of stereo-video and evaluating monitoring options for the Southern Bluefin Tuna Fishery ranch sector. Fisheries Research and Development Corporation report 2008/44, 46 pp

Pienaar LV, Thomson JA (1969) Allometric weight-length regression model. J Fish Res Board Can 26:123–131

Pix4D (2017) Pix4Dmapper. https://pix4d.com/. Accessed 27 Oct 2017

Pollefeys M, Koch R, Vergauwen M, Van Gool L (2000) Automated reconstruction of 3D scenes from sequences of images. ISPRS J Photogramm Remote Sens 55(4):251–267

Pollio J (1971) Underwater mapping with photography and sonar. Photogramm Eng 37(9):955–968

Pollio J (1972) Remote underwater systems on towed vehicles. Photogramm Eng 38(10):1002–1008

Rende FS, Irving AD, Lagudi A, Bruno F, Scalise S, Cappa P, Montefalcone M, Bacci T, Penna M, Trabucco B, Di Mento R, Cicero AM (2015) Pilot application of 3d underwater imaging techniques for mapping *Posidonia oceanica (L.)* delile meadows. Int Arch Photogramm Remote Sens Spat Inf Sci XL-5/W5:177–181. https://doi.org/10.5194/isprsarchives-XL-5-W5-177-2015

Rosen S, Jörgensen T, Hammersland-White D, Holst JC (2013) DeepVision: a stereo camera system provides highly accurate counts and lengths of fish passing inside a trawl. Can J Fish Aquat Sci 70(10):1456–1467

Santana-Garcon J, Newman SJ, Harvey ES (2014) Development and validation of a mid-water baited stereo-video technique for investigating pelagic fish assemblages. J Exp Mar Biol Ecol 452:82–90

Santos MN, Gaspar MB, Vasconcelos P, Monteiro CC (2002) Weight-length relationships for 50 selected fish species of the Algarve coast (southern Portugal). Fish Res 59(1–2):289–295

Schewe H, Moncreiff E, Gruendig L (1996) Improvement of fish farm pen design using computational structural modelling and large-scale underwater photogrammetry. Int Arch Photogramm Remote Sens 31(B5):524–529

Sedlazeck A, Koch R (2012) Perspective and non-perspective camera models in underwater imaging—overview and error analysis. In: Dellaert F, Frahm J-M, Pollefeys M, Leal-Taixé L, Rosenhahn B (eds) Outdoor and large-scale real-world scene analysis: 15th international workshop on theoretical foundations of computer vision, Dagstuhl Castle, Germany, 26 June–1 July 2011, Lecture notes in computer science, vol 7474. Springer, Berlin, pp 212–242

Seiler J, Williams A, Barrett N (2012) Assessing size, abundance and habitat preferences of the Ocean Perch *Helicolenus percoides* using a AUV-borne stereo camera system. Fish Res 129:64–72

Shieh ACR, Petrell RJ (1998) Measurement of fish size in Atlantic salmon (*salmo salar* l.) cages using stereographic video techniques. Aquac Eng 17(1):29–43

Shortis MR (2012) Multi-lens, multi-camera calibration of Sony Alpha NEX 5 digital cameras. In: Proceedings on CD-ROM, GSR_2 Geospatial Science Research Symposium, RMIT University, 10–12 December 2012, paper 30, 11 pp

Shortis MR (2015) Calibration techniques for accurate measurements by underwater camera systems. Sensors 15(12):30810–30826. https://doi.org/10.3390/s151229831

Shortis MR, Beyer HA (1997) Calibration stability of the Kodak DCS420 and 460 cameras. In: El-Hakim SF (ed) Videometrics IV, vol 3174. SPIE, pp 94–105

Shortis MR, Seager JW (2014) A practical target recognition system for close range photogrammetry. Photogramm Rec 29(147):337–355

Shortis MR, Clarke TA, Robson S (1995) Practical testing of the precision and accuracy of target image centring algorithms. In: El-Hakim SF (ed) Videometrics IV, vol 2598. SPIE, pp 65–76

Shortis MR, Miller S, Harvey ES, Robson S (2000) An analysis of the calibration stability and measurement accuracy of an underwater stereo-video system used for shellfish surveys. Geomatics Res Australas 73:1–24

Shortis MR, Harvey ES, Abdo DA (2009a) A review of underwater stereo-image measurement for marine biology and ecology applications. In: Gibson RN, Atkinson RJA, Gordon JDM (eds) Oceanography and marine biology: an annual review, vol 47. CRC Press, Boca Raton, pp 257–292

Shortis MR, Seager JW, Williams A, Barker BA, Sherlock M (2009b) Using stereo-video for deep water benthic habitat surveys. Mar Technol Soc J42(4):28–37

Skarlatos D, Demesticha S, Kiparissi S (2012) An 'open' method for 3d modelling and mapping in underwater archaeological sites. Int J Herit Digit Era 1(1):1–24

Telem G, Filin S (2010) Photogrammetric modeling of underwater environments. ISPRS J Photogramm Remote Sens 65(5):433–444

Underwater Photography Guide (2017) Underwater Digital Cameras. http://www.uwphotographyguide.com/digital-underwater-cameras. Accessed 27 Oct 2017

Van Damme T (2015) Computer vision photogrammetry for underwater archaeological site recording in a low-visibility environment. Int Arch Photogramm Remote Sens Spat Inf Sci XL-5/W5:231–238. https://doi.org/10.5194/isprsarchives-XL-5-W5-231-2015

Watson DL, Anderson MJ, Kendrick GA, Nardi K, Harvey ES (2009) Effects of protection from fishing on the lengths of targeted and non-targeted fish species at the Houtman Abrolhos Islands. West Aust Mar Ecol Prog Ser 384:241–249

Wehkamp M, Fischer P (2014) A practical guide to the use of consumer-level still cameras for precise stereogrammetric in situ assessments in aquatic environments. Underw Technol 32:111–128. https://doi.org/10.3723/ut.32.111

Williamson JE (1936) Twenty years under the sea. Ralph T. Hale & Company, Boston

Zhang Z (2000) A flexible new technique for camera calibration. IEEE Trans PAMI 22(11):1330–1334

Ziemann H, El-Hakim SF (1983) On the definition of lens distortion reference data with odd-powered polynomials. Can Surveyor 37(3):135–143

Check for updates

Legacy Data in 3D: The Cape Andreas Survey (1969–1970) and *Santo António de Tanná* Expeditions (1978–1979)

3

Jeremy Green

Abstract

This chapter explores the significance of legacy data as a source of new information and the possibility of extracting new information from sources of information that were recovered before the advent of computers and the digital revolution. Since then, much of the emphasis has been directed towards gathering new information and there has been little emphasis on records that date back over 50 years. This chapter examines two examples: the first the Cape Andreas Expedition in Cyprus 1969–1970 and the other the *Santo António de Tanná* excavation 1977–1980. Both case studies are examined for the elements of photography that can be used to extract new information and how data, in the future, can be best be collected to suit these developments.

Keywords

Cape Andreas · Cyprus · Kenya · Legacy data · Portuguese frigate · *Santo António de Tanná* · Shipwreck survey

3.1 Introduction

This chapter underlines the significance of legacy data as an important source of new information. The legacy data described in this chapter was collected in the late 1960s and 1970s. This was a time before desktop computers and GPS, when underwater cameras were just becoming more available and the underwater archaeological world was in its infancy. It is interesting to remember that, in those days, locating underwater archaeological sites was exceedingly difficult. Position could only be determined close to shore where land transits were the most reliable method and to some extent still are today, although they suffer from a lack of permanency. Additionally, where a survey track was required, horizontal sextant angles was the cheapest, although by far the most difficult method to utilize. Once out of sight of land, there was nothing available to the archaeologist, other than various offshore commercial electronic positioning systems, such as HiFix and MiniRanger; well beyond the budget of most archaeological projects. Surveying underwater archaeological sites was also difficult. Essentially, survey work relied on trilateration or simple offset surveys using tape measures and there was almost no possibility to work in 3D as the only available calculating systems available, at least in the early 1970s, was the slide-rule and log tables. Photography in the field was also difficult. Film cameras could only take up to 36 pictures before they required reloading; processing and printing in the field was difficult, as a dark room with processing facilities and an enlarger were required. This was the environment where the legacy data described in this chapter was collected.

This chapter deals with two projects that the author was involved in and which have been selected to illustrate the processing of legacy data. The first project was at Cape Andreas, Cyprus, which was the first archaeological project the author directed. The objectives of this project were based on the author's previous experience working with George Bass at Yassıada in Turkey and later with Michael Katzev on the Kyrenia excavation. After Cape Andreas, the author came to the Western Australian Museum and conducted the excavation of the Dutch East India shipwreck *Batavia*. The processing of the legacy data from that shipwreck is the subject of a PhD thesis and will not be discussed here (McAllister 2018). In 1978–1979 the photographic survey of the Portuguese frigate *Santo António de Tanná*, wrecked in Mombasa harbour in 1697 was undertaken. The two projects will be discussed in more detail below; however, some background to the two projects is required. As the primary objective of Cape Andreas work was to locate and survey

J. Green (✉)
Department of Maritime Archaeology, WA Museum, Fremantle, WA, Australia
e-mail: jeremy.green@museum.wa.gov.au

J. K. McCarthy et al. (eds.), *3D Recording and Interpretation for Maritime Archaeology*, Coastal Research Library 31,
https://doi.org/10.1007/978-3-030-03635-5_3

underwater archaeological sites, it presented an opportunity to investigate and explore new techniques and technology. At that time the underwater swim-line survey technique had only recently been developed, and the *Admiralty Manual of Hydrographic Surveying* (Hydrographer of the Navy 1965) provided information on maritime survey techniques. An experimental underwater theodolite was constructed to try and improve underwater site surveying. Photographic techniques were investigated using the Nikonos camera with refraction-corrected lens, which had only just become available. Bass et al. (1967) had developed an underwater photomosaic system at Cape Gelidonya and Williams (1969) had published *Simple Photogrammetery*, which introduced a range of photogrammetric techniques that could be applied underwater. With this range of techniques, the Cape Andreas project was undertaken.

The Mombasa survey, on the other hand, was a much more specific project. The hull of the ship was uncovered during the two seasons of excavation, and the objective was to record this in order to produce a site plan. By the late 1970s, technology had progressed. Programmable calculators were available; the Nikonos camera now had a 20 mm underwater-corrected lens and the author had worked in Australia to develop a stereo-bar photo tower to record sites. These techniques were used to record the complex hull structure of the *Santo António de Tanná*.

As it turned out. both projects subsequently provided an opportunity to reassess the data. With the advent of computers, Geographical Information Systems (GIS), satellite imagery and programs that allowed the data to be reprocessed, the subject of this chapter turns to examine the data collection methodology, the reprocessing of the data and the outcomes. While much has been published on the recent use of underwater photogrammetry with digital cameras, the author has found no references to published work on retrospective or legacy photogrammetric analysis for maritime archaeology. This is surprising as it is an area with huge potential. This is now beginning to be recognized the field of archaeology (Wallace 2017) and palaeontology (Falkingham et al. 2014; Lallensack et al. 2015).

3.2 Cape Andreas Expeditions

In 1969 and 1970, the Oxford University Research Laboratory for Archaeology conducted two underwater archaeological survey expeditions to Cape Andreas, Cyprus (Fig. 3.1), to record underwater archaeological material including shipwreck sites and anchors. The sites were found using a swim-line search technique, and they were then surveyed and photographed. The results were the subject of two publications (Green 1969, 1971b). This material has lain dormant and only recently, with the advent of a number of computer-related techniques, has now been reassessed. The positions of the sites, although accurately recorded on topographical maps at the time, did not have geographical coordinates, making it almost impossible to relocate them in the future. Using the original data, it has been possible, with the use of satellite imagery and the Esri ArcGIS program, to precisely locate all the sites and attribute approximate geographical coordinates (latitude and longitude) to them, ensuring the possibility of relocation in the future (Fig. 3.2).

The possibility of revisiting the data for these sites is due to the fact that both of these early maritime archaeological expeditions featured experiments in underwater photogrammetric techniques, which at that time were in their infancy. The expeditions used the relatively new underwater Nikonos 35 mm camera with a 27 mm water corrected lens to create photomosaics and to record sites and objects. The photographic data has now been reprocessed using *Agisoft PhotoScan/Metashape* and has resulted in some remarkable 3D plans of the sites.

The author had been involved in the Cyprus Archaeological Underwater Survey Expedition (CAUSE) that had visited the Cape in 1967, with a team from The University Museum, Pennsylvania and the Oxford University Research Laboratory for Archaeology (Green et al. 1967), and as the area seemed to be promising for a future survey it was selected for the project. The main objective of the Cape Andreas expeditions was to survey the seabed around the Cape and Khlides Islands for wreck sites and other archaeological material. As the water clarity around the Cape often produced visibility of around 70 m, the survey planned to use a swim-line technique with divers swimming at a depth of around 20 m visually searching the seabed up to a depth of 50 m. The divers were spaced at regular intervals on a line so that adjacent divers could see the same area, thus ensuring the seabed was systematically searched.

As there were no detailed hydrographic charts of the Cape Andreas area, the first priority of the 1969 expedition was to produce a detailed chart of the Cape delineating the 50 m contour. To do this an echo sounder was used to measure the depth and the position of the survey vessel was recorded using horizontal sextant angles to stations on the islands and Cape. As there were no survey points on the chain of islands extending from the Cape, the most detailed plan, at that time, was an aerial photograph. Therefore, the survey work had to start from scratch. Using a theodolite, a series of prominent survey stations on the islands were established that could be seen from the sea. Once established, the survey vessel made a series of runs perpendicular to the shore recording the track of the vessel with the horizontal sextants. Each sextant 'fix' was marked on the sonar paper trace and subsequently the data transferred to the plan. This enabled an accurate plan of the depth con-

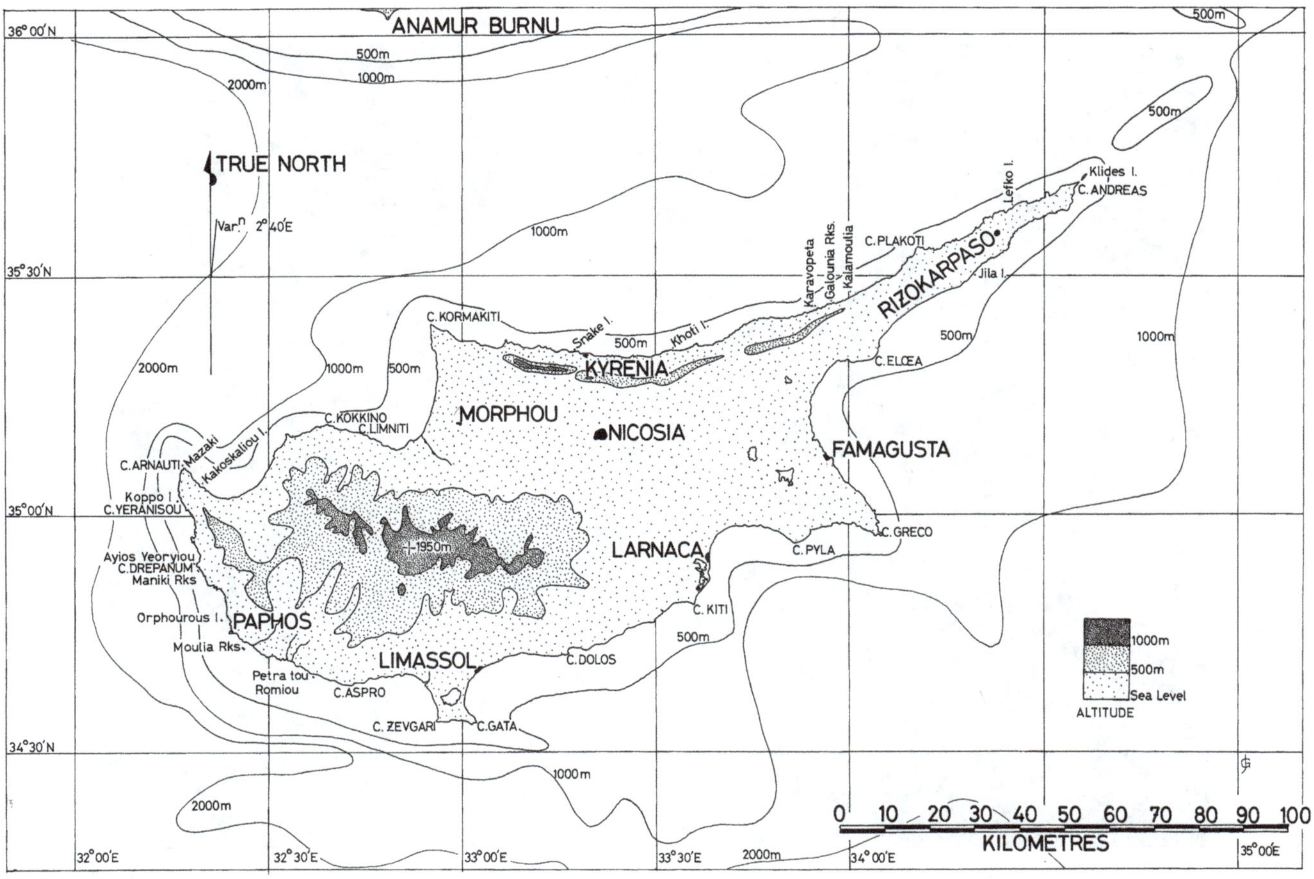

Fig. 3.1 Map of Cyprus showing Cape Andreas and the Khlides Islands

tours around the Cape and an estimate of the swim-line survey work that needed to be undertaken (see Fig. 3.3).

Once the vessel survey was completed, the swim-line surveys were undertaken, once again using horizontal sextant angles to plot the positions of the swim-lines. Different swim-line techniques were used in 1969 and 1970 and the results are shown in Fig. 3.4. Once a site was located, it was photographed and surveyed. At the large wreck sites, photographs were taken in order to create a photomosaic. To do this thin platted ski rope (selected because of its low stretch) marked at metre intervals, was laid out along the long axis of the site. This was used as a scale and to help the photographer ensure that the site was adequately covered. It was, by coincidence, this technique proved to be the most successful in processing the legacy data. The film was developed on-site. Images were printed and then manually laid up to create a photomosaic.

From the results of the 2 years surveys a large quantity of information was obtained from the swim-line work; this material was divided into three categories:

1. Wreck sites with ceramics, including material that may possibly be jettison;
2. Anchor sites-areas where anchors were closely associated; and
3. Individual anchors.

3.2.1 Wreck Sites with Ceramics

A total of ten pottery sites were located; some sites are little more than objects from spillage or jettison (Sites 1, 14 and 18). Sites 12 and 16; Sites 10 and 14; and Sites 17 and 24 had material that appears to be interrelated and it is difficult to decide whether the sites represent separate or associated events.

Site 12, on the north side of the island No. 4, is clearly a wreck site. It consists of an area approximately 20 × 15 m containing numerous heavily concreted Corinthian-style roof-tiles and cover-tiles. Figure 3.5 shows a hand-laid up photomosaic of the site and Fig. 3.6 shows a drawing of the distribution of the sherds.

Site 16, a few metres to the south of islands Nos 4 and 5, consisted of a scattered collection of concreted sherds of amphorae and tiles, together with a small concentration of small bowls and plates, many of which were intact. The tile sherds to the west of Tag 3 may represent material that has

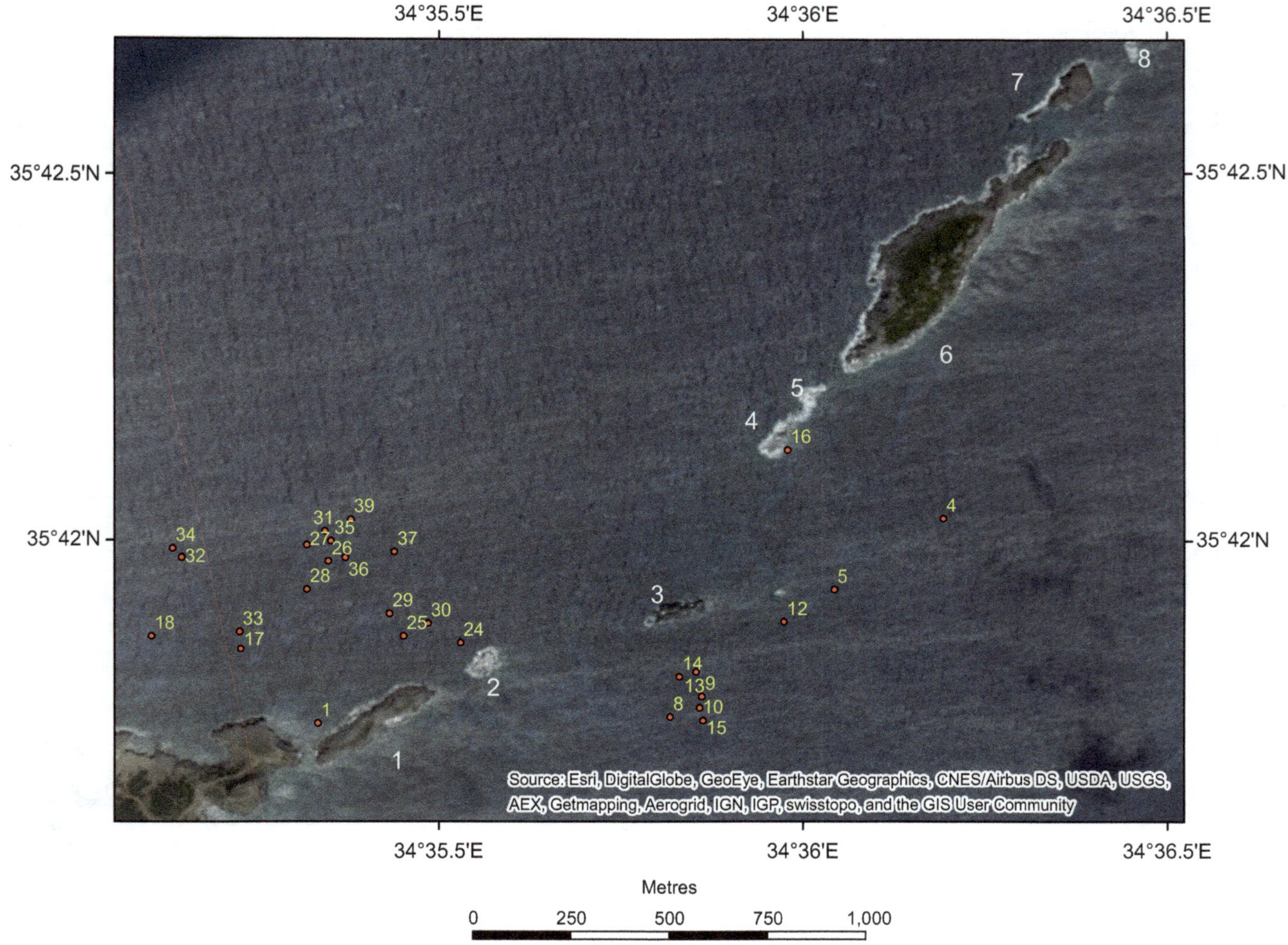

Fig. 3.2 Plan of Cape Andreas showing sites and latitude and longitude grid created from GIS

been washed over from the tile wreck, Site 12. It is difficult to establish if the two sites are associated and why such a large number (*c.* 25) of fine ware pottery objects should be concentrated, relatively undamaged, in such a small area.

Site 28 was located a week before the end of 1970 expedition. The superficial material lying at a depth of 20 m are Corinthian-style roof-tiles and cover-tiles (Fig. 3.7). The regular stacking indicated that the site was intact and represents the surface layer of cargo of a ship buried in a soft sand seabed. Two areas of tiles were noted. The larger area consisted of about 60 roof-tiles arranged in 4 rows, together with 8 cover-tiles; a further layer can be identified underneath these. The smaller area consists of about 15 tiles (Green 1971a).

Sites 17 and 24 lie around the north side of the second island; Site 24 was located by CAUSE in 1967 and was further investigated in 1969 (see Green 1969, Figs. 11 and 13).

Site 10 was located in 1969; it lies to the north of the rocks between the third and fourth islands. It is one of the most difficult sites to analyse, as the material is spread out over a large rocky area, 50 × 20 m. A variety of amphora sherds of different types and periods have been noted and recorded (Green 1969, Figs. 7 and 8). The team members constructed a large-scale mosaic of the site in order to try to produce a detailed plan. The initial impression is that this site represents spillage or jettison from several periods, rather than from several wrecks.

Site 1 consists of several looped handles and flat amphora bases concreted to the rocks around the southwest corner of the first island. In view of the small number of amphorae, this site can only represent jettison or spillage.

Site 18 consists of looped handles and pointed feet of amphorae scattered over an area of 300 sq. m. It is situated to the north of the small group of rocks off the south side of the Cape. In view of the considerable amount of pottery (and beer bottles) in the area it is likely that this has been used as a lee shore by ships since antiquity. The site therefore probably represents jettison of damaged amphorae from a ship sheltering and waiting for favourable winds.

Site 14 lies on the south side of the rocks mentioned with reference to Site 10. The site consists of a few tiles and amphorae, possibly jettison or spillage. It is surprising that out of the ten sites described, five consist mainly of looped-handle

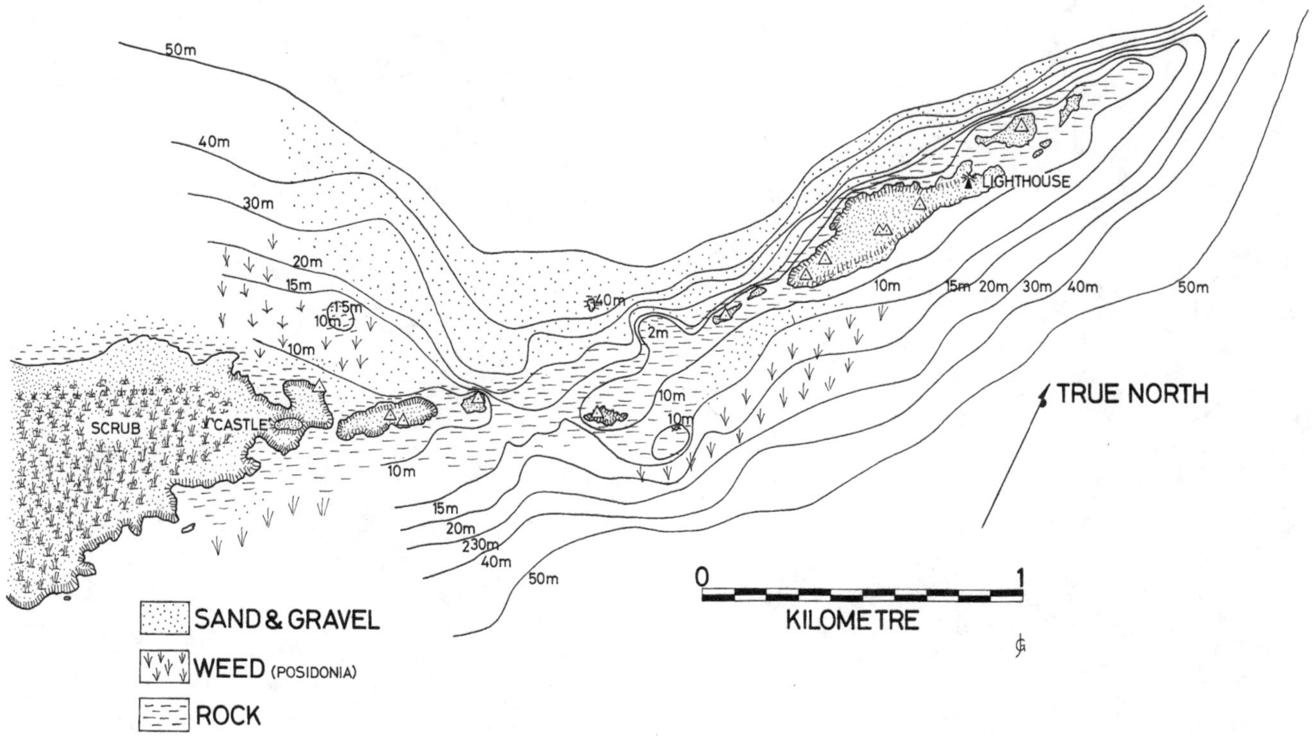

Fig. 3.3 Hydrograph chart of Cape Andreas produced from survey conducted in 1969

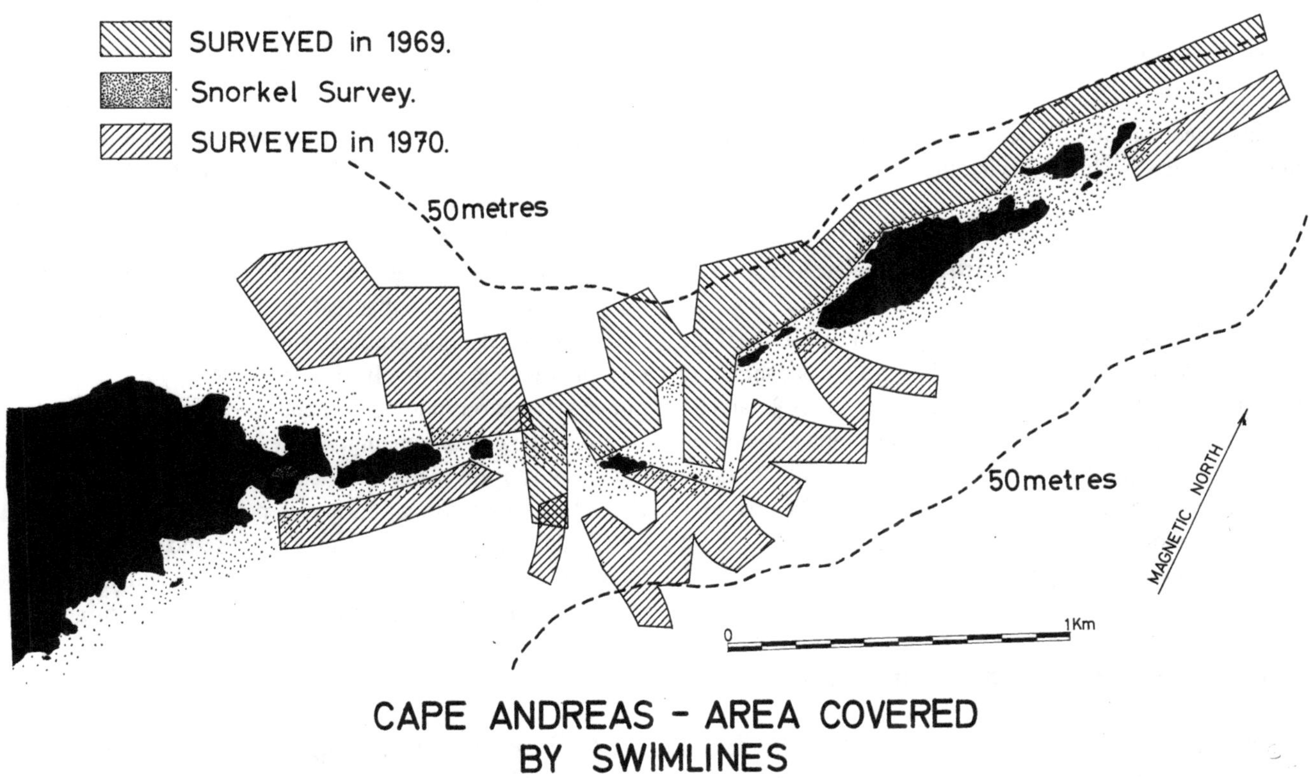

Fig. 3.4 Swim-line surveys of 1969 and 1970

Fig. 3.5 Photomosaic of Site 12 produced in 1970

amphorae and four are tile sites. Sites 12 and 28 are clearly complete wrecks of ships carrying tiles as a cargo.

3.2.2 Anchor Sites and Individual Anchors

Four anchor sites were recorded on the north side of the Cape Andreas. Two of the sites (23 and 26) were areas containing a large number of different types of anchors. Sites 23 and 26 were located at the point where the gently upward-sloping sand seabed changes to a steep rock cliff face. Site 23 has a total of 28 anchors: 18 iron, 8 lead and 2 stone. The positions of a total of about 50 anchors were recorded, but only about half were recorded photogrammetrically.

3.2.3 Reworking the Legacy Survey Data

As mentioned, in the late 1960s, surveying was limited to optical systems. As was typical in those days, relative position was accurate, but absolute position, in normal circumstances, was almost impossible to obtain. When accurate GPS first became available, most hydrographic charts needed to be corrected to conform to the absolute information. The plans produced in the 1960s, although accurate, therefore, could not be given precise latitude and longitude or be applied with any certainty to accurate modern maps or charts and could only be applied to large-scale Admiralty charts.

Using ArcGIS it was possible to georeference the plans produced in the 1960s that had used the outline features on the early aerial photograph. Using the World Map in Arc GIS, it was then relatively simple to identify coastline features on the aerial photograph and the World Map and thus complete the georeferencing. As the survey stations had been transferred to the plans these could then be located on the GIS. With this information it was possible to place all the survey data from the 1960s on the GIS and attribute them relatively accurate latitude and longitude coordinates. Thus, all the sites now have geographic coordinates with an estimated position accuracy of about ±5 m (see Fig. 3.2).

3.2.4 Reworking the Legacy Photographic Data

A total of 69 black and white 35 mm films were taken during the expeditions representing about 1700 images. The images were assessed for suitability for processing using *Agisoft*

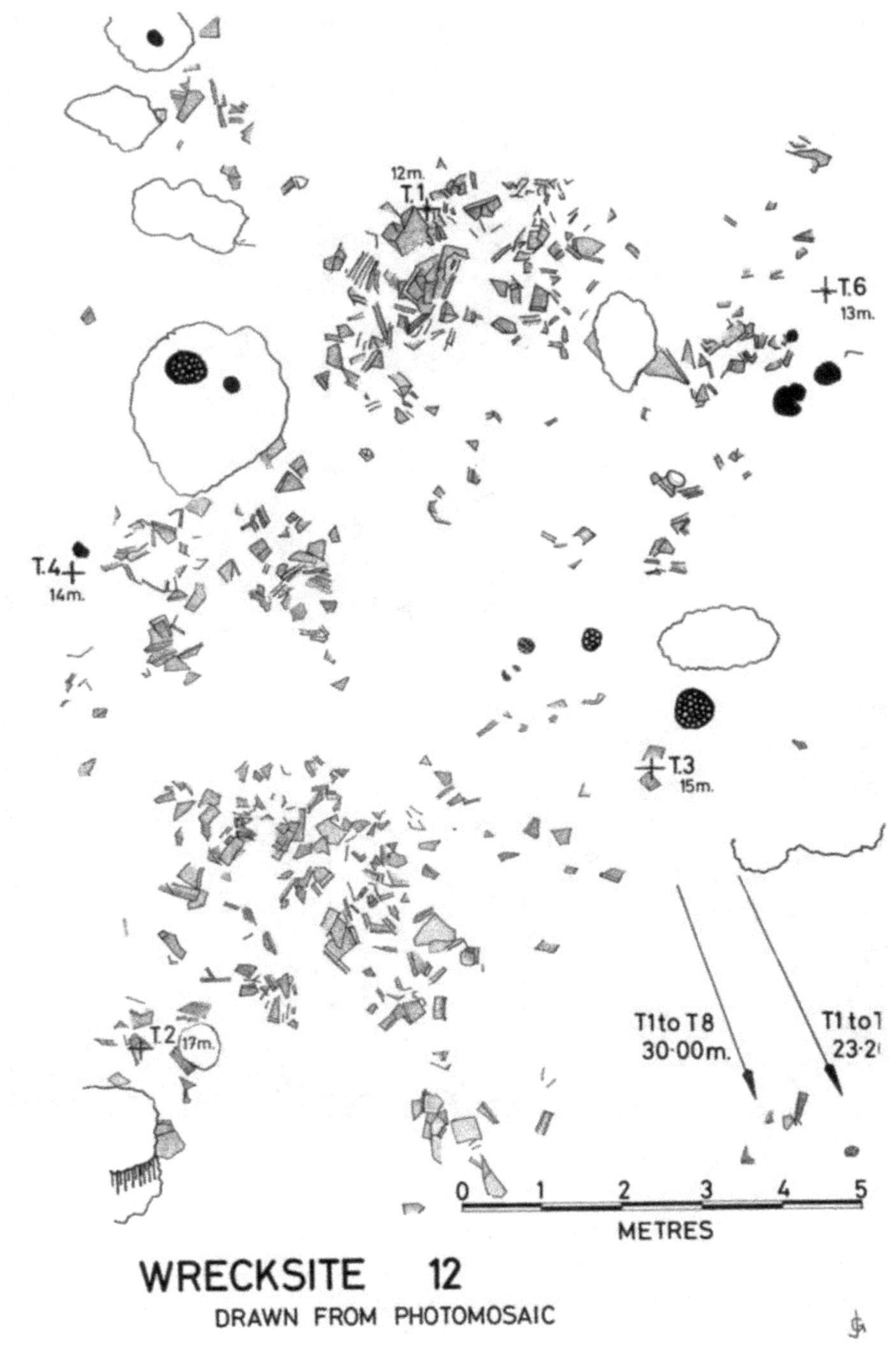

Fig. 3.6 Drawing taken from the photomosaic showing the ceramic material

PhotoScan/Metashape software. Initially, photos that were obviously unsuitable were rejected. This left photos that were mosaics of large sites and groups of photos of single or multiple objects such as anchors or ceramics.

3.2.5 Agisoft PhotoScan/Metashape

The images from Site 12 were selected first. Any image with a grid frame was rejected, as the grid frames were moved around the site and, thus, made the alignment for *Agisoft PhotoScan/Metashape* difficult (the option of masking the frames in *Agisoft PhotoScan/Metashape* was decided to be unnecessarily burdensome due to the large number of photographs). There were 99 images in the data set that had originally been used to construct the photomosaic. These were run through a high-end workstation with 4 X7560 Intel® Xeon 2.26 GHz CPUs and 512 GB of RAM. The *Agisoft PhotoScan/Metashape* settings were at the highest possible resolution. The alignment took 20 min resulting initially with 25 cameras out of the 99 aligning, giving 54,347 tie points and a 3D model with around 9.5 million faces

Fig. 3.7 Site 28

Fig. 3.8 PhotoScan/Metashape ortho-photograph of Site 12

(Fig. 3.8). The 25-camera chunk was then isolated and the program was re-run with the remaining 74 images. The second process aligned a further 23 cameras with 40,246 points, taking 1 h to build dense cloud and an hour to create the mesh (Fig. 3.8).

A similar method was used on Site 10 producing a good quality 3D image of the site (Fig. 3.9). Attempts to produce 3D images of the anchors, however, were generally unsuccessful partially because there were not enough photographs from different angles and in general the photographs had a 3D grid frame included in the view that disrupted the processing.

The well-known problem with *Agisoft PhotoScan/ Metashape* is that the program is a 'black box' and running the program on the same set of data produces different results on different occasions. In addition, there are many settings that can produce slightly different results. For the Cape Andreas material, a number of different models were produced. In general, the photomosaic runs without grid frames were the most successful in converting to 3D visualizations, however, the masking feature in *Agisoft PhotoScan* is yet to be tested on this material.

3.3 The *Santo António de Tanná* Shipwreck

In 1978 and 1979, photographic recording of the Portuguese wreck of the *Santo António de Tanná* (1697) was undertaken (Piercy 1976, 1977, 1978a, b, 1979a, b, 1981 & Sassoon 1982). The project, under the auspices of the National Museums of Kenya and the Institute for Nautical Archaeology, involved the excavation of the ship that lay about 50 m from shore under the walls of Fort Jesus, Mombasa, Kenya. The frigate *Santo António de Tanná* had been dispatched by the

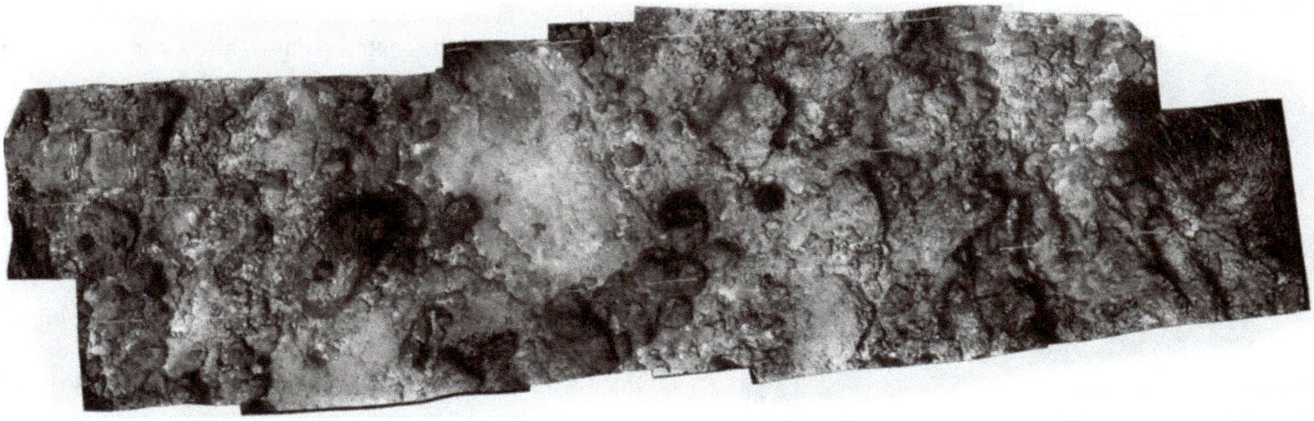

Fig. 3.9 Site 10 mosaic PhotoScan/Metashape

Portuguese Viceroy in Goa to relive the fort that was under siege by the Omanis. On arrival in 1697 the General anchored the frigate in front of the fort and was informed that all the Portuguese in the fort were dead and that about 25 Swahili men and 60 women were left defending the fort. The fort was immediately relieved, however, some time later the vessel broke its moorings, drifted onto the shore and sank (Fraga 2007; Killman 1974; Killman and Bentley-Buckle 1972). The fort finally fell to the Omanis in 1698 after a 3.5 year siege.

During the excavation of the site in 1978 and 1979 photographic recording was undertaken during periods of good visibility (Green 1978). This situation corresponded with the High Water Spring Tides, which brought clear oceanic water into the river that normally had low visibility (*c*. 2–3 m compared with 10–15 m during the High Water Springs).

The surveying techniques used to record the structure of the ship uncovered during the second season were based on the experience of the first season and were devised to meet the rather peculiar conditions of the site. Since it was required to produce detailed 3D plans, both photogrammetry and standard measurement recordings were used. The nature of the site, however, produced limitations in the application of both techniques. Poor visibility, except at high spring tides, precluded the constant use of photographic recording. Likewise, poor visibility and tidal currents made tape measurements unreliable and it was difficult to establish an accurate baseline for recording purposes. Additional problems were encountered because of the peculiar orientation of the ship, which lay on a steep slope with its bow inclined 20° down the slope and with a lateral tilt of 54° to port. The keelson was twisted along its exposed length, and there was evidence at the scarf joint that the stern section, including the keelson, had moved in relationship to the bow. These distortions have been extremely difficult to sort out because of the unusual orientation and the lack of a useful datum, such as the base of the keel to work from.

The overall shape of the structure was recorded by measuring profiles at 1 m intervals across the hull of the ship at right angles to the keelson. A stereo-photogrammetric survey over the whole of the inside of the ship was made so that detailed information of the internal structure could be recorded. Control points were put in place and surveyed so that the photogrammetric survey could be related to the profiles and incorporated in the overall plans. Detailed measurements were also made of the keelson, which served as the base line for the survey. Fraga (2007) produced a plan of the site based on the tape measurements taken in 1978–1979 and related this to the contemporary seventeenth and eighteenth-century Portuguese naval architecture texts of Lavanha (1610) and Oliveira (1578–1581).

3.3.1 Profile Recording

The profiles were measured at 1 m intervals along the keelson using a circular 0.75 m diameter, 360° protractor, graduated in half degrees and mounted on a bar measuring 1 m in length (Fig. 3.10). The bar was clamped to the upper (starboard) side of the keelson so that the plane of the protractor was at right angles to the keelson. A pin mounted at the centre of the protractor acted as a swivel for a 0.5 m length of thin string with a stirrup at the end, through which a 10 m survey tape was threaded. The angle and the true distance were recorded against the record number. Using this technique, it was possible to work in poor visibility (*c*. 1 m) even when the two operators were out of sight of each other. The overall method was fast and reasonably accurate. In a 61 min dive, with experience, it was possible to do two 6 m profiles consisting of a total of 80 readings. Using a small hand calculator (Hewlett Packard 25), the polar co-ordinates were

Fig. 3.10 Circular protractor used to record profiles

converted to rectangular (or x and y) co-ordinates, which greatly facilitated the plotting of data. The results, when converted to rectangular co-ordinates, were plotted on graph paper. The accuracy of the technique was basically governed by the size of the protractor and was about ±2%.

3.3.2 Trilateration Survey

The survey of the control points was carried out by trilateration. The upper starboard edge of the keelson was selected as the baseline. The control consisted of tags driven into the keelson and port and starboard extremities of the site at 2 m intervals. A 2 m rod was clamped against, and at right angles to the upper edge of the keelson opposite each of the keelson controls. Measurements were made to the three nearest controls on both the port and starboard sides of the site, from the base of the rod and at the keelson, and the mark 2 m above the keelson. Thus, for example, if the rod was at the 4 m keelson mark, 16 measurements were made from the base and top of the rod, 6 to the port 2, 4 and 6 marks, 6 to the starboard 2, 4 and 6 marks and 2 to the keelson 2 mark and 2 to the keelson 6 mark. The offsets of the control marks on the keelson to the rod were measured, and the angle of the rod to the true vertical was measured using a carpenter's level. With this information, it was, at the time, still not possible to calculate the 3D coordinates of all the control points. This was to come later, but at the time we were aware that there were ways to do this and rather futile attempts were made using the programmable calculator.

3.3.3 Photographic Recording

Two Nikonos III cameras were mounted 0.5 m apart on an aluminium stereo bar. The cameras were adjustable and provided with screws so that the vertical and horizontal tilt of the camera could be adjusted. Two targets, with viewing holes through their centres were mounted 0.5 m apart on a levelled bar at about 5 m from the levelled stereo bar. Using plane mirrors in place of the lenses, the stereo bar was adjusted so that the image of the target in the mirror, when viewed through the corresponding target, coincided with the centre of the optical axis of the fixed camera. This enabled the cameras to be adjusted so that the optical axes were accurately parallel and perpendicular to the stereo bar. The stereo bar was then mounted on the photo tower to be used underwater and the cameras remained on the bar for the whole of the survey. At the end of each underwater session, the bar was removed and taken ashore, but the cameras remained on the bar and the film could be extracted from the camera without disturbing the camera positions.

The photo tower consisted of a 2 m^2 base graduated in 0.1 m intervals, with stays supporting a 2 m bar, 1.88 m vertically above the base (Fig. 3.11). The bar was constructed so that, when the stereo bar was fitted to it, the optical axes of the camera lenses lay on the centre line of the base square, equidistant about the centre of the grid. Fine adjustments were made using the mirror system to set the photographic plane of the stereo bar parallel to the plane of the grid frame. With this arrangement, there was a photographic overlay over the whole of the 2 m grid frame, the 15 mm lens had, in fact a focal length in water of 20.8 mm.

Table tennis balls on 0.5 m of white string were attached to the mid-points of each side of the grid frame. As these floated upright in water, it was possible to determine the orientation of the camera plane to the true vertical. The orientation of each pair of stereo photographs could be determined for reference purposes by extending the line of the strings to the nadir point. The line joining the nadir point to the principal point of the photograph, gave the direction of the true vertical, and the ratio of the length of this line to the effective focal length of the photograph gave the tangent of the angle of the true vertical to the camera axis. In practice, however, the small current that was always present prevented this from being effective.

Fig. 3.11 Phototower on site showing slope of site

Photographic exposures were made by moving the tower at 1 m intervals, thus ensuring good end-lap between the stereo pairs although it was difficult to ensure that the adjacent runs had good overlap. Using two plastic buckets for buoyancy the photo tower could be moved around quite easily by one person, although for accurate positioning two people were required. Great care had to be taken not to stir up sediment, particularly as most photography was carried out at high water when there was little current. Considerable variation in the quality of photographs was noted over the three High-Water Spring Tide periods when this photogrammetric coverage was carried out. This was due partially to the variation in the quantity of suspended matter in the water and also to the light level caused by the effects of clouds and the time of day. In many cases the coverage was a compromise, particularly as the best visibility conditions, corresponding to the highest tides, fell just after dawn or just before sunset, when the light levels were too low for good results. Under normal circumstances, using FP4 rated at 400 ASA and given twice normal development in D76, exposures ranged from f2.8 at 1/30 to f5.6 at 1/60.

Considerable planning was required in preparation for the High-Water Spring Tide periods when this type of photogrammetry was possible. The timbers had to be cleaned of any silt, sand and other material that collected as a natural result of the muddy water and stray discharge from the airlifts. If possible, airlifting was terminated about an hour before high-water, and the timbers were then brushed to remove the fine silt. No other diving was carried out during the photographic runs to prevent disturbance of the fine silt in other areas on the site. The photomosaic was created by printing the images and then gluing them down on a paste board (Fig. 3.12).

3.3.4 Agisoft PhotoScan/Metashape

The most significant problem with processing the *Santo António de Tanná* material in *Agisoft PhotoScan/Metashape* has been the presence of the photo tower and table tennis balls. Because the tower was placed on the interior surface of the hull, the orientation of the tower, in relation to the Cartesian coordinates of the hull of the ship, was different in each photo-pair. *Agisoft PhotoScan/Metashape*, thus, has the problem that in each photo there is a tower frame in exactly the same position and a view of the hull in a random orientation. Work on the material has proceeded over the years since the introduction of *Agisoft PhotoScan/Metashape* in 2010 and results have slowly improved. A major breakthrough occurred thanks to a series of photographs taken using the stereo bar by itself without the tower. On one particularly clear day, the bar was swum at a high altitude over the site and a series of stereo photographic pairs were recorded. These were run through *Agisoft PhotoScan/Metashape* and a good 3D model was obtained (Fig. 3.13), however because of the height, the resolution is not particularly good.

Currently a project at Curtin University's HIVE has just been completed where the photographs and survey information were combined to develop a high-resolution orthophotograph (Fig. 3.14) and a DEM (Fig. 3.15). Using the masking technique in *Agisoft PhotoScan/Metashape*, the tower and the table tennis balls were removed, and a high-quality mesh has been achieved, although with some holes in the coverage (see Shaw 2018).

In addition, the stereo pairs were processed to create individual models that were stitched together to produce a different approach to obtaining a 3D model. Essentially the objective of the project is to discover a method of processing

Fig. 3.12 Handmade photomosaic

Fig. 3.13 Perspective view of 3D model created in PhotoScan/Metashape

stereo photographic coverage to produce a 3D model. This will have enormous implications for legacy photography and a method of reassessing excavations.

3.4 Conclusions

The ultimate question is what does the 3D visualization of legacy data do for the archaeologist? It is obvious that visualization of a site in 3D is interesting and has a considerable significance in presenting the underwater archaeological world to the public. It remains less clear, however, what the implications are for the archaeological world. One significant issue is the ability to obtain an orthomosaic of a site, which compared with the hand-produced photomosaic—made by laying up paper prints of images and matching them—is a significant improvement in accuracy. Working with 2D prints of a site with any significant 3D component is always a compromise. The ability to produce an othomosaic and, thus, create a plan that is geometrically correct is

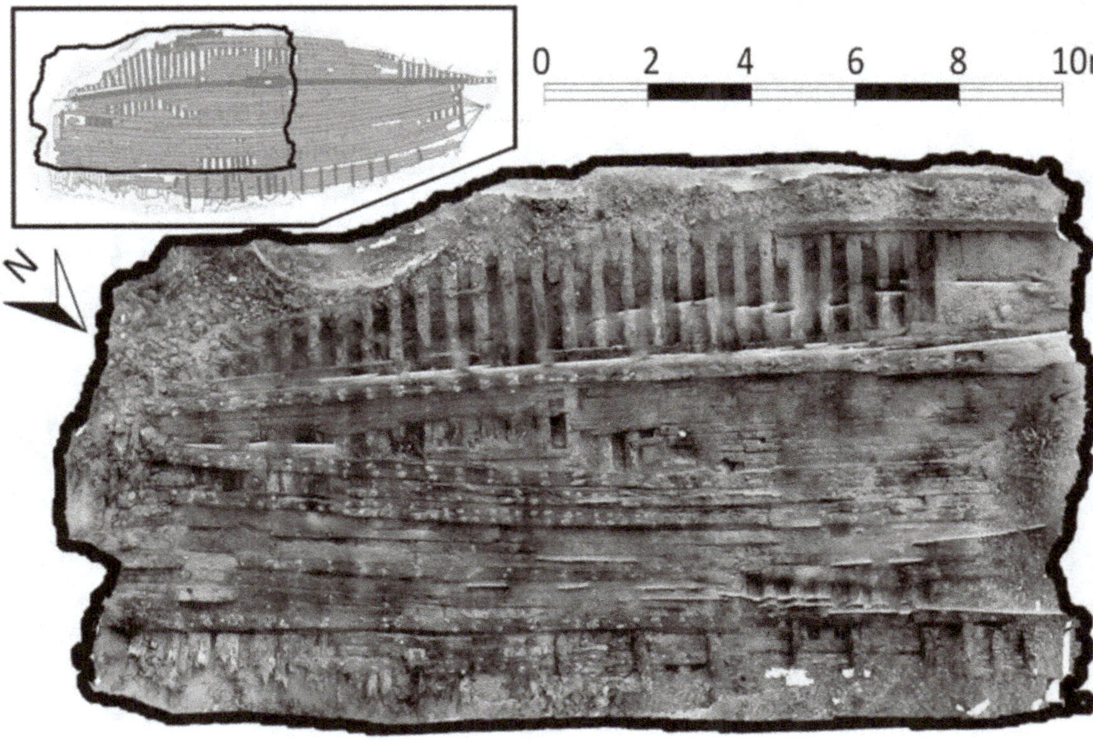

Fig. 3.14 High resolution orthophotograph created in PhotoScan/Metashape by HIVE

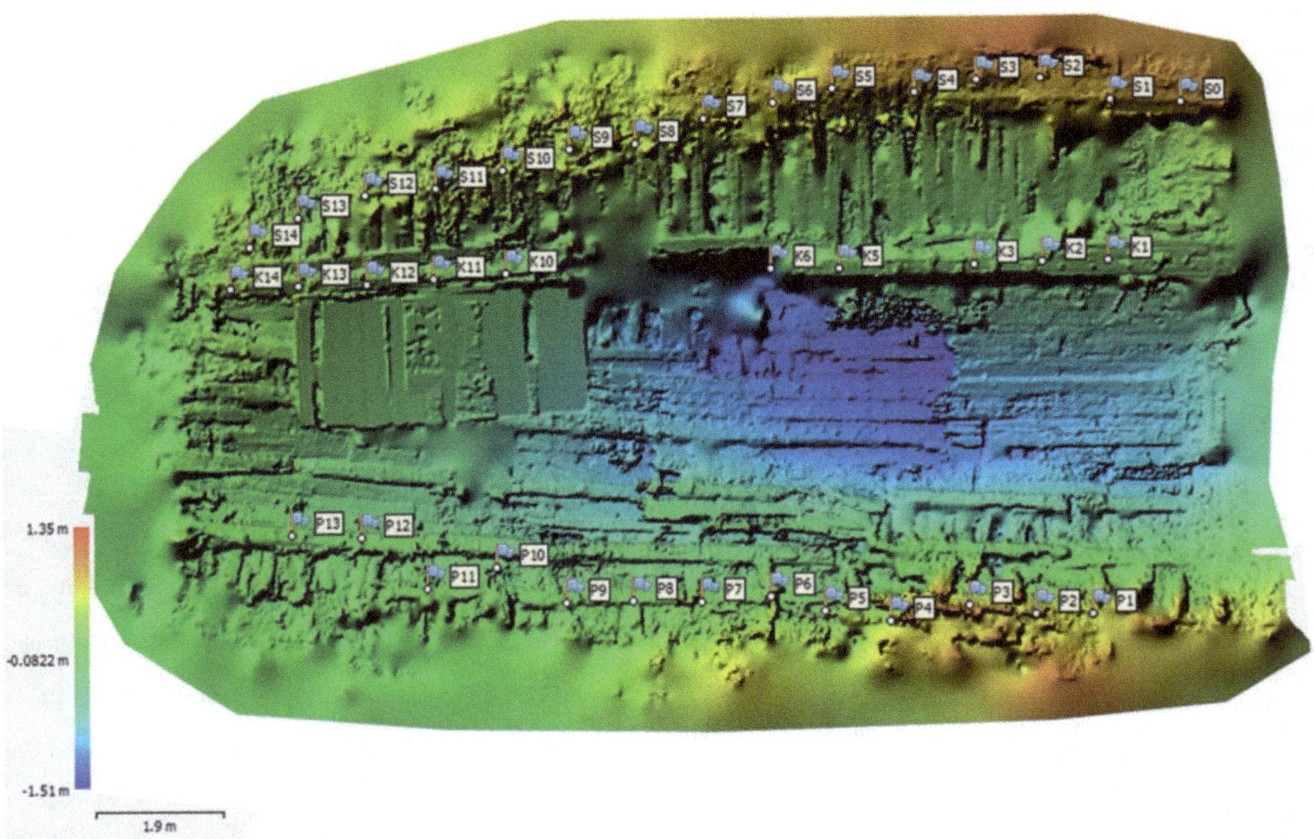

Fig. 3.15 A Digital Elevation Model (DEM) showing control points produced by HIVE

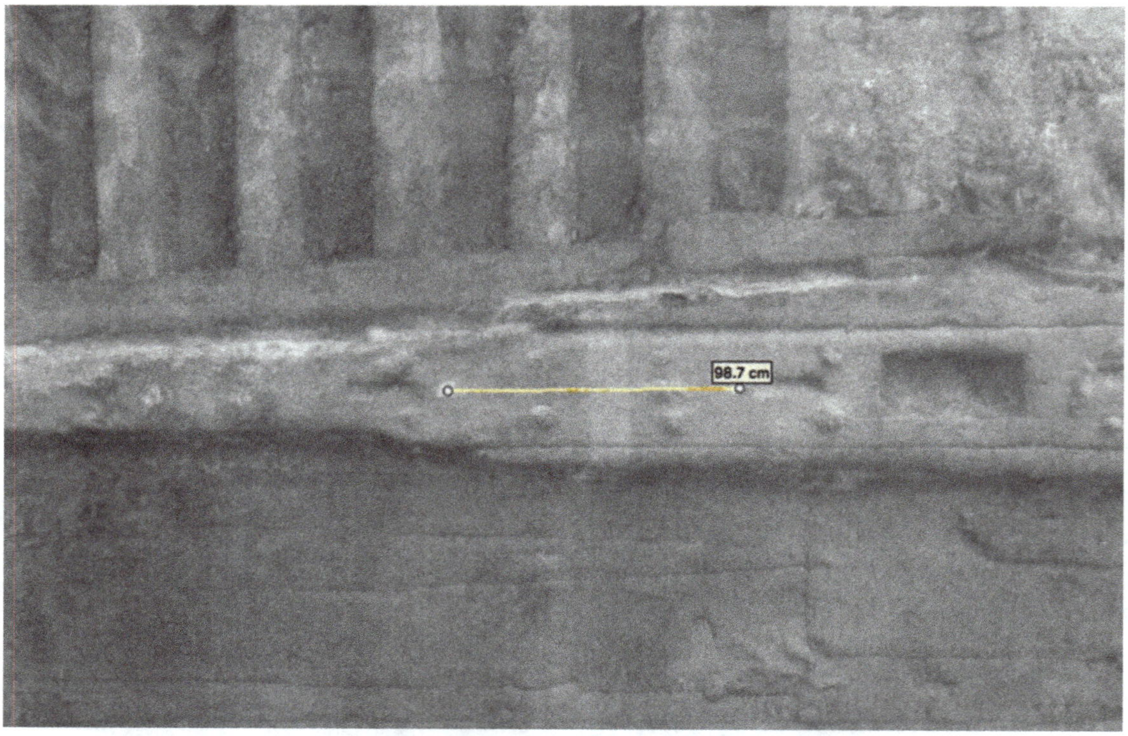

Fig. 3.16 Detail of PhotoScan/Metashape model of *Santo António de Tanná* keelson showing 3D measurement between two control points

significant in the interpretation of the site, particularly as most site photographs have scales included. The orthophotograph can easily be scaled as the survey lines are marked in metre intervals thus providing an overall site scale. It is thus possible to make a count, catalogue and measure the artefacts on the site, something that would be almost impossible with the paper-based photomosaic.

The question of 3D measurement of a site is more complicated. In the case of Cape Andreas, the sites were relatively flat so 3D measurements were less important. The situation with the *Santo António de Tanná* is much more interesting. The 3D model of the site is surprisingly detailed and enables almost any measurement from the site to be obtained. For example, the control points on the keelson, 1 m apart is shown in Fig. 3.16 as 98.7 cm, this, considering that the photography was taken 40 years ago, is quite remarkable. This particular aspect of accuracy in a 3D model is currently under further investigation and will be the subject of a later report.

Acknowledgements The author would like to acknowledge the information supplied by Lachlan Shaw, the Curtin University intern, who worked on the *Santo António de Tanná* project as part of a grant from the Australian Research Council Linkage Grant *Shipwrecks of the Roaring 40s* (LP130100137). Also, Andrew Woods, Petra Helmholtz, David Belton and Joshua Hollick from Curtin University HIVE who supervised this project. I would like to thank Robin Piercy, the director of the *Santo António de Tanná* project for providing the photographic record of the site. Last, thanks goes to Patrick Baker and the Cape Andreas team for their help and support.

References

Bass GF, Throckmorton P, Taylor JP, Hennessy JB, Shulman AR, Buchholz H-G (1967) Cape Gelidonya: a Bronze Age shipwreck. Trans Am Philos Soc 57(8):1–177. https://doi.org/10.2307/1005978

Falkingham PL, Bates KT, Farrow JO (2014) Historical photogrammetry: bird's paluxy river dinosaur chase sequence digitally reconstructed as it was prior to excavation 70 years ago. PLoS One 9(4):e93247

Fraga TM (2007) Santo Antonio de Tanná: story and reconstruction. PhD dissertation, Texas A&M University

Green JN (1969) The research laboratory for archaeology Cape Andreas expedition 1969. Research Laboratory for Archaeology, Oxford

Green JN (1971a) Cape Andreas: a survey of a tile wreck. Prospezioni Archaeologiche 6:141–178

Green JN (1971b) An underwater archaeological survey of Cape Andreas, Cyprus 1969–1970: a preliminary report. In: Blackman DJ (ed) Maritime Archaeology: proceedings of the 23 Symposium 1971 of the Colston Research Society held in the University of Bristol, 4–8 April. Butterworths, London

Green JN (1978) The survey procedure. Int J Naut Archaeol 7(4):311–314

Green JN, Hall ET, Katzev ML (1967) Survey of a Greek shipwreck off Kyrenia. Cyprus Archaeometry 10:47–56

Hydrographer of the Navy (1965) Admiralty manual of hydrographic surveying, vol 1. Hydrographer of the Navy, London

Killman J (1974) Fort Jesus a Portuguese fortress on the E. African coast. Clarendon Press, Oxford

Killman J, Bentley-Buckle AW (1972) A Portuguese wreck off Mombasa, Kenya. Int J Naut Archaeol 1(1):153–157. https://doi.org/10.1111/j.1095-9270.1972.tb00687.x

Lallensack JN, Sander PM, Knötschke N, Wings O (2015) Dinosaur tracks from the Langenberg Quarry (Late Jurassic, Germany) reconstructed with historical photogrammetry: evidence for large theropods soon after insular dwarfism. Palaeontol Electron 18(2):1–34

Lavanha JB (1610) Livro Primeiro da Architectura Naval. Original edition, Lisboa, 1610

McAllister M 2018 'Seeing is Believing': investigating the influence of photogrammetric digital 3D modelling of underwater shipwreck sites on archaeological interpretation. PhD dissertation, University of Western Australia

Oliveira F (1578–1581) O Livro da Fábrica das Naus. Lisbon

Piercy RCM (1976) The Mombasa shipwreck. Inst Naut Archaeol Newsl 3(3):1–5

Piercy RCM (1977) Mombasa wreck excavation: preliminary report, 1977. Int J Naut Archaeol 6(4):331–347. https://doi.org/10.1111/j.1095-9270.1977.tb01033.x

Piercy RCM (1978a) Mombasa wreck excavation: second preliminary report, 1978. Int J Naut Archaeol 7(4):301–319. https://doi.org/10.1111/j.1095-9270.1978.tb01080.x

Piercy RCM (1978b) The 1978 season at Mombasa. Inst Naut Archaeol Newsl 5(4):1–5

Piercy RCM (1979a) Mombasa wreck excavation: third preliminary report, 1979. Int J Naut Archaeol 8(4):303–309. https://doi.org/10.1111/j.1095-9270.1979.tb01135.x

Piercy RCM (1979b) Mombasa. Inst Naut Archaeol Newsl 6(3):8

Piercy RCM (1981) Mombasa wreck excavation: fourth preliminary report, 1980. Int J Naut Archaeol 10(2):109–118. https://doi.org/10.1111/j.1095-9270.1981.tb00020.x

Sassoon H (1982) The sinking of the *Santo Antonio de Tanna* in Mombasa harbour. Paideuma 28:101–108

Shaw LR (2018) The photogrammetric analysis of the *Santo Antonio de Tanna*. http://museum.wa.gov.au/maritime-archaeology-db/maritime-reports/photogrammetric-analysis-santo-antonio-de-tanna. Accessed 6 Aug 2018

Wallace CAB (2017) Retrospective photogrammetry in Greek archaeology. Stud Digit Herit 1(2):607–626

Williams JCC (1969) Simple photogrammetry: plan-making from small-camera photographs taken in the air, on the ground or underwater. Academic, London

4 Systematic Photogrammetric Recording of the Gnalić Shipwreck Hull Remains and Artefacts

Irena Radić Rossi, Jose Casabán, Kotaro Yamafune, Rodrigo Torres, and Katarina Batur

Abstract

In September 1967 an important shipwreck site was discovered near the islet of Gnalić in Northern Dalmatia (Croatia). It immediately raised significant interest in the scientific community and the broader public. Due to logistical and financial issues, the excavation ceased after five short-term rescue research campaigns, over a total duration of 54 working days. Renewed interest in the site, particularly the hull remains, resulted in reviving the project after 45 years. The trial campaign, carried out in 2012, had a positive outcome, and the excavation has continued annually in a systematic way. The nature of the site demanded significant effort to document the excavated areas. Considering all the temporal restrictions caused by various reasons, photogrammetry proved to be an extremely helpful and efficient tool.

Keywords

Artefacts · Croatia · Gnalić shipwreck · Hull remains · Photogrammetry · Underwater recording

4.1 Introduction

During the past decade, photogrammetry has rapidly developed from a sophisticated skill practiced by a small group of devoted experts with appropriate equipment and special software, to a broadly available tool, which can be undertaken with few restrictions regarding educational level, professional background or virtual-modelling experience. Applying the photogrammetric process throughout a demanding ongoing underwater archaeological excavation has allowed the authors to experience the benefits and limitations of the photogrammetric recording of cargo, equipment and hull remains, and to exploit its positive features.

Since its beginning, the Gnalić project team consisted of experts with significant underwater research experience, young researchers devoted to the application of new technologies, and enthusiastic students and volunteer divers. This diverse group conducted a series of experiments which targeted the development of an efficient photogrammetric recording system. The goals were to produce a seamless integration into traditional photographic recording, improved monitoring of the excavation process, and ultimately the production of enhanced images and material for public outreach. After five consecutive years of site experience, the authors report the outcome of this operation, in order to share experience and recommend best practices for the scientific community, to enhance the recording and processing steps from the perspective of the various levels of end users, and ultimately to improve underwater photogrammetric recording results.

It is important to emphasize that the implementation of the photogrammetric recording process was not the consequence of pre-planned systematic activity on the shipwreck site, but resulted from spontaneous positive collaboration between multiple team members with various expertise. They invested their time and effort into finding optimum solutions within the framework of a project with an extremely limited budget, fully exploiting the advantages of the

I. Radić Rossi (✉) · K. Batur
Department of Archaeology, University of Zadar, Zadar, Croatia
e-mail: irradic@unizd.hr

J. Casabán
Institute of Nautical Archaeology, College Station, TX, USA
e-mail: jlcasaban@tamu.edu

K. Yamafune
A.P.P.A.R.A.T.U.S. LLC, Yonago City, Totori Prefecture, Japan

R. Torres
Centro de Investigaciones del Patrimonio Costero (CIPAC-CURE), Maldonado, Uruguay
e-mail: rodrigo.torres@cure.edu.uy

J. K. McCarthy et al. (eds.), *3D Recording and Interpretation for Maritime Archaeology*, Coastal Research Library 31,
https://doi.org/10.1007/978-3-030-03635-5_4

development of broadly available equipment and software coupled with the creative atmosphere in the field. After 5 years, photogrammetry became one of the indispensable components of our recording methodology, not replacing but complementing the traditional organization of the underwater research.

4.2 The Shipwreck of Gnalić

The shipwreck of Gnalić, was officially found in 1967. Local divers relocated the site in the early 1960s, but it was only revealed to local authorities in 1967. It is one of the most important sixteenth-century shipwreck sites discovered to date. Besides the variety of cargo containers and cargo of various origins, precisely dated guns from the Alberghetti workshop, and the well-preserved portion of the hull, its importance is reflected in hundreds of archival documents, which clearly define its cultural, historical, social, economic and political context.

The shipwreck belongs to a group of sites in Croatian waters which were partly salvaged in the past. Unfortunately, for many wrecks found in the decades following the 1960s, both public opinion and responsible institutions considered a set of short rescue campaigns sufficient. However, the outcome of the Gnalić project, revived after 45 years, has clearly demonstrated the opposite. The importance and history of the ship also exceeded all expectations.

4.2.1 History of Research

The initial report of this important discovery led to immediate action by Ivo Petricioli, professor at the Department of Archaeology and Art History of the Faculty of Humanities and Social Sciences in Zadar, to rescue the ship's cargo. Three rescue campaigns were conducted in 1967 and 1968 (Petricioli and Uranija 1970), with an additional two campaigns in 1972 and 1973 (Petricioli 1981; Božulić 2013; Radić Rossi et al. 2016). In 1973, the Italian art historian Astone Gasparetto (1973) proposed an identification of the ship based on the archival research conducted at the State Archive of Venice.[1] According to Gasparetto, the shipwreck remains corresponded to that of *Gagiana* (*Gaiana*, or *Gagliana*), sunk 'in the waters of Murter' or 'in the waters of Biograd (Zara Vecchia)' in autumn 1583.

After years of neglect, Zdenko Brusić attempted to restart the excavation in 1996, but his attempt was unsuccessful due to administrative issues. Nevertheless, his attempt resulted in a comprehensive summary of what was known from the previous underwater research based on the old documentation (Brusić 2006: 78, Fig. 2). Then, in 2004, the range of glass from the ship's cargo was published (Lazar and Wilmott 2006), and a colourful overview of the most attractive finds, targeted at the general public and fund raising (Mileusnić 2004).

Finally, a partial excavation of the shipwreck took place in 2012, providing a unique opportunity to verify the archaeological potential of the site, which proved decisively high (Radić Rossi et al. 2013).[2] Systematic research began in 2013, and through October 2017 has encompassed approximately 200 m^2 of ship remains and seabed examination.

4.2.2 The Ship

For decades, knowledge about the ship relied mostly on the cargo items, a modest selection of underwater sketches and photographs, several recovered elements of the hull preserved in the Local Heritage Museum of Biograd na Moru (Beltrame 2006), and Gasparetto's proposed identification. This situation has drastically changed since 2012 through the systematic examination of both archaeological and historical sources.

4.2.2.1 Historical Documents

The identification of the vessel proposed by Gasparetto in 1973 has been fully confirmed by the archival research conducted in parallel with the renewed excavation efforts (Radić Rossi et al. 2013).[3] The heavily loaded merchantman sunk at the islet of Gnalić in early November 1583 was in fact *Gagliana grossa*—a merchantman with a capacity of 1200 Venetian barrels (Ven. *botti*), i.e. around 700 tons (Lane 1934; Tucci 1967), and an estimated length of 35–40 m.

The ship was built in Venice in 1567–1569, and successfully launched. The Ottomans captured the vessel near Valona, in Albania, during the War of Cyprus (1570–1573), and it remained in Ottoman hands for 10 years. In 1581, it was sold in Pera, Constantinople, to the Christian merchant Odoardo da Gagliano. Following the usual trading route between Venice and Constantinople, the ship sunk in 1583 near the islet of Gnalić, in northern Dalmatia, while carrying a valuable cargo for the Sultan Murat III (Radić Rossi et al.

[1] Although Gasparetto examined just the notarial archive of Catti out of over 50–60 notaries active in Venice at the time (personal information M. Bondioli), he managed to trace important information. Gasparetto based his choice on the work of Tenenti (1959).

[2] The support was provided by the Ministry of Culture of the Republic of Croatia, through the engagement of Josip Belamarić and Zlatko Uzelac. The Center for Maritime Archaeology and Conservation (CMAC) of Texas A&M University and the City of Biograd na Moru have also supported the project.

[3] Since 2012 Mariangela Nicolardi and Mauro Bondioli have conducted systematic research in the State Archive of Venice, which started from Gasparetto's presumption, and has confirmed the identification of the ship multiple times (Radić Rossi et al. 2013: 75–88).

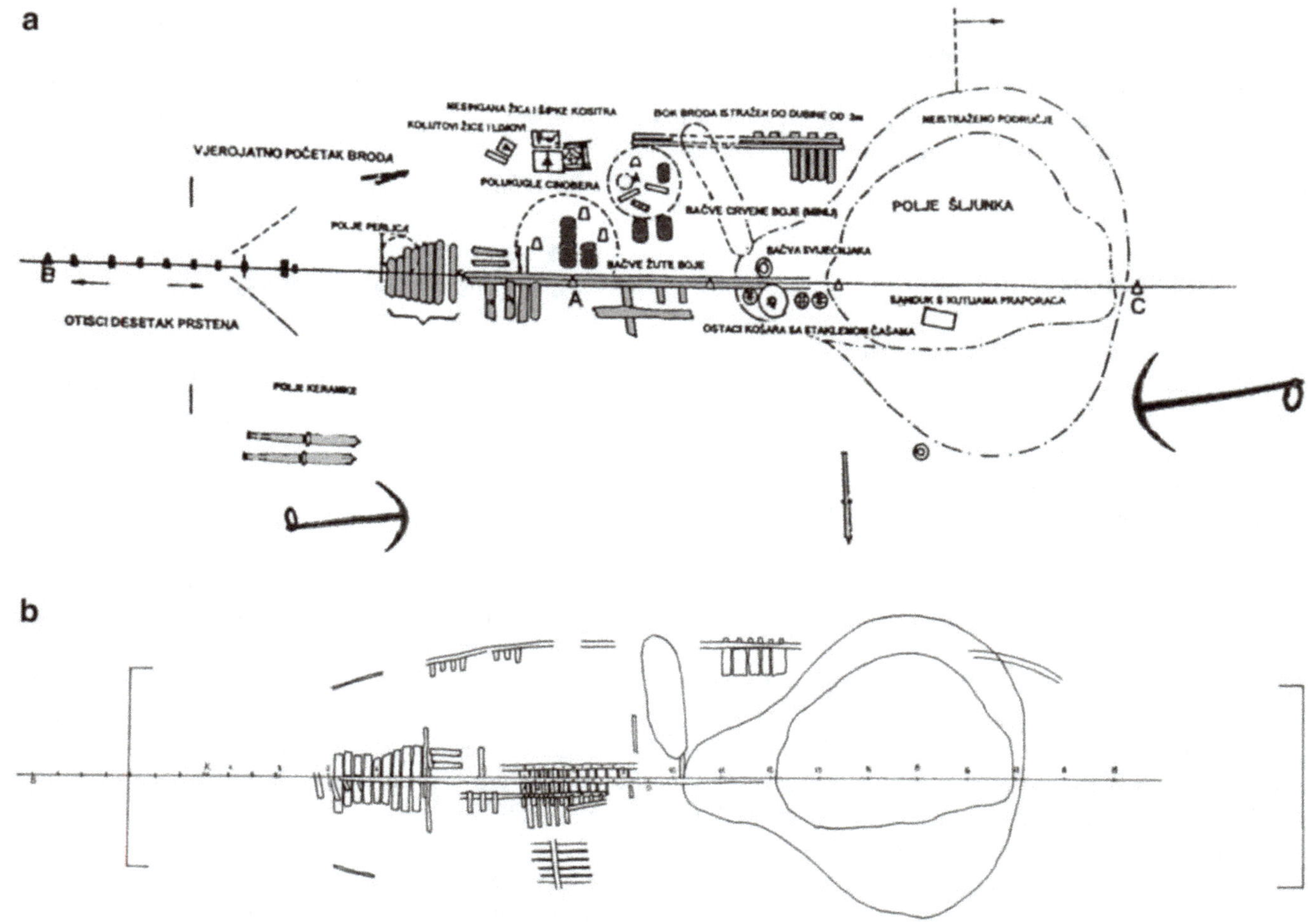

Fig. 4.1 (**a** and **b**) The two sketches indicating (**a**) the main groups of finds noticed during the rescue operations 1967–1973 and 1996; and (**b**) the wooden elements of the hull noticed in the trenches, and the surface layer (Z. Brusić), with the position of the hull remains shown bottom right on drawing (**b**)

2013: 86). The most precious part of the cargo was salvaged between December 1583 and February 1584, and the remaining part and the ship's hull lay undisturbed on the bottom for almost 400 years.

4.2.2.2 Archaeological Sources

In the 1990s, Zdenko Brusić reviewed the documentation from the previous Gnalić shipwreck interventions, and consolidated all of the archaeological information available from the plans, sketches and reports (Fig. 4.1a, b). Based on the presumption that the ship lay on its keel, with its bow to the west, and considering the armament items previously recovered, he suggested the interpretation of the ship as a heavy merchant galley (Brusić 2006: 80). Gasparetto, however, had already challenged this mind by 1973, and in 2006 Carlo Beltrame, relying on information provided by Mauro Bondioli and earlier images of the site, determined that the ship sunk at Gnalić was not a galley but a round ship.[4]

The recent archaeological work has also added more detail to our understanding of the remains of the Gnalić shipwreck. Excavation in the western part of the site allowed for the examination of exposed elements of the hull, which led to the conclusion that the keel area should be identified along the northern extremity of the site. The missing deadwood in the stern area had left a gap that was tentatively identified as a big crack, but this interpretation was corrected after excavation of the broader surface and correct identification of the keel.

The area covered by the surface finds, estimated from sidescan sonar and sub bottom profiler survey results,[5] measures approximately 15 × 60 m, with a maximum thickness of 1.5 m of sediment above the hull (1350 m^2 in total). The six excavation campaigns to date, including the test campaign carried out in 2012, lasted in total 330 days, with 200 m^2 of the site exposed. The complete recovery of the

[4]On the other hand, Beltrame (2006: 93) concluded that the ship was preserved below the waterline, which turned out to be incorrect.

[5]The survey was executed by the Department of Geology of the University of Patras, Greece, under the direction of George Papatheodorou.

artefacts and detailed cleaning of the hull has been completed for 140 m²—the excavation of the remaining surface area will be carried into future archaeological campaigns.

The Gnalić shipwreck documentation encompasses both traditional recording procedures and photogrammetric recording of the excavation progress. In accordance with the aim of this publication, the following text presents the experience of the photogrammetric recording of the Gnalić shipwreck excavation from 2012 to 2016 through the description of data acquisition and processing, and it discusses the advantages and limitations as experienced by experts with various responsibilities within the research team.

4.3 Systematic Photogrammetric Recording of Site and Finds

Considering the importance of the Gnalić shipwreck, the project's demanding underwater research conditions, limited financial resources and consequently small time frame, and the ever-present threat of losing information, the recording process had to be extremely efficient. The documentation process focused on:

1. Developing an accurate site plan, based on multi-layered information;
2. Recording the advancement of the excavation during each campaign;
3. Combining the results of each excavation campaign;
4. Continuously mapping the spatial distribution of finds before recovery;
5. Accurate recording of the hull;
6. Documenting material for future scientific presentation of the research; and
7. Producing material for public promotion of the project.

The initial Gnalić site recording process in 2012 followed the traditional system of tagging, measuring, drawing, photographing and video recording. Relying on the direct experience of the photogrammetric recording of the Late Roman shipwreck of Pakoštane, Croatia, carried out within the framework of an international project directed by the University of Zadar and *Centre National de la Recherche Scientifique* (CNRS)—Centre Camille Jullian, France,[6] and its contribution to the documentation process (Dumas 2012), it was apparent that photogrammetry should not be omitted from the organization of the underwater work. Even with that prior experience, it was challenging to ensure everything was available to guarantee the correct execution of the operation. The first attempt of photogrammetric assessment of the site, during the 2012 trial campaign, however, was extremely encouraging and resulted in photogrammetry becoming an essential part of the recording procedure.

The rapid development of the software, which became readily available and increasingly user friendly, combined with the effort in developing and testing the recording system, resulted in what is reported in the following text.

4.3.1 Trial Campaign 2012

The Gnalić shipwreck excavation restarted in 2012 in the form of a trial campaign, which lasted just 10 working days.[7] The team consisted of 15 divers, who each spent 30 min twice a day on the seabed. The area chosen for the trial excavation was a transversal cross section of the ship in the western part of the site—an area that had been previously excavated. This choice was influenced by the need to check the state of the preservation of the hull in an area that had been exposed previously and where all artefacts already would have been recovered.

It would be easy to focus on inadequate documentation inherited from the past projects, and the un-systematic assessment of the excavated areas—instead, it should be stressed that the first researchers did an excellent job of preserving the site and its finds for the future. They worked under completely different conditions, without any experience, logistics and expertise in underwater archaeological excavation. Yet, they demonstrated in fieldwork reports the highest level of awareness of the importance of the shipwreck site and a strong desire to systematically study all of its components (Radić Rossi et al. 2013: 70, 73). Therefore, everything accomplished in the restarted Gnalić project carries out respect for what was done in the past and is an attempt to realize the dream of those pioneering researchers.

At the beginning of the campaign, a metal grid, composed of seven squares each 2 × 2 m, was positioned across the site, and the excavation started simultaneously from its northern and southern edge. It should be stressed that it is usually thought that the main function of the grid is accurate documentation. This may have been true during the early development of underwater recording, but today its main function certainly exceeds documentation issues. Based on extensive fieldwork experience, with much effort invested in the training of students and amateur scientific divers, two main purposes significantly justify the positioning of the solid grid

[6] The photogrammetry of the Pakoštane shipwreck was carried out by Vincent Dumas and Philipe Grosscaux.

[7] The comprehensive duration of the campaign was 30 days, but it encompassed the assessment of the old finds and documentation in the Local Heritage Museum in Biograd na Moru. The operation was co-directed by the Department of Archaeology of the University of Zadar, represented by Irena Radić Rossi, and the Nautical Archaeology Program of Texas A&M University, represented by Filipe Castro.

Fig. 4.2 The photogrammetric model of the part of the hull exposed in 2012 at the northern edge of the excavation area (P. Drap)

over large, delicate surfaces. First, the grid provides solid support for divers—they can rest on it while working without moving finds or disturbing the site regardless of the task they are undertaking (excavating, photographing, filming, sampling, recovering artefacts, etc.). Another important function of the grid is the orientation of the diver, i.e. reducing to a minimum any possible confusion of their assigned work area.

It was observed in 2012, however, that the grid presents a serious obstacle to accurate photogrammetric recording if the target area exceeds the surface of a single square. Therefore, during all the photogrammetric recording operations in the following years the grid was removed from the excavated areas before capturing photos, and repositioned after the operations were completed.

The trial photogrammetric recording in 2012 was led by Pierre Drap from CNRS' Laboratory of Information Systems and Technology (LSIS) he and his team joined the excavation for 3 days. The more complex methodology and sophisticated computer programs applied at that time are no longer in use (P. Drap, personal communication). This initial photogrammetric recording was mainly targeted to demonstrating what can be relatively quickly and easily obtained through the systematic photographic recording of the excavated area (Fig. 4.2), or bigger areas in relation to the distribution of the surface finds (Fig. 4.3).

As discussed, the main goal of the trial excavation campaign was to check the presence of the cargo items and the state of preservation of the wooden elements of the hull in order to verify the need to restart the project. Therefore, it did not focus on the accurate cleaning and recording of the hull structure, which in any case would have been impossible in just ten working days. Thus, photogrammetry was the best solution for quick data collection and visualization of the situation on the seabed.

4.3.2 Research Campaign 2013

During the 2013 excavation campaign, a team led by José L. Casabán conducted the 3D photogrammetric recording of the remains of the Gnalić shipwreck covering a total surface of 300 m^2 at an average depth of 25 m (Fig. 4.4). The methods applied included the trilateration of a network of control points and the photographic coverage of the shipwreck remains (Casabán et al. 2014). The images were later aligned using *Agisoft PhotoScan/Metashape* to produce a dense point cloud and a mesh, representing the

Fig. 4.3 Photogrammetric model of the situation on the western half of the site in 2012, comprising a mound of ballast stones, a row of barrels filled with intensely red hematite powder, and exposed wood (P. Drap)

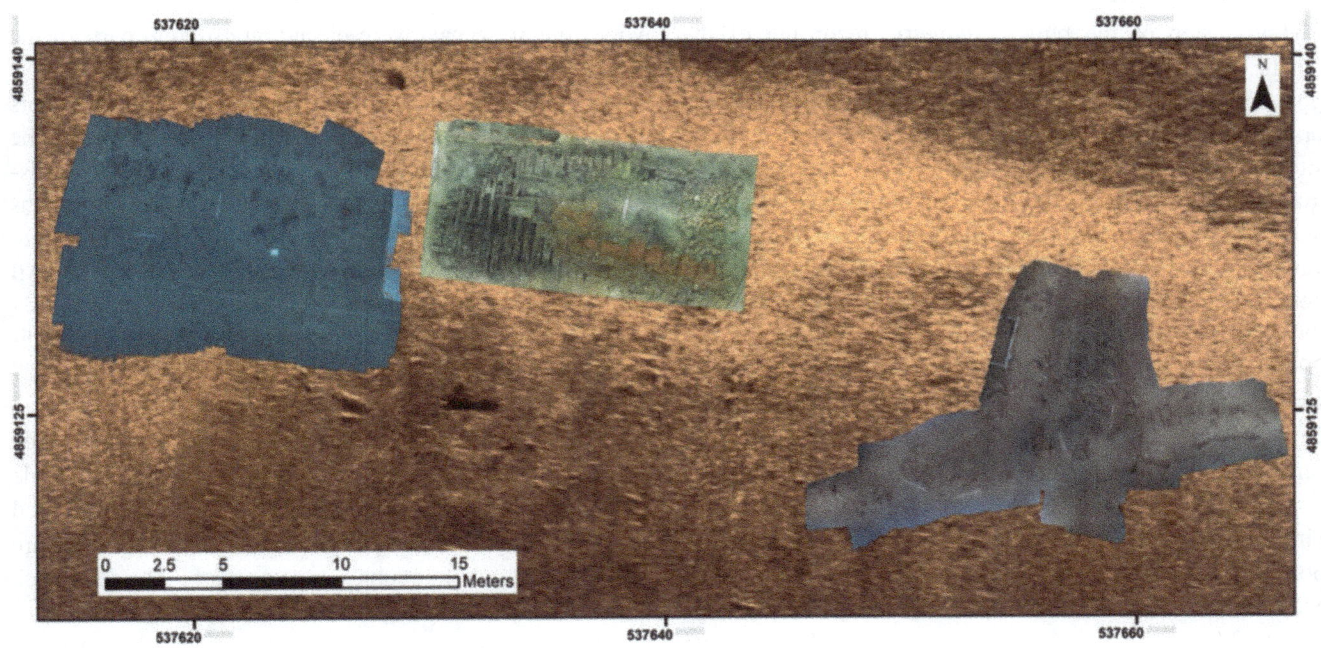

Fig. 4.4 Site area documented photogrammetrically, 2013 Gnalić Research Campaign (J.L. Casabán)

surface of the shipwreck including its artefacts and hull remains. The model was georeferenced using a network of control points and, finally, a texture was added to the model based on the images. An orthophoto of the shipwreck was then generated from the photogrammetric model and imported into *AutoCAD Map 3D* to trace a site plan, while longitudinal and transversal sections were extracted from the photogrammetric model. The orthophoto plans and sections were integrated into a GIS database to perform spatial analysis.

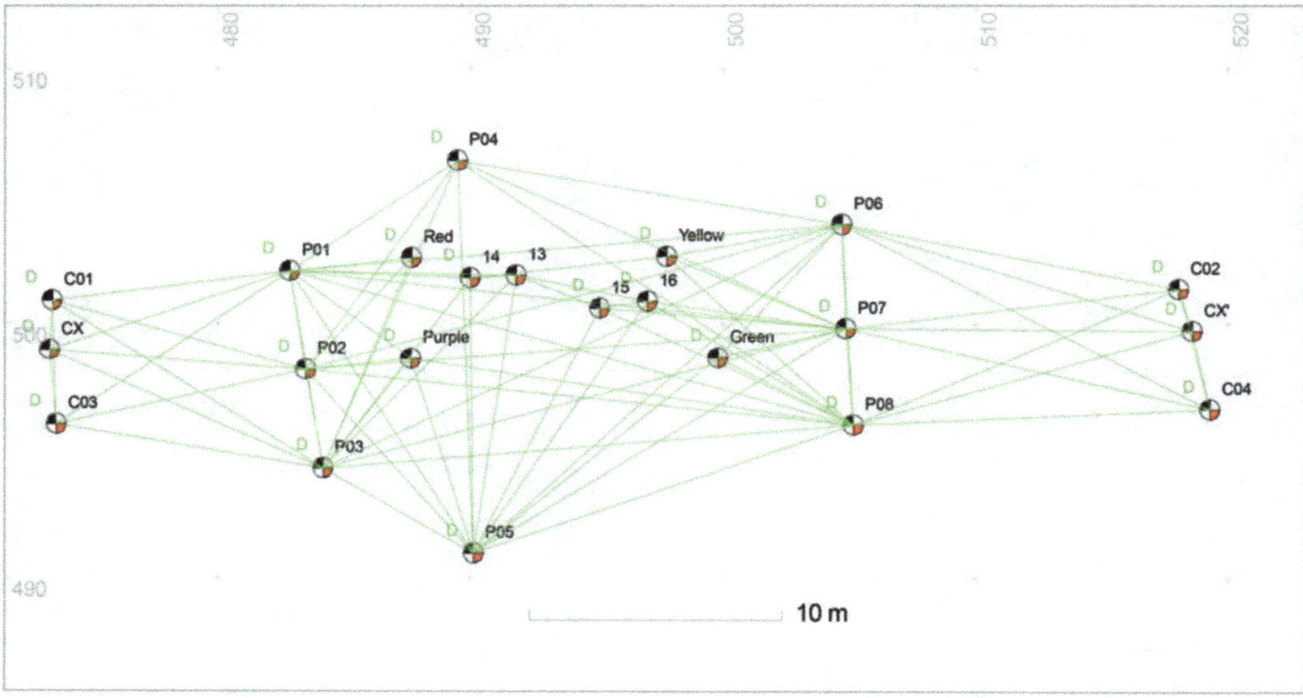

Fig. 4.5 Control points network, 2013 Gnalić Research Campaign (J.L. Casabán)

4.3.2.1 Control Points and Multi-image Coverage of the Site

The first step of the photogrammetric recording of the Gnalić shipwreck was the creation of a network of control points to georeference the photogrammetric model and the orthophoto generated from the multi-image coverage of the site. The georeferencing system, which was applied for the multi-image coverage of the Gnalić shipwreck, was pioneered at the Institute of Nautical Archaeology (INA) excavations of the Classical shipwreck at Tektaş Burnu (Turkey) between 1999 and 2001 (Green et al. 2002: 284, 288–290), the Archaic Greek shipwreck at Pabuç Burnu (Turkey) in 2002–2003 (Polzer 2004: 3–11), and the Phoenician shipwreck at Bajo de la Campana (Spain) in 2007–2011 (Polzer and Casaban 2012: 12–14). This method is based on a series of control points evenly distributed over the site, which also appear in the photographic coverage of the mapping area. The x, y and z coordinates of each point are obtained through 3D trilateration or 'Direct Survey Measurement' (DSM), and depth measurements (Bowens 2008: 127–128). In other words, the distances between the points are measured using measuring tapes, while their depths are determined with diving computers. Then, all the linear data acquired in this way is processed on a laptop using *Site Recorder SE*, a software program which calculates the 3D position of each point based on the linear measurements and depths. The RMS residual for network of control points produced with *Site Recorder SE* for the 2013 photogrammetric recording of the Gnalić shipwreck was 15 mm, using a distance and depth adjustments of 40 mm and 100 mm (Fig. 4.5). In addition to the control point network, four scale bars were placed on different parts of the mapping area to provide extra measuring references. Finally, six additional distance and depth measurements were taken in specific areas of the site to determine the accuracy of the resulting photogrammetric model. The distances were taken with a measuring tape while the depths were measured using the same dive computer (Suunto Vyper) employed for the network of control points. All these extra measurements were added to the final processing of the photogrammetric model to strengthen the precision of the control point network.

After the control points were positioned on the site, a multi-image photographic survey of the site was conducted ensuring that all the site features, control points, and scale bars showed in the photographs taken. The images required a 60% vertical overlap and 80% horizontal overlap to ensure an optimum photographic coverage of the shipwreck remains.[8] The multi-image coverage of the different areas of the shipwreck was conducted in 2013 using a Nikon D200 DSLR camera equipped with a single manual strobe light that required calibration under water at the beginning of each dive. A single diver carried out the photographic coverage of the site, following parallel and transversal transects to pro-

[8] These overlapping percentages for the images are recommended in Section 2 of the Shooting Process Planning suggested in the website of Agisoft (2018).

Fig. 4.6 Orthomosaic of the hull remains, 2013 Gnalić Research Campaign (J.L. Casabán)

duce enough overlap between the images. Additional photographs of particular areas of the site were taken from different angles to ensure that each part of the shipwreck was visible from at least two camera locations, and to minimize blind spots which could hinder analysis of archaeological features. A few of these blind spots still occurred, however, since the photographic coverage was mainly conducted perpendicularly to the surface of the hull remains and the time limitations of the project did not allow additional photos to be taken from different angles in all cases. Despite this inconvenience, it was still possible to produce an accurate and detailed orthophoto of the hull remains of the Gnalić shipwreck. The multi-image coverage of the excavation area was conducted several times during the archaeological season to document the different stages of the archaeological work.

The images taken following this method were then processed using *Adobe Photoshop* to improve their quality by manipulating the image settings such as white balance, exposure, contrast, brightness, and clarity. The image corrections were intended to ensure the best quality of the resulting photogrammetric model and the orthomosaic since the visibility conditions on site were not always ideal, and varied from 1 day to the next depending on the currents and the excavation work.

4.3.2.2 Image Processing, 3D Model, and Orthophoto Generation

After corrections, the images were processed following general workflow tasks in *Agisoft PhotoScan/Metashape*, until the mesh model of the shipwreck was created. At this stage, the control points that appeared in the photos were plotted manually on the model using the program tools. Then, the x-, y-, and z-coordinates of each control point were loaded into *Agisoft PhotoScan/Metashape*. This allowed for the georeferencing and optimization of the photogrammetric model using all the coordinates and measurements taken previously, providing a method to check the accuracy of the model. Finally, a texture based on the multi-image coverage of the site was added to the model and a 1:1 orthomosaic of the site was generated (Fig. 4.6).

The georeferenced orthomosaic was imported into *AutoCAD Map 3D*, a software package which combines the Computer Aid Design (CAD) tools and the main data formats used in Geographic Information Systems (GIS) to trace a 2D site plan directly from the orthomosaic in real scale (Fig. 4.7). In addition, several longitudinal and transversal sections of the hull remains were obtained from the photogrammetric 3D model of the site using software packages such as *Rhinoceros* or *Autodesk Maya*.

Fig. 4.7 Tracing of the hull remains of the Gnalić ship based on the 2013 orthomosaic (J.L. Casabán)

4.3.2.3 GIS Analysis

Finally, both the georeferenced orthophotos and site plans were integrated into ESRI *ArcGIS* to manage and to analyse the photogrammetric data combined with other types of information generated in any archaeological project. In this case, the GIS database is used as a tool to produce more rapidly different types of site plans that a trained nautical archaeologist will use to interpret the hull remains (see Steffy 1994: 191–250). In order to analyse the photogrammetric information provided in the orthophotos, the archaeological features and hull components represented in the site plans generated from the orthophotos would be linked to the hull catalogue compiled during the excavation. The catalogue includes the descriptions of the different hull timbers, their dimensions, types of wood, and any other observation and interpretation made by the archaeologists during the excavation of the hull remains.

The GIS database would be used then to perform different types of analyses in order to understand the site formation sequence based on the spatial distribution of hull timbers and related artefacts. In addition, different site plans of the site will be generated based on the data gathered in the timber catalogue. The criteria employed to produce the different set of plans would include the type of hull components (stem, keel, sternpost, keelson, frames, planking, ceiling planking, and other components), dimensions, types of wood used for the construction of the hull, visible scarfs and butts, and other structural components. Separate plans showing the location of scattered small finds, such as nails, will also be produced because they could provide information about hull components that were not preserved.

The analysis of the information provided in the different sets of plans generated with ESRI *ArcGIS* tools will be used by nautical archaeologists to understand the design and construction sequence of the vessel, including hull modifications or repairs occurred during its operational life, using reverse engineering. This information was later combined with the hull lines obtained from the mesh of the photogrammetric model in order to produce the lines drawings of the hull and, finally, the construction drawings.

4.3.3 Research Campaign 2014

In 2014, the photogrammetric recording on the Gnalić shipwreck site continued. This year the recording team, led by Kotaro Yamafune and Rodrigo Torres, experimented with slightly different methods.

4.3.3.1 Local Coordinate System

At the start of the season, a new local coordinate system was established using control network around the site. While a local coordinate system had been established in 2013, it had been difficult to create secured control points of trilateration because of the shortage of time.

It was decided to create a local control network, in order to make sure all the recording throughout the season could be related to the same x, y, z datum, providing a 4D

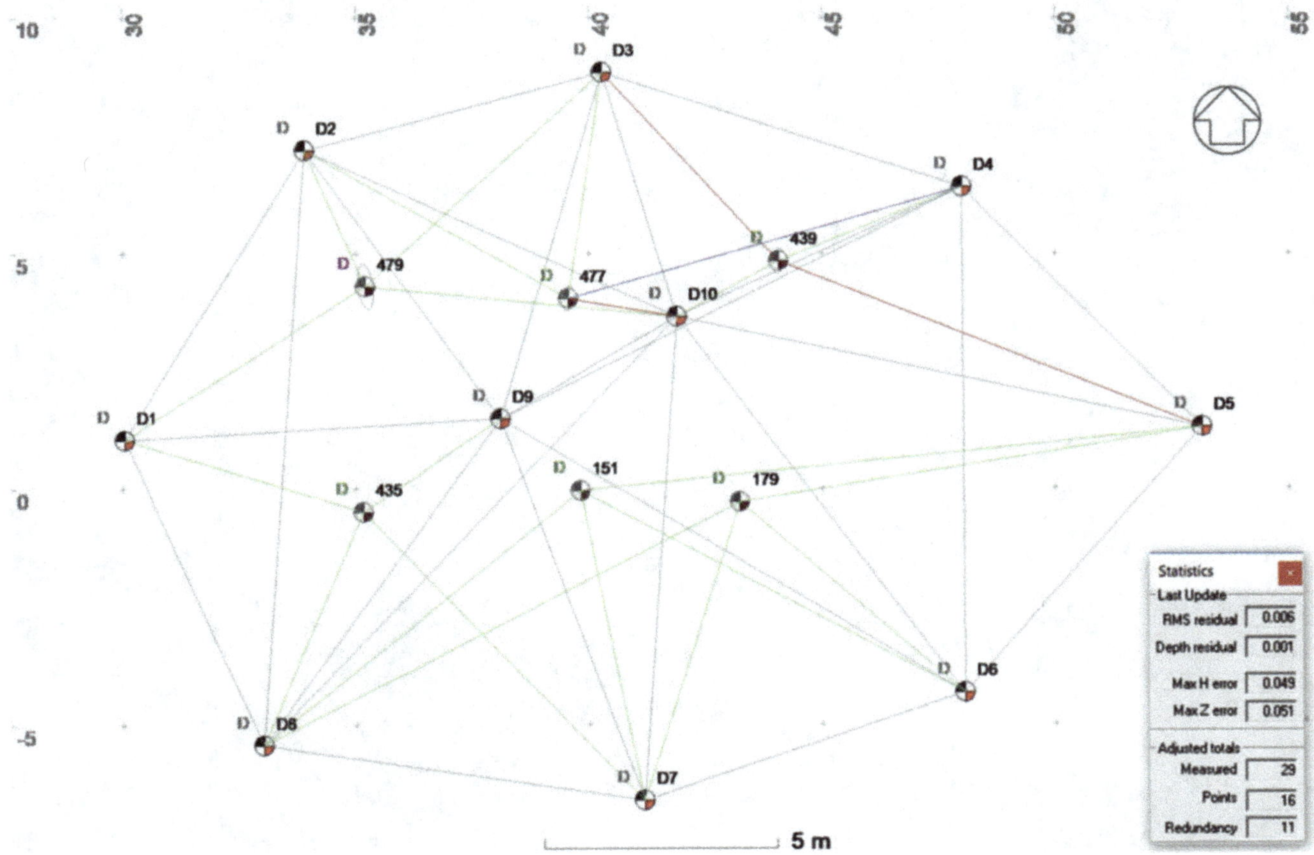

Fig. 4.8 Local coordinate network of the Gnalić shipwreck site (R. Torres, K. Yamafune)

recording strategy (x, y, z, t) which could accommodate photogrammetry, but also triangulation and grid offset plotting, if necessary. Moreover, once the local coordinate system was re-established in 2014 (Yamafune et al. 2016), it was possible to apply it to 2013 photogrammetric models using/creating the common points, such as tags of timbers that had been retained for both field seasons. The extracted x, y, z, coordinates from the common points of 2014 models were subsequently applied to the 2013 model.

The method used to establish a local coordinate system, according to conventional DSM-techniques (Atkinson et al. 1988; Rule 1989) has been published by Yamafune (2017). Regular fibreglass measuring tapes were employed for distance measurements and a diving computer for depths (10^{-3} and 10^{-1} reading resolutions, respectively). Depth measurements of all control points were taken as quickly as possible, with the same dive computer (UWATEC Aladin Ultra) to keep consistency and minimize the effect of tidal variation. Depth readings at the reference datum (D1) were recorded before and after each DSM measurement session, to function as a vertical datum and reference for tide correction throughout the excavation (Fig. 4.8).

Once the statistical ('Best Fit') adjustment provided by *Site Recorder 4 SE* (RMS 0.006 m, in this case) was considered satisfactory, a report was exported in .txt format into *MS Excel*. A clean x, y, z spreadsheet was then produced, with coordinates from *Site Recorder 4 SE* adjustment. Since this adjustment produces a network which is correctly scaled but not georeferenced, however, the spreadsheet was transferred to *ESRI ArcGIS 10.1* for spatial orientation and coordinates transformation.

To do so, two control points which run across the ship's axis were chosen (in this specific case, D1 and D5) to act as the excavation baseline. Then a temporary tight reel line from D1 to D5 was laid, and careful magnetic compass bearing was taken. With the help of the Internet (NOAA 2018) the magnetic deviation was calculated based on the inserted date and site coordinates, in order to provide models and orthophoto site plans with true north.

When the network was locked and transformed, six more internal secondary control points (SCP; 12bit water-proofed *Agisoft PhotoScan/Metashape* coded targets) were added in the system. These SCP were then trilaterated to fit into the primary control network, and x, y, z coordinates were derived for them. The SCP coordinates were used to spatially reference all partial photogrammetry models produced through-

Fig. 4.9 Composite 3D digital model of the state of the excavation by the end of the 2014 research campaign. This model was composed of five separate photogrammetric models whose base-photos were taken on five different days (K. Yamafune and R. Torres)

out the season, allowing for the overlapping, matching, and scaling of the models and orthophotos. Scale bars were also used to calibrate scaling on each recording/photo-shooting session. Here is where the 4D excavation control should start, plotting artefacts, layers, 3D models, etc. It is important to highlight that the strategy was conceived to allow for in-field processing and on-the-fly excavation feedback.

4.3.3.2 Composite Models

The second experimental approach tested on the 2014 campaign was the development of 'composite' photogrammetric models. Operational experience found that the calculation power of the project computer was limited; this meant that when a photogrammetric model of the entire shipwreck site, or a larger area, was created, the available detail in the 3D model deteriorated due to operator reduction of the number of meshes and resolution of textures, in order for the computer to complete the processing. Therefore, to acquire the best results for accurate archaeological information, a large site had to be separated into smaller areas. In other words, if 3D models of the site were created as separate pieces, the number of meshes and the resolution of the 3D models of the entire site could be higher. Moreover, if those 3D model pieces were created separately under the local coordinate system, these pieces could then be merged automatically in other 3D modelling software which has better rendering power. In 2014, the final composite 3D model of the campaign was created from five different model pieces, with each of those pieces created from photogrammetry performed on a different day. These five model pieces were then exported into *Autodesk Maya* to compose the separate models into one single model (Fig. 4.9). After the 2014 campaign a CG animation was created, and uploaded to YouTube for dissemination. Nevertheless, it is worth noting that since the August 2015, *Photoscan v1.2* added a new function called 'Build tiled model.' This new function allows computing power to be concentrated on a small area specified by the size of a 'bounding box' for building mesh and texture within. In other words, once dense cloud data of the entire site is created, high resolution models of smaller area can be created within the 3D model of the site.

4.3.3.3 GIS Analysis

The third noteworthy approach tested during the 2014 field season was the improved application of GIS with photogrammetry, and its integration into the excavation workflow.

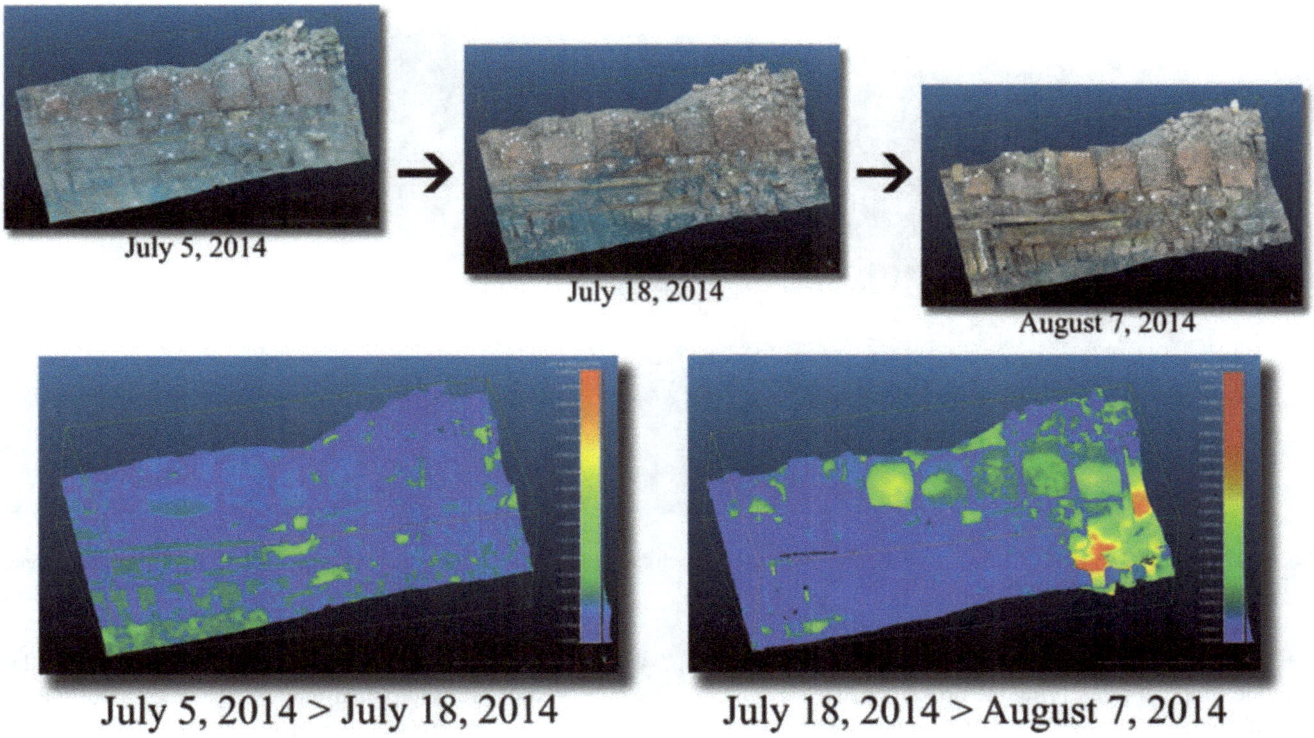

Fig. 4.10 Points based deviation analysis in *CloudCompare*. Differences between two different data set were displayed in colors (K. Yamafune)

Today, this has become a common application of photogrammetry in archaeological projects; however, it was still rare in 2014. The GIS software *ESRI ArcGIS* was used as an interactive map, updated every day using the photogrammetric orthomosaic and information gathered by the excavators. The main dataset imported into the GIS software as a base map was an orthophotomosaic generated by *Agisoft PhotoScan/Metashape*. Since the 3D photogrammetric models had already been created under the local coordinate system, an orthomosaic generated from georeferenced 3D models already contained georeferenced information. Whenever a new orthomosaic was generated and imported into ESRI *ArcGIS*, the maps of the shipwreck site were updated. Additionally, information on artefacts recovered from the site was digitally catalogued, and then linked to the GIS database. Moreover, these site plans and other information were printed out on waterproof paper, which archaeologists brought underwater to execute their assigned tasks faster and more efficiently. This served as a georeferenced database for the project, and it helped the entire excavation process by providing up-to-date information throughout the campaign.

4.3.3.4 Points-Based Deviation Analysis

The final experimental method applied during the 2014 campaign was 'points-based deviation analysis.' This concept takes advantage of software such as *CloudCompare*, which can compare two different point cloud data sets, and generate differences between the two different data sets, showing the results/calculations empirically as quantitative data, or graphically as differences in colours. In 2014, this type of analysis was applied to the repetitive photogrammetric recording of the same areas throughout the 2-month campaign. The dense points cloud data of the area, which contained large barrels and smaller casks, were exported into *CloudCompare* and the deviation analysis was applied (Fig. 4.10). The software displayed 'excavated areas' in colour and provided differences in quantitative data (for instances, how much additional surface had been excavated). Based on the results, it is suggested that stratigraphic recording and analysis could be aided by application of 'deviation analysis'. Additionally, this application could be used for multi-year site monitoring; in other words, the 'deviation analysis' based on photogrammetry could track changes on underwater sites over time, and could be used for site management plans.

4.3.4 Research Campaigns 2015 and 2016

In 2015 and 2016 systematic photogrammetric recording continued in the same manner. During the two campaigns, the excavation went on along the keel, the exposed ballast, and the area south of the big barrels filled with intense-red hematite. Extremities of the preserved keel area were

Fig. 4.11 (**a** and **b**) Composite photogrammetric 3D model of the research campaigns 2015 and 2016, and the respective orthomosaic (K. Yamafune, R. Torres, S. Govorčin, D. Gorički)

reached, but the excavation continued in both directions, with the scope of identifying the broken posts.

Unfortunately, the location of the control points placed during the 2014 campaign slightly changed for various reasons. This meant that it was not possible to use these control points to add new reference points for newly exposed areas for photogrammetric recording. For this reason, different methods were applied to create the local coordinate system for 3D photogrammetric models of newly exposed shipwreck areas (Fig. 4.11a, b).

In order to apply the new method, capturing photos for photogrammetry covered slightly larger areas than necessary. Once photogrammetric models of each area were created with intentionally wider capturing areas, the photogrammetric models of the same areas from 2014 campaign were opened. Then, in *Agisoft PhotoScan/Metashape*, markers were created on exact mutual points on all models; since the 2014 models were already georeferenced, or contained the local coordinate system, the 2014 coordinate system could then be applied to the 2015 and 2016 models. Therefore the 2015 and 2016 photogrammetric models were georeferenced based on the coordinate system of the 2014 campaign. However, this method may be less reliable once excavated areas are extended further. In any case, the new coordinate system was essential for photogrammetric recording in following campaigns.

4.3.5 Mapping the Area of Archaeological Interest in 2017

The 2017 field season continued, following the same research and documentation procedures, in the southern and central area of the hull. According to the system, recently elaborated by Yamafune (2017), the photogrammetric mapping of the whole area of archaeological interest was successfully executed during six consecutive dives, covering the area of 60 × 20 m (Fig. 4.12).

This new methodology does not require direct measurements of control points that was applied during 2013 and 2014 campaigns. The scale bars were placed on the mapping area to scale constrain created 3D models and to allow the application of a 3D CAD software, such as *3D Rhinoceros,* to create a local coordinate system.

In order to produce this photogrammetric model, the site was divided into two half-areas, 30 × 20 m each, following the requirements of one photo shooting session. Nine scale bars (five 1-m scale bars and four 0.5-m scale bars), and three coded targets, acting as reference points, were positioned in each half-area. After the successful creation of one half-area model, *Agisoft Photoscan/Metashape* calculated the residual error of 0.0045 m, based on 0.7-m markers/scale bars, created in *Agisoft PhotoScan/Metashape* software.

Once the initial model was created, however, it was possible to apply the 'Optimize camera position' command on *Agisoft PhotoScan/Metashape,* in order to fix distortions in the created model. As a result, the residual error of the Gnalić model became 0.0003 m (0.3 mm) in 0.7 m (Fig. 4.13). This theoretically indicates that the possible positional error from one end of the site to the other (around 60 m) is approximately 2.7 cm.

Once the precision of the model was considered satisfactory, the depth measurements of three reference points were taken using a dive computer. When the photogrammetric model was created, scaled, its camera positions optimized, and re-processed, distances between three control points were calculated within *Agisoft PhotoScan/Metashape* software. Then, a triangle was created in *Rhinoceros* 3D CAD

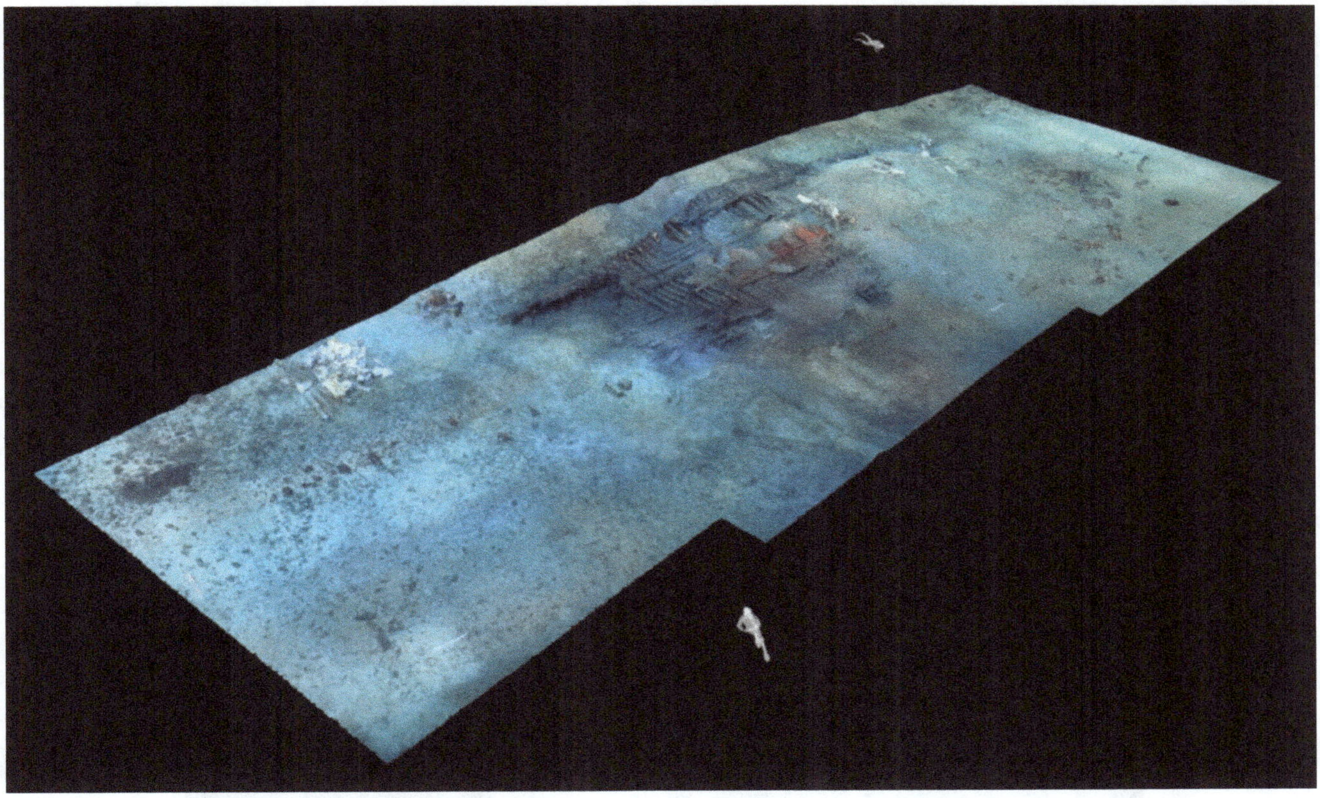

Fig. 4.12 Photogrammetric model of the entire Gnalić shipwreck site. The area measures approximately 60 × 20 m. The model includes furthermost stern (rudder pintles?) and bow (grapnel anchor chain?) concretions (K. Yamafune)

software, and its position was adjusted based on depths of three reference points. Finally, *Rhinoceros* provided x, y, and z coordinates of the adjusted position of these local reference points, which were applied to the photogrammetric models in *Agisoft PhotoScan/Metashape* (Fig. 4.14).

This new methodology has been applied on various underwater archaeological sites, with successful results. It is a great advantage that it does not require much preparatory work to create an accurate local coordinate system for photogrammetric recording, such as DSM. For instance, for recording the Gnalić Shipwreck site in its entirety, one dive to places scale bars and three reference points, and two dives for photography on the western half-area were necessary; followed by one dive to move the scale bars and position the reference points in the eastern half-area, and two more dives to complete the photography. This means that it required just six dives (25 min each), or 150 min of one diver's time, to cover 60 × 20 m surface, while maintaining accuracy of 2.7 cm residual errors over the whole length of the site (i.e. for 1 m long object possible error is 0.4 mm).

4.4 Timber and Artefact Recording

The process of systematic photogrammetric recording also encompasses recovered artefacts, and elements of the ship's hull. The main issue in the photogrammetric recording of such relatively small objects is that photographs have to be taken from all the directions, i.e. cover all the surfaces. In other words, if photogrammetry was performed on an artefact in the same manner as it is performed on an archaeological site, then the side on which the artefact rests could not be modelled properly, because it is hidden from the camera.

To solve this problem, in *Agisoft PhotoScan/Metashape* a process known as 'masking' can be used to proceed correctly with the photogrammetric recording. The program has various masking methods, yet during the 2016 Gnalić

a

Scale Bars	Distance (m)	Accuracy (m)	Error (m)
☒ point 1 – point 2	0.700000	0.000100	0.000071
☒ point 3 – point 4	0.700000	0.000100	0.000010
☒ point 5 – point 6	0.700000	0.000100	0.000459
☒ point 7 – point 8	0.700000	0.000100	-0.000530
☒ point 9 – point 10	0.700000	0.000100	-0.000175
☒ point 11 – point 12	0.350000	0.000100	0.000274
☒ point 13 – point 14	0.350000	0.000100	0.000407
☒ point 15 – point 16	0.350000	0.000100	-0.000224
☒ point 17 – point 18	0.350000	0.000100	-0.000123
Total Error			
Control scale bars			0.000304

b

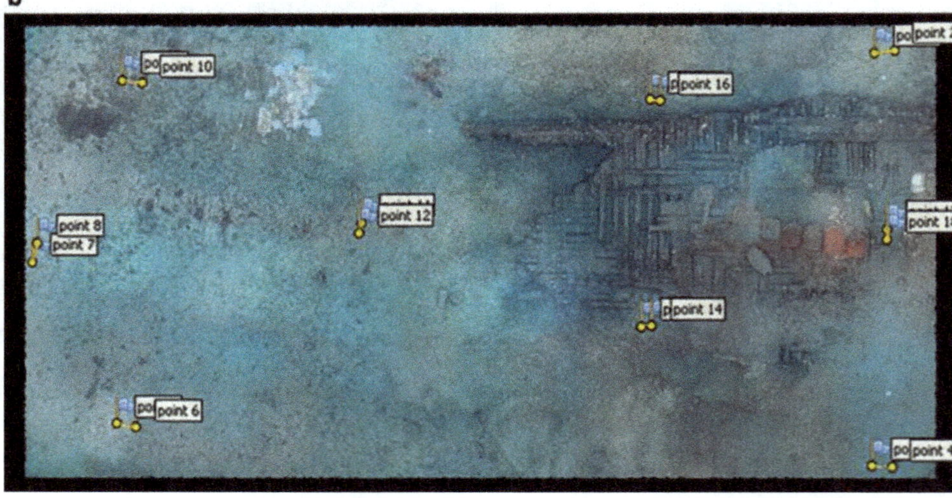

Fig. 4.13 After optimizing camera positions in *Agisoft PhotoScan/Metashape*, the residual error of 0.7-m scale bars is 0.0003 m (0.3 mm)

research campaign the 'masking from model' method was exploited. First, two or more 3D models of a single object were created, taking care that each time the object rested on a different side. After scaling the created models by using scale-bars placed around the object, the ground plane (all meshes of the 3D model except meshes of the object itself) was erased. Then 'masking' was applied to photos using the 'masking from model' method. After 'masks' were successfully created on all the original photos, they were gathered under one 'chunk', and the regular photogrammetric workflow of *Agisoft PhotoScan/Metashape* proceeded, exploiting the created 'masks.' As a result of the 'masking' procedures, photogrammetric 3D digital models of several recovered timbers, the bottom part of the bilge pump and a probable pintle concretion of the ship's rudder were successfully created.

The purpose of photogrammetric modelling of ship timbers was to create a 3D record of each recovered timber for timber catalogues. According to nautical archaeological methodology, to fully understand ship's structure it is important to understand and record all the dimensions, position and types of fastenings and scarves, tool marks, and so on. Therefore, recording the elements of the hull in as much detail as possible is an important task (Steffy 1994). Creating timber and artefact catalogues, however, is usually labour intensive and requires significant experience. For this reason, the photogrammetric 3D models were conceived to facilitate and speed up the operation. Once a 3D model of a ship-timber or other object was created, orthomosaics were generated using six different projection planes (top view, bottom view, front view, back view, right view, and left view). Next, these orthomosaics of the six different projection planes were aligned in series in

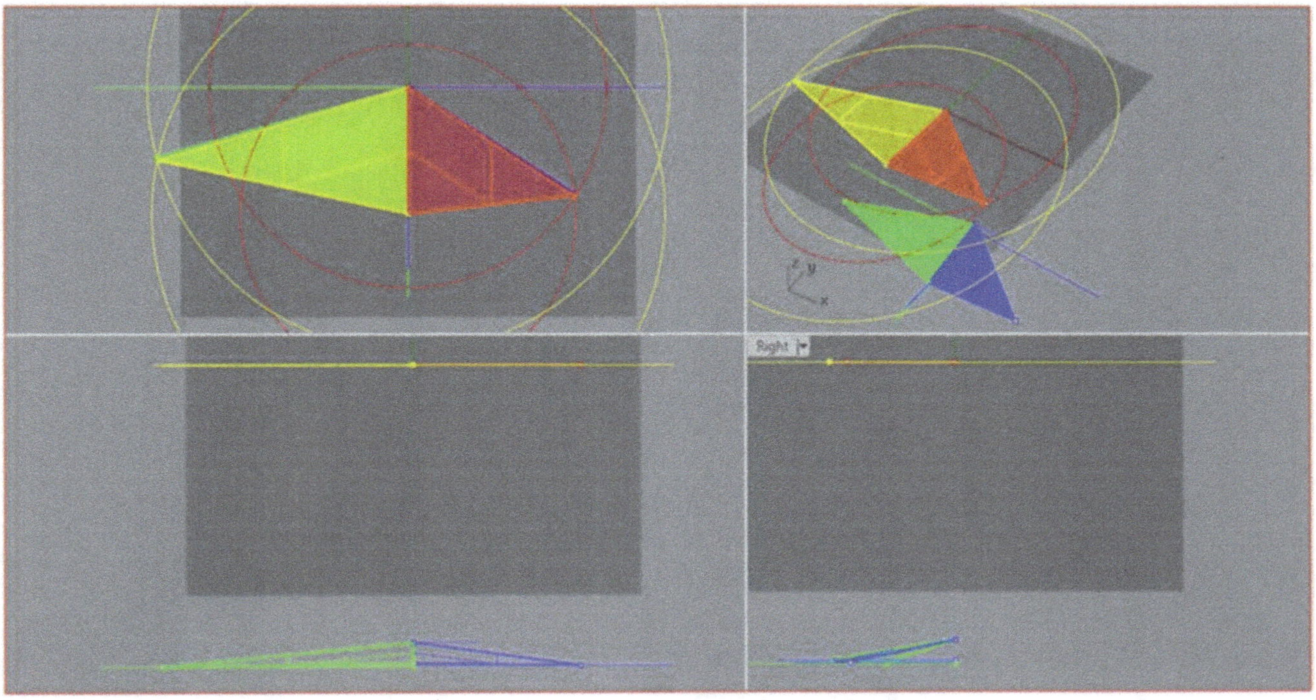

Fig. 4.14 Screenshot image of creating a local coordinate system for Gnalić Shipwreck site. The yellow triangle was created based on the distances between reference points on the western half-area in *Agisoft PhotoScan/Metashape*, and a red triangle was created in the same way for the eastern half-area. The green triangle indicates positions of reference points on the western half-area of the site, and a blue triangle indicates reference points on the eastern half-area. Position/rotation of the green and the blue triangles were adjusted by depth measurements

Adobe Photoshop and converted into the artefact drawing (Yamafune 2016).

The main advantage of using photogrammetry to generate the basic timber and artefact drawings is its efficiency in terms of required recording time. Capturing the data to create a 3D model is fast, requiring only the time to photograph the object. After this, the artefact can quickly be returned to suitable conservation conditions, or to site, while post-processing to create the 3D model is carried out. For instance, during the 2016 campaign, the bottom part of the ship's bilge pump was recovered in order to record its structure in both 3D digital model format and traditional 2D artefact drawing (Fig. 4.15). The pump was found in the starboard side of the ship, surrounded by planks forming a triangular structure, which could be interpreted as the pump well. Notches on the sides on the foot valve had the function of fixing the pump between two frames, while the upper part of the pump was not preserved. After recording the exact position of the bilge pump inside the pump well, it was taken to the conservation laboratory to conduct detailed documentation, using photogrammetry. Traditional manual recording of the artefact by 2D drawing would have certainly required much more time, causing a longer exposure of the waterlogged artefact to drying conditions. In this case of quick photogrammetric recording automatically transferred into a 2D drawing (with accuracy checked through the observations and direct measurements), the bilge pump was rapidly returned to the site to be preserved in situ until conditions for conservation can be assured. In summary, using photogrammetry and the 'masking' method, the team successfully acquired necessary archaeological data on wooden structures of the ship with minimum damage.

Another interesting element that was recorded by photogrammetry was a probable pintle concretion, recovered from the area of the stern. The area where it was recovered consisted of concreted objects which might have been pintle and gudgeon, the ship's elements connecting the rudder to the transom and sternpost. The pintle was attached to the rudder, although the wooden part either disintegrated or remains rest below the sediment. The method applied to record the presumed pintle was the same as for the timbers and bilge pump (Fig. 4.16), the only difference being that *Agisoft PhotoScan/Metashape* control points were used instead of scale bars. As at present it was not possible to arrange for proper conservation treatment, after recording, the concretion was returned to the site.

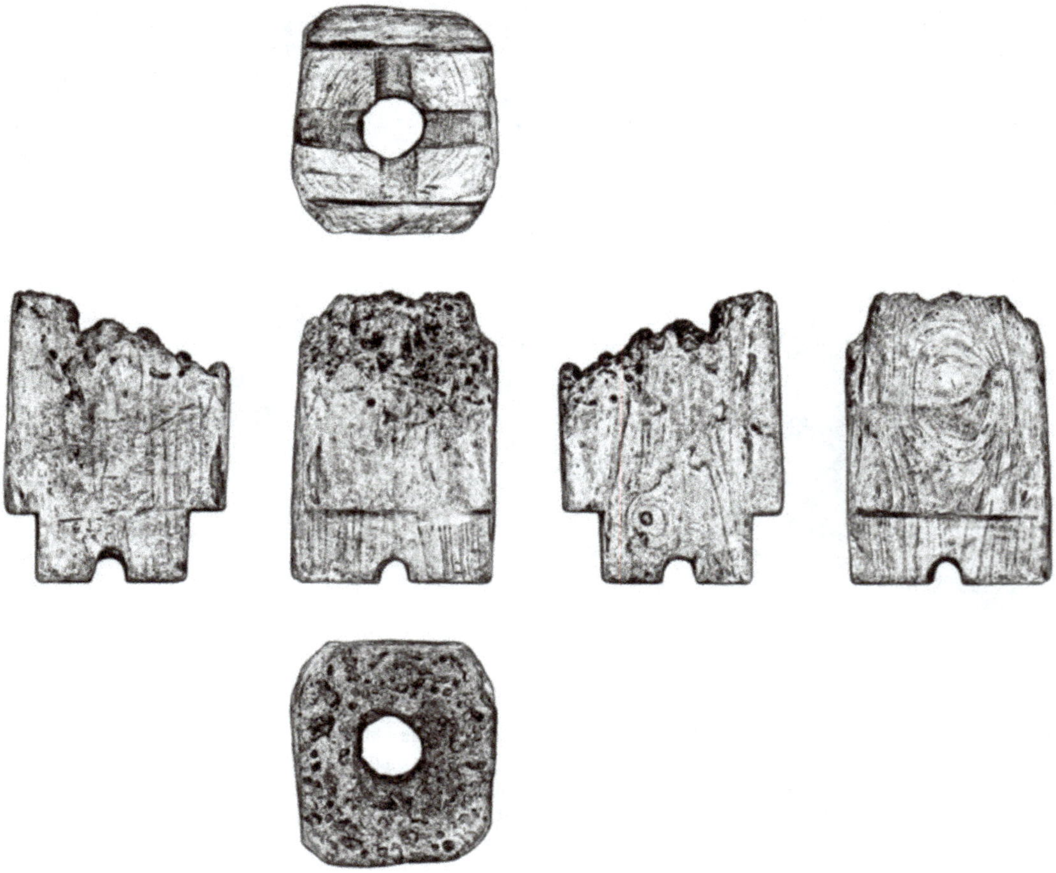

Fig. 4.15 Artefact drawing of the bottom of the bilge pump structure of Gnalić shipwreck. Orthomosaics of six different projection plans were generated from the 3D photogrammetric model, and then these orthomosaics were aligned and converted into artefact drawings (K. Yamafune, K. Batur)

4.5 Virtual Reality Application

After the 2014 excavation season, the application VR GNALIĆ was created in order to exploit the results of the photogrammetric record, by integrating the images into a format suitable for exploration on a computer, or with a VR viewer, for example the HTC VIVE headset. The experiment was conducted by Ervin Šilić and his team from the Novena Digital Multimedia Studio in Zagreb, Croatia.

In the coordinate system of the virtual space of the application, basic orientation points were defined for the integration of all the photogrammetric models in the virtual world. The positioning of newly produced models is designed as an automated process. Opening the application recalls the defined models and presents them in a virtual environment in which it is possible to move them (Fig. 4.17a–c). The virtual environment was created with the *Unity* multiplatform game engine, and the *ASP.NET* web framework was used for the Content Management System.

Besides moving around the site, the interactive component allows the user to open or hide each excavation phase. It is also possible to select the timbers which were fully recorded, and to examine them in detail. On selection, each timber appears in a separate space in the form of a 3D model, which can be examined from all sides, and cut along any of the three axes in order to obtain cross-sections in various positions (Fig. 4.17a–c).

VR GNALIĆ can be exploited for presenting the research results to the wider public in an attractive and exciting way but could also be useful to permit discussion among scientists, who could 'visit' the site, and integrate their suggestions or comments, even if they could not be physically present during the excavation campaigns.

Ongoing technological development is rapidly increasing the potential of such exploitation of 3D models of sites and finds, and there is no doubt that it this technology could be widely exploited for the promotion of underwater archaeological projects, and the justification of their importance. They can provide ideal complementary material for temporary exhibitions or museum displays, either in real or virtual form (Fig. 4.18).

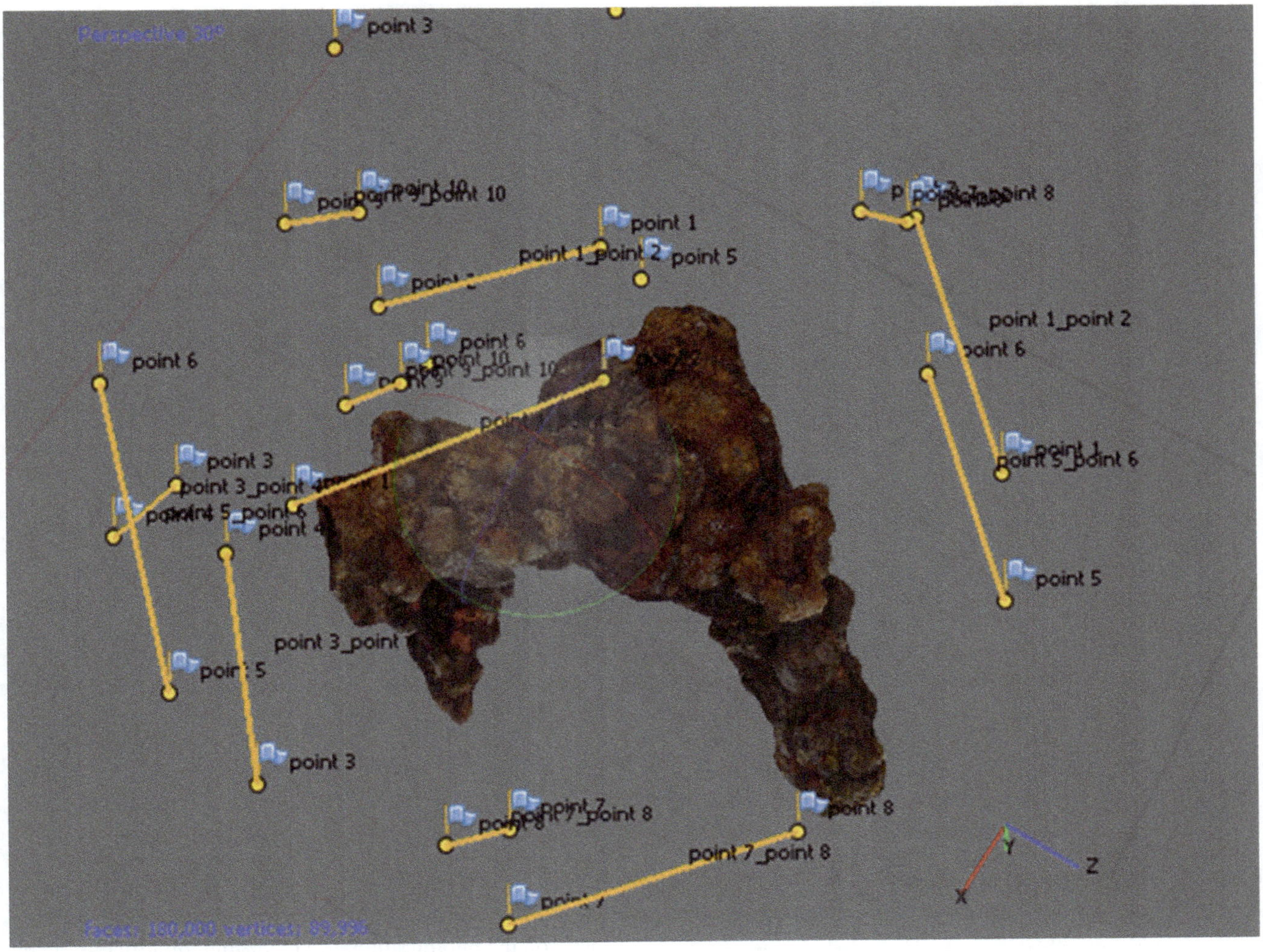

Fig. 4.16 Probable pintle concretion, as recorded (K. Batur)

4.6 Automation of the Underwater Recording Process

In the framework of the *Breaking the Surface* field workshop on underwater robotics and applications (held in Biograd na Moru, Croatia, October 2016), the Autonomous Underwater Vehicle GIRONA 500 was employed to record the state of research of the Gnalić shipwreck site (Gracias et al. 2013). The work was executed by the research team from the Computer Vision and Robotics Research Institute of the University of Girona, under the direction of Pere Ridao (Ridao and Gracias 2017). It was conceived to demonstrate the state-of-the art application of underwater robotics for rapid high-resolution mapping of shipwreck sites.

The AUV was programmed to survey the shipwreck at multiple altitudes, and the data collected was used to build 2D photomosaics and 3D optical reconstructions with 1 × 1 mm pixel resolution, as well as to develop topological panoramic maps, which were made available during the same field workshop (BTS 2016). The team from Girona had previously performed detailed AUV mapping of the *La Lune* shipwreck (Gracias et al. 2013) and some shipwreck sites along the Catalan coast (e.g. Hurtós et al. 2014). The experimental mapping of the Gnalić shipwreck had the most positive outcome and demonstrated the potential of the automated recording process, considered as complementing rather than replacing the work of the divers. The comparison of the accuracy of the automatically generated photos, and consequently the orthomosaics and 3D models, with the results of the photogrammetric recording executed by the diving team is in progress. This analysis will serve to improve the automated process, which could contribute greatly to recording the advancement of the excavation, and would be essential for documenting deep-water sites.

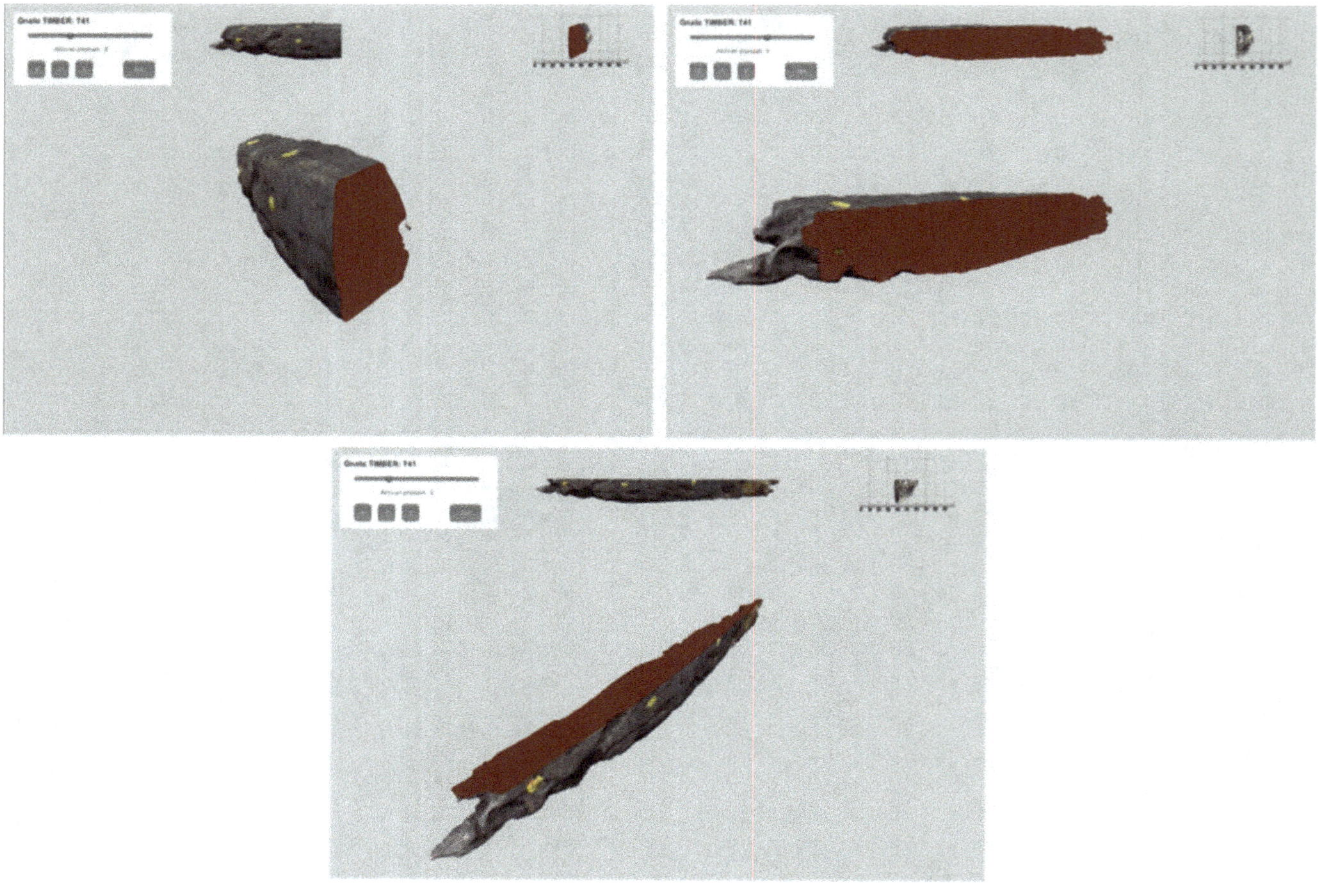

Fig. 4.17 (**a–c**) Obtaining cross-sections of the recorded timbers along the x, y and z axes (Novena Ltd.)

4.7 Conclusions

Despite all the advantages, after six years of intense underwater photogrammetric recording experience on an extremely demanding shipwreck site, the authors conclude that photogrammetry is not an absolute, sole recording system, that supersedes all others. It certainly helps in quick and precise recording of artefacts and structures, providing data that could be used in various formats and for various purposes. On the other hand, its accuracy of geometry and resolution of texture has limitations. Therefore, detailed data have to be recorded manually in order to fully understand shipwreck sites. Moreover, it does not define the relationships between the recorded elements, as it only records the visible 'surface' data.

Photogrammetry is a recording technique, and, as such, it assists the research procedure. This means that researchers must know how to properly excavate and document the site, recover and conserve the archaeological finds etc. An unfortunate recent trend in this discipline is an increasing number of excavation campaigns in which archaeologists focus mostly on the photogrammetric recording, blindly relying on its data. It is a well-known fact that underwater archaeological campaigns require good preparation, organization, excavation, site and artefact conservation, and publication of the results. Photogrammetric recording cannot replace any of these phases, but could help us in their execution.

As long as the archaeologists that work in the underwater environment, however, understand the advantages and limitations of photogrammetric recording, it can be fully integrated into the workflow of research campaigns. Once it functions properly as an integral part of the project, it can greatly help the archaeological research, by generating an accurate record through a relatively cheap and time-saving process, and subsequently provide attractive material for the public promotion of the project, educational purposes and museum display. It can also help in a 3D visualization of the advancement of the excavation, something that is nearly impossible or extremely time-consuming with the traditional recording process. Photogrammetry is still often called 'innovative.' In fact, photogrammetry is now a common, indispensable tool in the process of recording the underwater archaeological sites and finds. Although tech-

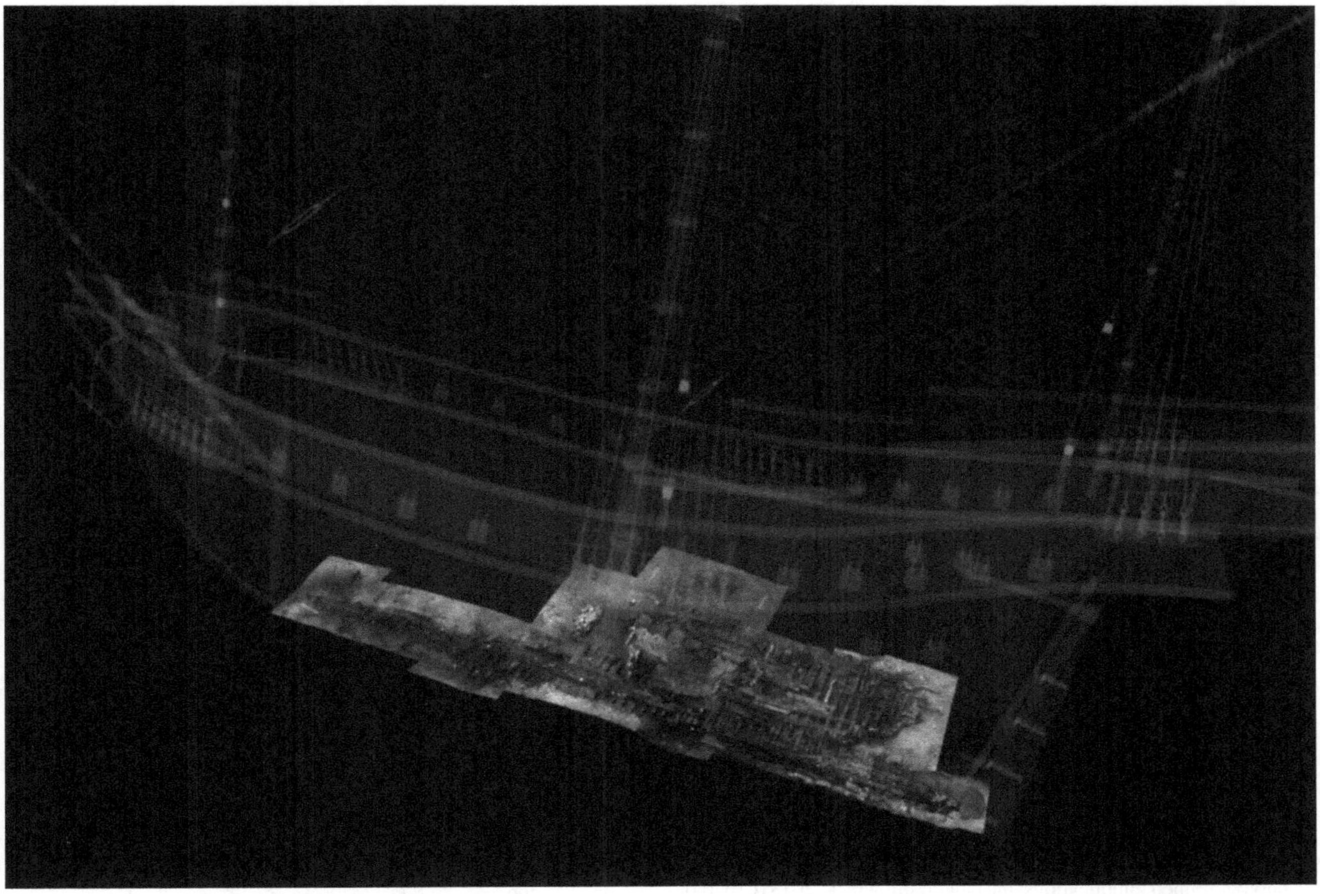

Fig. 4.18 Position of the excavated part of the hull within the representation of a sixteenth-century ship (K. Yamafune)

nological upgrades continue to improve the hardware, software and accuracy of photogrammetric recording, the true innovation lies in avoiding overreliance on photogrammetry, but instead striving to understand its drawbacks and limitations. Simply performing photogrammetry as the end goal of an underwater excavation is not sufficient. Innovations, such as the Gnalić shipwreck photogrammetry discussed in this article continue to adapt and refine the processes for recording underwater archaeological sites and their excavations, while examining and understanding the capabilities and shortcomings of these digital methods.

Acknowledgements We express our gratitude to the Ministry of Culture of the Republic of Croatia, the Croatian Science Foundation (Archaeology of Adriatic Shipbuilding and Seafaring Project [AdriaS], IP-09-2014-8211), the City of Biograd na Moru, the Municipality of Tkon, the Nautical Archaeology Program of Texas A&M University, the Institute of Nautical Archaeology, the Institute for Maritime Heritage ARS NAUTICA, and the Local Heritage Museum of Biograd na Moru for their financial and logistical support of the Gnalić shipwreck excavation. Special thanks to the Association for Promotion of Underwater Archaeology (FUWA) from Koblenz, Germany, which invested a lot of money and human power into supporting the excavation since 2013, thus significantly extending the excavation seasons. We are also grateful to Filipe Castro from the Ship Reconstruction Laboratory of Texas A&M University for having supported the project in the period from 2012 to 2014, including the work on the photogrammetric recording of the site; to Dave Ruff, PhD candidate from the same institution, whose help in editing the English text was a precious contribution to the finalization of this and many other articles, and to Pierre Drap and his small CNRS team, which executed the first photogrammetric recording of the Gnalić shipwreck remains. Credit for the outcome of the systematic photogrammetric recording is shared by Sebastian Govorčin, Javier Rodrigez Pandozi, Dražen Gorički, Mirko Belošević, and Danijel and Ranko Frka. The outstanding quality of the photographic coverage of the Gnalić site is a direct result of the quality of the team members who performed the photographic coverage of the researched areas, providing the basis for producing the 3D models. Regarding the underwater photogrammetry, we would like to point out the great contribution of Pere Ridao and the team of the Computer Vision and Robotics Research Institute of the University of Girona, in the hopes that the fruitful initial experiment with AUV-based photogrammetric recording of the Gnalić shipwreck executed in 2016 will continue to develop in the future. We are also grateful for the support of Ervin Šilić and his team from the Novena Digital Media Studio, and look forward to continued collaboration. We would especially like to thank Mariangela Nicolardi and Mauro Bondioli for their most patient and long-lasting examination of the documents in the State Archive of Venice regarding the ship's context, its personal story and the hull structure, as well as all the other participants in the fieldwork and post-excavation research, whose list includes several hundreds of names from all over the world. Finally, we express our gratitude to the anonymous referees for their comments, corrections and suggestions, that improved the content of the article, and to Miranda Richardson for excellent feedback regarding structural and stylistic aspects of the text.

References

Agisoft PhotoScan (2018) http://www.agisoft.com/pdf/tips_and_tricks/Image%20Capture%20Tips%20-%20Equipment%20and%20Shooting%20Scenarios.pdf. Accessed 8 Aug 2018

Atkinson K, Duncan A, Green J (1988) The application of a least squares adjustment program to underwater archaeology survey. Int J Naut Archaeol 17(2):119–131. https://doi.org/10.1111/j.1095-9270.1988.tb00631.x

Beltrame C (2006) Osservazioni preliminari sullo scafo e l'equipaggiamento della nave di Gnalić. In: Guštin M, Gelichi S, Spindler K (eds) The heritage of Serenissima: the presentation of the architectural and archaeological remains of the Venetian Republic, Proceedings of the international conference Izola, Venezia, 4–9 November 2005, pp 93–95. Annales Mediterranea, Koper

Bowens A (ed) (2008) Underwater archaeology: the NAS guide to principles and practice. Wiley-Blackwell, Portsmouth

Božulić G (2013) Zbirka "Teret potopljenog broda iz 16. stoljeća" Zavičajnog muzeja Biograd na Moru/The "cargo of a 16th century sunken ship" collection of the Biograd na Moru Heritage Museum. In: Filep A, Jurdana E, Pandžić A (eds) Gnalić; Blago potonulog broda iz 16. stoljeća/Gnalić—Treasure of a 16th century sunken ship, exhibition catalogue. Croatian History Museum, Zagreb, pp 37–49

Brusić Z (2006) Tre naufragi del XVII o XVIII secolo lungo la costa adriatica orientale. In: Guštin M, Gelichi S, Spindler K (eds) The heritage of Serenissima: the presentation of the architectural and archaeological remains of the Venetian Republic, Proceedings of the international conference Izola, Venezia, 4–9 November 2005, pp 77–83. Annales Mediterranea, Koper

BTS (2016) Breaking the Surface 2016, Girona 500 Tutorial (Data Analysis). http://bts.fer.hr/. Accessed 15 Aug 2018

Casabán J, Radić Rossi I, Yamafune K, Castro F (2014) Underwater photogrammetry applications: the Gnalić shipwreck, 2013 (Croatia). In: Abstracts of the IKUWA V conference, Cartagena, Spain, 14–18 October 2014

Dumas V (2012) La photogrammétrie numérique appliquée à l'architecture navale: le cas de l'èpave de Pakoštane (Annexe). In: Boetto G, Radić Rossi I, Marlier S, Brusić Z (eds) L'épave de Pakoštane, Croatie (fin IVe–début Ve siècle apr. j.-c.). Archaeonautica 17:105–151 (Annexe:143–145)

Gasparetto A (1973) The Gnalić wreck: identification of the ship. J Glass Stud 15:79–84

Gracias N, Ridao P, Garcia R, Escartíny J, L' Hour M, Cibecchini F, Campos R, Carreras M, Ribas D, Palomeras N, Magi L, Palomer A, Nicosevici T, Prados R, Hegedüsz R, Neumann L, De Filippox F, Mallios A (2013) Mapping the moon: using a lightweight AUV to survey the site of the 17th century ship 'La Lune'. OCEANS, Bergen, Norway, 10–14 June 2013, MTS/IEEE. https://doi.org/10.1109/OCEANS-Bergen.2013.6608142

Green J, Matthews S, Turanli T (2002) Underwater archaeological surveying using PhotoModeler, VirtualMapper: different applications for different problems. Int J Naut Archaeol 31(2):283–292. https://doi.org/10.1006/ijna.2002.1041

Holt P (2010) Site Recorder 4 exercise book. Version 1.5, June 2011. 3H Consulting Ltd. http://www.3hconsulting.com/Downloads/SiteRecorder4ExerciseBook.pdf. Accessed 19 May 2017

Hurtós N, Nagappa S, Palomeras N, Salvi J (2014) Real-time mosaicing with two-dimensional forward-looking sonar. In: 2014 IEEE international conference on Robotics and Automation (ICRA), Hong Kong, pp 601–606. https://doi.org/10.1109/ICRA.2014.6906916

Lane F (1934) Venetian ships and shipbuilders of the Renaissance. The John Hopkins Press, Baltimore

Lazar I, Willmott H (2006) The glass from the Gnalić wreck. Annales Mediterranea, Koper

Mileusnić Z (ed) (2004) The Venetian shipwreck at Gnalić. Annales Mediterranea, Koper

NOAA (2018) Magnetic field calculators. https://www.ngdc.noaa.gov/geomag-web/#declination. Accessed 14 Aug 2018

Petricioli S (1981) Deset godina rada na hidroarheološkom nalazu kod Gnalića. Godišnjak zaštite spomenika culture. Hrvatske 6(7):37–45

Petricioli S, Uranija V (eds) (1970) Brod kod Gnalića - naše najbogatije hidroarheološko nalazište, Vrulje—Glasilo Narodnog muzeja u Zadru 1

Polzer M (2004) An Archaic laced hull in the Aegean: the 2003 excavation and study of Pabuç Burnu ship remains. INA Quaterly 31(3):3–11

Polzer M, Casaban J (2012) Photogrammetry: a legacy of innovation reaching back to Yassıada. INA Quarterly 39(1/2):13–17

Radić Rossi I, Bondioli M, Nicolardi M, Brusić Z, Čoralić L, Vieira de Castro F (2013) Brodolom kod Gnalića—Ogledalo renesansne Europe/the shipwreck of Gnalić—mirror of Renaissance Europe. In: Filep A, Jurdana E, Pandžić A (eds) Gnalić: Blago potonulog broda iz 16. stoljeća/Gnalić—treasure of a 16th century sunken ship, exhibition catalogue. Croatian History Museum, Zagreb, pp 65–95

Radić Rossi I, Nicolardi M, Batur K (2016) The Gnalić shipwreck: microcosm of the Late Renaissance world. In: Davison D, Gaffney V, Miracle P, Sofaer J (eds) Croatia at the Crossroads: a consideration of archaeological and historical connectivity. Archaeopress Archaeology, Oxford, pp 23–248

Ridao P, Gracias N (2017) The methodology and the results of the AUV (Girona 500 survey of the present state of the Gnalić shipwreck site. In: Abstracts In Poseidon's realm XXII: international conference on underwater archaeology (We're all in the same boat—the social importance of ships, rafts and ferries), Koblenz, 17–19 March 2017

Rule N (1989) The Direct Survey Method (DSM) of underwater survey, and its application underwater. Int J Naut Archaeol 18(2):157–162. https://doi.org/10.1111/j.1095-9270.1989.tb00187.x

Steffy JR (1994) Wooden ship building and the interpretation of shipwrecks. Texas A&M University Press, College Station

Tenenti A (1959) Naufrages, corsaires et assurances maritimes à Venise, 1592–1609. SEVPEN, Paris

Tucci U (1967) Un problema di metrologia navale: la botte veneziana. Studi veneziani IX:201–246

Yamafune K (2016) Using computer vision photogrammetry (Agisoft Photoscan) to record and analyze underwater shipwreck sites. PhD dissertation, Texas A&M University

Yamafune K (2017) A methodology for accurate and quick photogrammetric recording of underwater cultural heritage. In: Proceedings of the 3rd Asia-Pacific Regional conference on Underwater Cultural Heritage, Hong Kong, 1, pp 517–537

Yamafune K, Torres R, Castro F (2016) Multi-Image photogrammetry to record and reconstruct underwater shipwreck sites. J Archaeol Method Theory 24(3):703–725. https://doi.org/10.1007/s10816-016-9283-1

5 Underwater Photogrammetric Recording at the Site of Anfeh, Lebanon

Lucy Semaan and Mohammed Saeed Salama

Abstract

This chapter considers the application of underwater photogrammetry to record and document the underwater cultural heritage at the site of Anfeh in North Lebanon. Although photogrammetry has become a standard procedure in the field of maritime archaeology worldwide, this is the first use of this recording method in the country. The research context is presented, followed by the methodology adopted according to the particularities of the site and then the results of work undertaken over two campaigns: one in 2016 and one in 2017. The main aims in this chapter are to demonstrate the advantages of a low-cost and time-effective method of documenting sites, where the funding prohibits the use of more expensive geophysical equipment. The application of multi-image photogrammetry as a recording technique at Anfeh has merit in providing global access to artefacts in their in situ context. The results generated from 3D data were particularly informative to the study of a substantial collection of anchors of different types and sizes, without removing them from their underwater context. By calculating volume from the 3D scan, an estimation of the weight of these could be thus achieved and will serve in future analysis of the vessels plying the maritime routes at Anfeh.

Keywords

Underwater archaeology · Maritime material culture · Anchors · Eastern Mediterranean

5.1 Introduction

Documentation is an intrinsic part of the archaeological process and presents innate challenges when research takes place in a submerged setting. Indeed, underwater archaeologists are faced with several constraints—not encountered in terrestrial archaeology—such as maximizing bottom time during each dive, the depth of the operations, as well as the ambient conditions of visibility, currents, and temperature. Underwater surveying methods using photography have developed substantially in the last few decades, and one such application, multi-image photogrammetry, has become a commonly used recording tool.

In lay terms, photogrammetry is defined as a process for taking measurements from photographs. However, multi-image photogrammetry and its related applications are much wider in scope as they breach the spatial and temporal limitations encountered when working under water, and also constitute a dissemination tool that fosters accessibility and hermeneutic discussions about the archaeological material. The nature, range, and development of photogrammetry go beyond the scope of this chapter and have been argued elsewhere (Balletti et al. 2016; Costa et al. 2015; Gawlik 2014; Green et al. 2002; Drap 2012; Polzer and Casaban 2012; McCarthy 2014; McCarthy and Benjamin 2014; Van Damme 2015; Yamafune et al. 2016; and the various chapters in this volume). Notably, McCarthy (2014, 176) stressed the potential of this technique as 'a practical and cost-effective method for accurate survey and as a tool for community engagement with heritage.' Also, from the perspective of applying such a technology to the documentation of shipwrecks (Balletti et al. 2015, 2016; Skarlatos et al. 2012, 2014) and underwa-

L. Semaan (✉)
Department of Archaeology and Museology, University of Balamand, Al Kurah, Lebanon
e-mail: lucy.semaan@balamand.edu.lb

M. S. Salama
Alexandria Centre for Maritime Archaeology & Underwater Cultural Heritage, Alexandria University, Alexandria, Egypt

J. K. McCarthy et al. (eds.), *3D Recording and Interpretation for Maritime Archaeology*, Coastal Research Library 31,
https://doi.org/10.1007/978-3-030-03635-5_5

ter sites (Bruno et al. 2015), and even for deep-sea underwater surveying (Drap et al. 2015), such a technology represents the opportunity to transpose the study of and accessibility to underwater cultural heritage from the marine environment onto dry land in a non-intrusive manner.

This chapter details the application of multi-image photogrammetry at Anfeh, Lebanon in an academic context and research into the maritime cultural landscape of Anfeh.[1] The impact of such a technology in Lebanon is enhanced by the fact that it is the first endeavour of its kind in the country. To the authors' knowledge at present, no other research project pertaining to underwater archaeology has applied 3D photogrammetric recording to the documentation of the Lebanese underwater cultural heritage.

5.1.1 Context of the Research

The coastal town of Anfeh is located approximately 70 km north of Beirut and 15 km south of Tripoli. It is bordered by the village of Chekka and the Barghoun River to the south; to the north, by the agricultural area of Hraishi and the village of Qalamoun; and to the east by the villages of Barghoun and Zakroun (Fig. 5.1). Anfeh extends westwards to the Mediterranean Sea by a 400-metre long and 120-wide peninsula (Fig. 5.2). The latter, Ras al-Qalaat, is roughly oriented on a NNW-SSE axis while rising some 14 m above mean sea level (MSL). Three rock-cut moats separate it from the mainland at its easternmost limit and are carved on a north-south axis roughly perpendicular to the peninsula (Fig. 5.3) (Chaaya 2016, pp. 281–282; Lawrence 1988, 28; Nordiguian and Voisin 1999, 381; Panayot-Haroun 2015, 399). The coastline north of Ras al-Qalaat is exposed as it offers no lee from the dominant SW winds. It consists of cliffs that drop in places onto a narrow rocky shore that is hazardous for seafaring. Closer to the peninsula, the coastline forms two large well-protected shallow bays—the Nhayreh Bay and the bay of Ras al-Safi—that offer natural havens (Fig. 5.4). Due to coastal urbanisation that started developing in the 1980s, both sides of the Nhayreh Bay are occupied by modern beach resorts. This leaves a narrow space in the bottom of the bay for the present-day modest fishermen's harbour. To the south of Ras al-Qalaat, the rocky shoreline is low-lying and consists of a small cove with an open bay that are suitable for anchoring and landing places when the northerly winds blow, being in the lee of the promontory.

Recent terrestrial excavation and survey work on the peninsula and its hinterland by the Department of Archaeology and Museology (DAM),[2] University of Balamand (UoB), suggests an occupation of the promontory of Ras al-Qalaat and of the modern town of Anfeh that extends most likely from the Early Bronze Age to the Ottoman period (Panayot-Haroun 2015, 2016). In May 2016 and in September 2017, DAM[3] led a one month detailed recording of underwater cultural material at the site through multi-image photogrammetry. This was preceded by a three week underwater visual survey undertaken in 2013 in the waters around Ras al-Qalaat and off the north and south coastal stretches of the modern town that aimed at assessing the underwater archaeological potential at Anfeh (Fig. 5.5) (Semaan 2016; Semaan et al. 2016).

Considering the logistical constraints in terms of time, and human and financial resources, the 2013 preliminary survey also laid the groundwork for the subsequent underwater photogrammetry surveys. The later surveys mapped submerged relevant archaeological features and artefacts, and generated accurate in situ 3D records of these, while reducing the time and cost of archaeological survey work. Indeed, this low-cost and user-friendly method for acquiring high-resolution datasets is well suited for low-budget research since, traditionally, underwater archaeological research projects are associated with high financial and logistical costs. At Anfeh, these detailed datasets constitute base-line knowledge of its seabed, which is hitherto understudied, and offers first-hand documentation for further research.

An innovative application has been the calculation of volume, and by extrapolation the weight, of in situ stone anchors based on the density of the material associated with each stone type (Table 5.3). Indeed, the standard practice of stone anchor recording or 'anchorology', set by pioneer Honor Frost in Mediterranean archaeology (Frost 1997), favours weight as a key typological parameter, which can reflect both the antiquity of anchors and the size of the carrying vessel. Underwater photogrammetry allows a much more accurate calculation of anchor weight, as recently demonstrated by Fulton et al. (2016). Moreover, the site's submerged archaeological material is vulnerable to looting—thus multi-image photogrammetry allows close monitoring and is a means to mitigate potential losses.

[1]This research project constitutes Semaan's 3-year post-doctoral fellowship (2015–2018) at UoB with a grant from the Honor Frost Foundation and UoB. Hence, all fieldwork related activities are financed by Semaan's research allowance.

[2]The Anfeh Project is led by Nadine Panayot-Haroun, Head of DAM.

[3]The teams were supervised by Lucy Semaan (UoB) and comprised maritime archaeologists Rupert Brandmeier (Ludwig-Maximilians-University), Clara Fuquen (University of Southampton), Menna-Allah Ibrahim, Ziad Morsy, Mohammad Saeed, and Maii Tarek (University of Alexandria), Enzo Cocca and Salvatore Colella (Università degli Studi di Napoli L'Orientale); archaeologist Hadi Choueiri; as well as Lebanese divers Hussein al-Hajj, Mario Kozaily, Serge Soued, Shadi Zein, and Elie Semaan who intermittently joined the team on a voluntary basis.

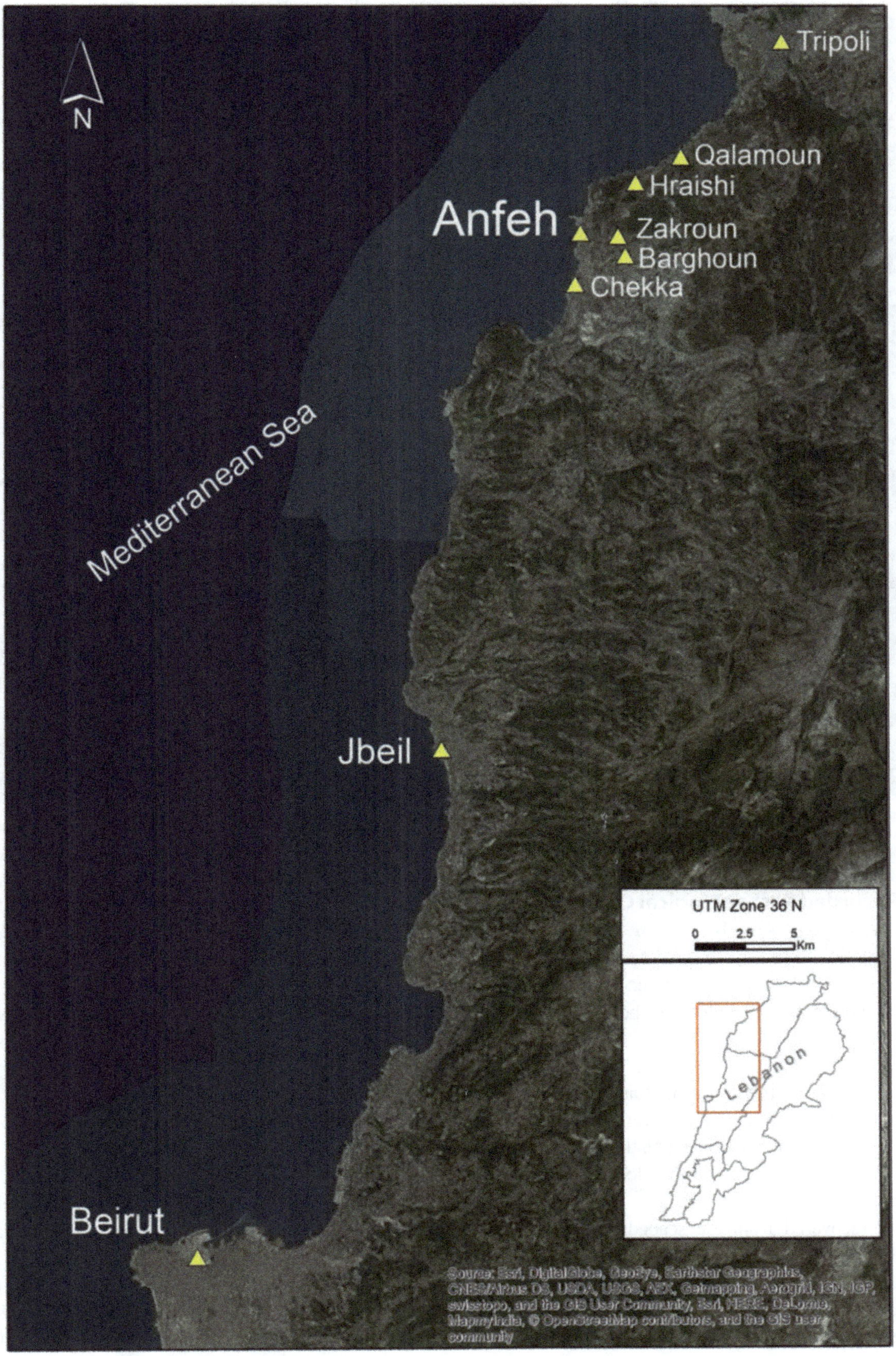

Fig. 5.1 A map of the location of Anfeh and neighbouring areas (C. Safadi)

Fig. 5.2 Aerial image of Ras al-Qalaat extending westwards into the Mediterranean Sea (R. Tanissa)

5.1.2 Recorded Archaeological Cultural Heritage at Anfeh

The targeted artefacts and features recorded using multi-image photogrammetry at Anfeh can be divided into four categories:

1. Onshore ramps or slipways: There are two slipways on the Ras al-Qalaat promontory that were recorded via photogrammetry and one further east on the southern coast of Ras al-Safi bay; while the fourth is located on the southern coast of Anfeh;
2. Anchors recovered from the seabed: 12 anchors were retrieved in 2013 during the underwater visual survey for the purpose of dating and typology. Another two were recovered in 2016. The 14 anchors were recorded via photogrammetry on land;
3. Underwater anchors: These 17 anchors were recorded in situ in 2013 and 2016 and were left under water. They either constitute isolated finds or anchors located in groups adjacent to each other. The team defined the groups and zones to be covered according to the organic distribution of these artefacts; and
4. Underwater masonry blocks: These were mainly recorded previously in 2013. As with the anchors, they were also found either as isolated finds or in larger clusters.

5.1.3 Methodology

The authors used the *Agisoft Photoscan/Metashape Professional edition®* software to record and map underwater material such as masonry blocks and anchors as well as coastal features, mainly the above-mentioned four slipways; and to produce orthogonal projections, sections, plans and drawings as supporting visual documentation in the framework of researching the maritime cultural landscape of Anfeh. One of the future goals is to combine the 3D rendering of anchors and masonry blocks with other digital techniques to produce a virtual tour of the modelled areas for the general public, espe-

Fig. 5.3 Aerial image of the three moats on the eastern side of Ras al-Qalaat (R. Tanissa)

cially the non-divers, so they gain an appreciation of the underwater cultural heritage at Anfeh (Fig. 5.6). These activities are part of a wider workflow that was inspired by Yamafune et al. (2016). As this still is a work in progress, this chapter will discuss the procedures of underwater photography, photogrammetry, and orthophotos illustrated in the workflow diagram.

5.2 Underwater Photography

5.2.1 Equipment

During the 2016 and 2017 seasons,[4] the team used four different underwater photography systems for data collection, in order to compare and optimise results (Table 5.1): (1) GoPro HERO4 Black Edition with its stock underwater housing; (2) Canon PowerShot G15 compact camera, with an underwater Fantasea housing (Fig. 5.7: to the right); (3) Canon EOS 70D DSRL equipped with a Canon 20-mm lens that provides a great depth-of-field while getting closer to the target. The camera was set in an IKELITE Underwater TTL housing, mounted on an aluminium tray with dual quick release handles, and with a modular 8 in. dome with 2.75-in. lens extension. The whole kit was completed with two DS-161 strobes and their light diffusers. The Ikelite housing had a hemispherical glass in front of the lens in order to minimize the refraction error due to the water/glass interface (Drap 2012, 115), and thus generate sharper image corners caused by the use of a wide angle lens (Fig. 5.7: to the left) (Yamafune et al. 2016, 7); and (4) Sony DSC-RX100 com-

[4]Images for the anchors, the masonry blocks in S7, S8 and U9 were captured in 2016 by Mohammad Saeed and Ziad Morsy, and the ones for the masonry blocks located in R10 and T8, T9 were captured in 2017 by Salvatore Colella and processed by Enzo Cocca (See Archaeological Survey Results below).

Fig. 5.4 Aerial image of Ras al-Qalaat and its north and south bays, taken in 1956. (©Directorate of Geographical Services, Ministry of Defence, Lebanon, 1/5000, Photo 494, Mission 1/100; Property of the Department of Archaeology and Museology at UoB)

pact camera set in a Nauticam Underwater housing working with an INON UWL-H100 wide conversion lens type 2.[5]

A high performance computer is usually required to run multi-image photogrammetry software as it analyses a substantial amount of images (McCarthy 2014, 177). Therefore, the team used a MSI GT80 Titan SLI high-end gaming laptop with a 2.70-GHz Intel Core i7-6820HQ CPU, 32 GB of RAM and Dual NVidia GeForce GTX video card, which successfully dealt with the large photogrammetric datasets.

5.2.2 Data Collection

First, the various targeted areas were marked and delimited: either visually using natural markers on the seabed as references for small surfaces, or with ropes and rods for larger surfaces. The features and artefacts were then cleaned of debris and algae and then tagged. A 1-m scale bar, or sometimes two, were placed next to objects. The scale bars were graded in black and white every 10 cm. Then, a set of clear and well-exposed images was taken of the targeted area from free positions, with a considerable overlap of over 50% between images. As Drap (2012, 114) argues: 'The key factor of this method is redundancy: each point of measured space must be seen in at least three photographs.' This is also done to optimize results in *Agisoft Photoscan/Metashape* as each image is compared to every other image. The settings of the cameras, and particularly the lens zoom, were not changed during data capture.

The series of photos were taken with an estimated field of view of 45–70°, and from different positions while swimming in a circular and/or a zigzag pattern. When the objects and their wider landscape were being recorded, the diver covered the limits of the landscape first. Indeed, several oblique photos were taken around each location/group of artefacts and features to provide a wider coverage. Subsequently and in a continuous mode, the diving photog-

[5]This system was used in 2017 by Salvatore Colella while the previous three systems were used in 2016.

Fig. 5.5 Bathymetry map of Ras al-Qalaat and Anfeh's coastline showing the grid system and the areas covered during fieldwork from 2013 to 2017 (Produced for the underwater visual survey at Anfeh by the Institute of Environment (UoB) and modified in GIS by C. Safadi and E. Cocca)

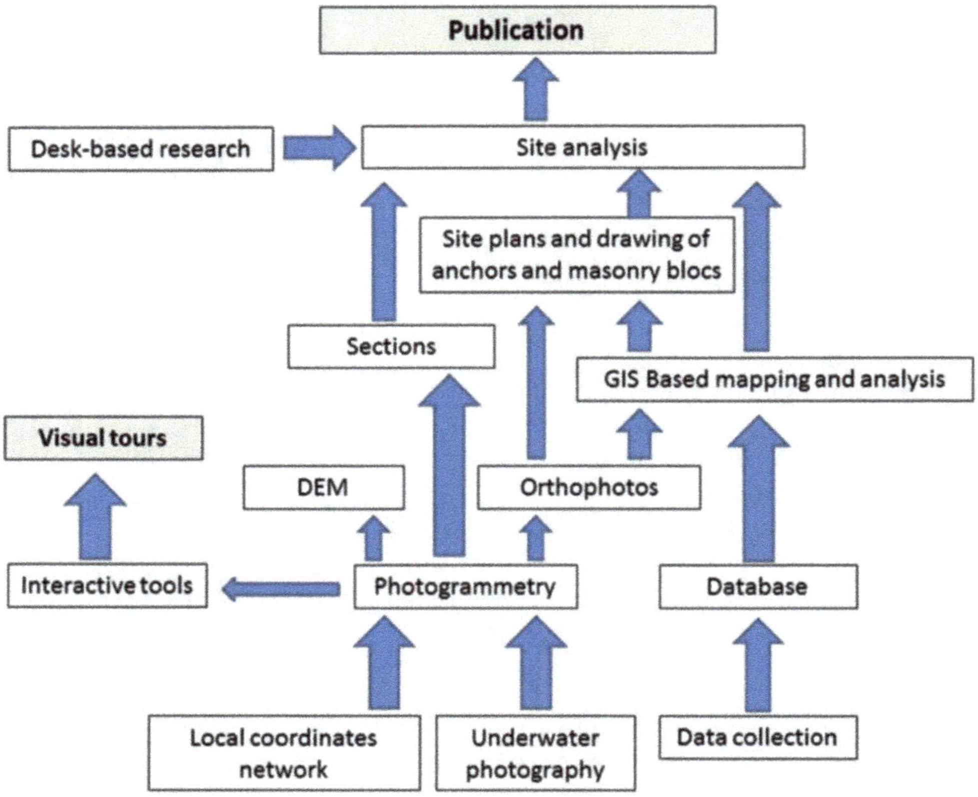

Fig. 5.6 Workflow for the multi-image photogrammetry applied to Anfeh's underwater cultural heritage following the methodology set by Yamafune et al. (2016) (L. Semaan modified from Yamafune et al. 2016:Figure 1)

Table 5.1 Summary table of the characteristics of each camera system

Camera systems	GoPro HERO4 black	Canon PowerShot G15	Canon EOS 70D DSRL	Sony RX-100
Image sensor	CMOS	CMOS 1/1.7″	APS-C 'Dual Pixel CMOS AF'	Exmor CMOS sensor 1″
Sensor size	1/2.3″	41.51 mm² (7.44 × 5.58 mm)	337.5 mm² (22.50 × 15.00 mm)	116.16 mm² (13.2 × 8.8 mm)
Resolution	12MP, 4000 × 3000 pixels	12MP, 4000 × 3000 pixels	20.2MP, 5472 × 3648 pixels	20.2 MP, 5472 × 3648 pixels
Focal length	Wide-angle lens	28–140 mm (equivalent at 35 mm to 6.1–30.5 mm F/1.8–2.8)	18–135 mm (equivalent at 35 mm to 29–216 mm, F/3.5–5.6)	10.4–37.1 mm (equivalent at 35 mm to 28–100 mm, F/1.8–4.9)
ISO range	6400 (max)	80–12800	100–12800 standard 25600 expanded	125–6400
Zoom ratio	–	5×	7.50×	3.6×
Aperture range	F/2.8	F/1.8–F/2.8	F/3.5–F/22 (wide) –F/5.6–F/36 (tele)	F/1.8 (wide)–F/4.9 (tele)
Shutter speed	2 s–1/8192 s	15 s–1/4000 s	30s–1/8000 s	30s–1/2000
Flash		Built-in	Built-in	Built-in
Supported operating system	Windows 7, 8, 8.1, Mac OS X 10.8	Windows XP, Vista, 7, 8, 10, Mac OS X	Windows XP, Vista, 7, 8, 10, Mac OS X	Windows XP, Vista,7, 8, 10, Mac OS X
Operating environment	Max depth: 60 m, max temperature: 50 °C	Temperature: 0°–40 °C Humidity: 10–90%	Temperature: 0°–40 °C Humidity: 0–85%	Temperature: 0°–40 °C
Dimensions	41 × 59 × 21/30 mm	107 × 76 × 40 mm (without housing)	139 × 104 × 79 mm (without housing and strobes)	101.6 × 58.16 × 1.41 mm (without housing)
Weight	88 g (152 g with housing)	352 g (without housing)	755 g (without lens dome, housing, and strobes)	213 g (body only)

Fig. 5.7 Two of the camera systems used: to the left, the Canon EOS 70D DSRL with the Ikelite housing and to the right the Canon PowerShot G15 with the Fantasea housing (L. Semaan)

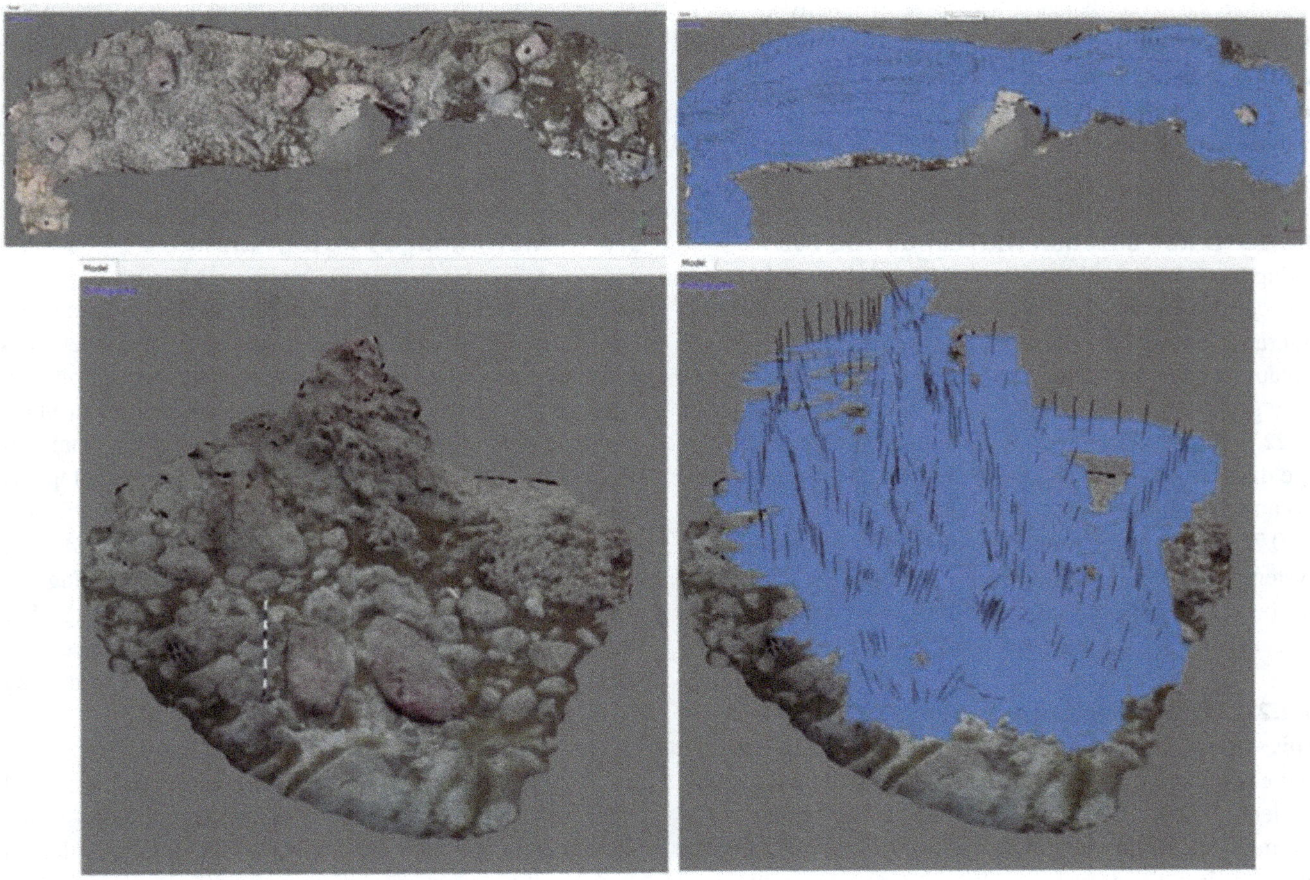

Fig. 5.8 Examples of the camera positions covering two groups of anchors (top: Group 2 and 3), and two well-weathered masonry blocs (bottom) (Photos: M. S. Salama)

rapher captured a series of pictures that covered the object up-close to a distance above object between a range of 0.5–1.5 m with a 360° capture around the object itself (Fig. 5.8).

During data acquisition the GoPro camera did not provide clear high-resolution images. The focal length of its lens is quite small, and as a wide angle fisheye, caused important distortions (Capra et al. 2015, 71).[6] The authors decided, therefore, to forego capturing images with this system and focused on the other two higher-end systems:

5.2.2.1 Data Collection with CanonG15

This system was light-weight, compact, and easy to use. Hence it did not cause major issues for the team during data acquisition. This system has a 28-mm equivalent lens (*f: lens focal length*) and a camera with sensor sizes of 7.6 mm (*S_x: sensor size in x*) by 5.7 mm (*S_y: sensor size in y*). Swimming on an average of 1 m (*d: distance from the object*) above the targeted sites, and considering the relational equation for producing good quality images $C_{x=} \frac{Sx \times d}{f}$ (Yamafune et al. 2016), the diver was able to cover a rectangular area of 0.27 m (*C_x: coverage in x*) by 0.20 m (*C_y: coverage in y*) on each shot. Perhaps the one downside of this system, however, was that the white balance had to be manually evaluated depending on the depth of the targeted area for JPEG acquisition. For example, one area extended between a depth of 4 and 10 m which complicated the finding of the right white balance underwater and optimising the quality of the images.

5.2.2.2 Data Collection with Canon EOS 70D

Using this system enabled the team to obtain clear shots with true colours and a good exposure of the site. Also, this configuration allowed for a wider coverage than the Canon G15, because a wide angle lens could be used, thus ensuring more time-saving swimming plans: with a 20-mm lens (*f*), a sensor of 22.5 (S_x) by 15 mm (S_y), and while swimming on an average distance of 1 m (*d*) above the targets, the diver was able to capture shots equating to a rectangular area of 1.12 m (C_x) by 0.75 m (C_y). The Canon EOS 70D with its Ikelite housing system, however, is quite bulky to manipulate, and the strobes would only capture about 300–600 photos in a single dive, depending on the flash power.

5.2.2.3 Data Collection with Sony DSC-RX100

This system was similar to the Canon G15 as it was also light and easily manoeuvred underwater. It captured photos with the lens at 12.18 mm (*f*) and a sensor size of 13.2 mm (S_x) by 8.8 mm (S_y). Swimming on an average of 1 m (*d*) above the targeted sites, the diver was able to cover a rectangular area of 1.08 m (C_x) by 0.72 m (C_y) respectively on each shot.

5.2.3 Image processing

Upon acquisition, the images were downloaded to the computer, sorted, and tagged. This was done using *Adobe Bridge*, which is a digital asset management software that can import and bulk rename batches of pictures, while granting easy access to metadata. Tagging pictures was done systematically, by filling the related fields of the software accordingly: 'Area code _Dive Log_ Artefact ID_Material/Subject_ Date_Photo number'. The Area code refers to the square individual number of the survey grid; the Dive Log is the number of the dive related to collecting the related picture; Artefact ID stands for the inventory name of the archaeological object; Material is the nature of the artefact photographed such as masonry, anchor, pottery, etc., and the subject refers to pictures showing people in action or the underwater environment for example; the date relates to the date when the pictures were taken; and the photo number is the serial number automatically generated by the camera used.

The images collected needed to be processed before they could be used to construct accurate 3D photo models. This was done through using *Adobe Lightroom 5* and *Photoshop CS5* to edit the colour balance, the contrast, and the brightness properties of the photos. The authors preferred not to greatly modify the colours of the images, in order to stay faithful to the actual state of the object and the landscape.

5.3 Multi-image Photogrammetry

Subsequent to processing, overlapping digital images taken from different viewpoints were loaded in one batch into *Agisoft Photoscan/Metashape*, which analyses each picture, automatically detects matching correlated features in unordered picture collections, and creates 3D models from still images. At Anfeh, this software was used to develop 3D models of the collected data, as it proved to be reliable and easy to use in its automation and interface. Also, *Agisoft Photoscan/Metashape* has the advantage of rendering the camera calibration and bundle adjustments obsolete. The MSI laptop used was able to process quite dense point clouds which were limited between a few hours to processing overnight.

The camera and housing system used in 2017, however, the Sony DSC-RX100, needed calibration. A plastic A4 checkered sheet was printed and plunged in the water. Subsequently and with the camera at the same focal length used for the underwater photogrammetry, team members Enzo Cocca and Salvatore Colella took photos in order to frame just the sheet in various settings. Generally, 10–15 photos were taken after each photogrammetry immersion. They were then processed with *Agisoft Lens* software in order to calibrate the camera system. Once this was done, the exported.xml file was used in *Agisoft Photoscan/Metashape* to correct the errors of the images. Below is a list of the calibration parameters used in *Agisoft Photoscan/Metashape* to align the images:

[6]This is, however, not always the case as explained by Van Damme (2015, p 236).

- fx, fy: the x and y dimensions of the focal length measured in pixels;
- cx, cy: coordinates of the main points, i.e. the coordinates of the optical axis intersection with the sensor plane;
- skew: skew transformation coefficient;
- k1, k2, k3, k4: radial distorsion coefficients; and
- p1, p2, p3, p4: tangential distortion coefficients.

There are several tasks required by *Agisoft Photoscan/Metashape* in order to load and process images, and produce 3D models (Agisoft 2016, 8). Below are the ones that were applied specifically during the multi-image photogrammetric process at Anfeh:

1. Loading photos into the software and aligning them which results in creating sparse point clouds and showing camera positions;
2. Building dense point cloud which is based on the estimated camera positions and the captured pictures;
3. Building mesh, that is, generating a 3D polygonal model that represents the object surface based on the dense point cloud (Agisoft 2016, v);
4. Generating texture and applying it to the mesh; and
5. Exporting orthomosaic, 3D PDF files, and Digital Elevation Models (DEM).

The general settings for processing the various 3D models are compiled in Table 5.2.

As previously mentioned, the nature of the recorded material at Anfeh varied between on-land and underwater material culture.[7] During the 2016 season a total of about 30,000 photos were captured, including a total of 23 dives dedicated to the task of underwater photogrammetry equating to a total of 24 h of shooting underwater. So far, the total days for processing data are equivalent to 65 days. Meanwhile during the 2017 season, a total of 1762 photos were captured during a total of three dives which equated to about 4 h of shooting underwater. The total time spent processing these was 8 h and 4 min.

Table 5.2 General settings followed for all 3D models

Align photos	Accuracy/high–medium
	Pair preselecting/generic
Build dense cloud	Quality/high–medium
	Depth filtering/mild–moderate
Build mesh	Polygon count/medium–low
	Interpolation/enabled
Build texture	Mapping mode/generic
	Blending mode/mosaic

[7]See 'Recorded Coastal and Underwater Cultural Heritage at Anfeh' section above.

5.3.1 Orthophotos

Subsequent to generating 3D models for each artefact or group of artefacts, orthophotos were extracted as 1:1 scaled top view photomosaics with *Agisoft Photoscan/Metashape*. These 2D images were captured in high quality and exported as JPEG formats. Decimated versions of these were also produced to facilitate their use in presentations and in other instances where highly accurate files were not needed. These are currently used as a basis for drawing and tracing 2D site plans in AutoCAD® software.

In the case of the isolated finds, the orthophotos were directly generated from the 3D model. In the case of the larger groups of finds—such as the different groups of anchors at the feet of the southern reef and the large area of masonry blocks on the north-eastern side of Ras al-Qalaat—however, the photomosaics were patched together from the different batches used in generating the photogrammetric models of the sites. In such instances, the batches were close enough to each other and there was sufficient overlap between each group. Therefore, the authors linked these batches by marking common points then aligning the chunks to each other according to these points.

5.3.2 Export Adobe 3D PDFs

The team was also able to able to export 3D models as 3D PDF format. One of the main advantages of generating such format files is the careful study of the material while reducing the time required for measurements underwater; as well as extrapolating measures and dimensions of artefacts taken at the office from the corresponding 3D PDF files (Tables 5.3 and 5.4).

5.4 Archaeological Survey Results

Since this chapter pertains mainly to underwater recording and photogrammetry, the related processes of recording submerged anchors and limestone masonry blocks lying on the seabed will be explained here. Indeed, the team was able to effectively obtain the accurate shapes, geometry, colour, measurements, and spatial distribution of the submerged artefacts.

5.4.1 Isolated Anchors

Several anchors were located during the 2013 and 2016 visual surveys, some of which were left in situ and recorded photographically for 3D photogrammetry. One example is a one-hole triangular anchor (AN.S12.004) of substantial

Table 5.3 Measurements of the anchors mentioned in the text according to their 3D PDF files

Anchor	Length (cm)	Width (cm)	Thickness (cm)	Hole diameter (cm)	Estimated weight (kg)
AN. S12 004	71.8	65.8	36	22	337.82
AN. S11. 273	35.7	20.8	5.8	4.6–4.1	8.56
AN. S11. 286	46.7	27.2	7	9.4	18.20
AN. S11. 287	46	31.3	4.9	12.3	13.98
AN. S11. 288	47.2	30.1	4.8	16.4	12.54
AN. S11. 274	42.5	31.5	11.5	5.3	32.70
AN. S11. 291	45	30	10.5	8.5	29.33
AN. S11. 292	77.8	58.5	30.8	17.2	287.33
AN. S11. 293	76	64.6	35	24.7	334.95
AN. T11. 294	49	51.95	14.5	20.8	69.08
AN. T11. 295	23.6	20.1	12.5	4	12.46
AN. T11. 296	67.2	63.3	33	22.9	273.86
AN. T11. 297	64.3	41.5	20	14	108.63
AN. T11. 298	73.8	47	15.5	12.4	112.08
AN. T11. 299	37.9	31	7.2	4.5	18.02
AN. T11. 300	50	30	7.9	8.3	24.67
AN. T10. 146	57.6	38.8	19.7	9.9	91.82

Table 5.4 Measurements of the masonry blocs mentioned in the text according to their 3D PDF files

Block	Length (cm)	Width (cm)	Depth (cm)
AN. S8. 188	157	133	94.3
AN. S8. 189	190	124	102
AN. S8. 190	132	87.6	81.2
AN. S8. 191	157	120	86.9
AN. S8. 192	177	127	91.9
AN. U9. 240	100	32.2	12.2

The measurements of AN.U9.240 are the ones taken in 2016 when the block was found half buried in sediments

dimensions (Table 5.3), that was located at a depth of 12 m and lodged under the feet of the southern reef of Ras al-Qalaat (Figs. 5.9 and 5.10).

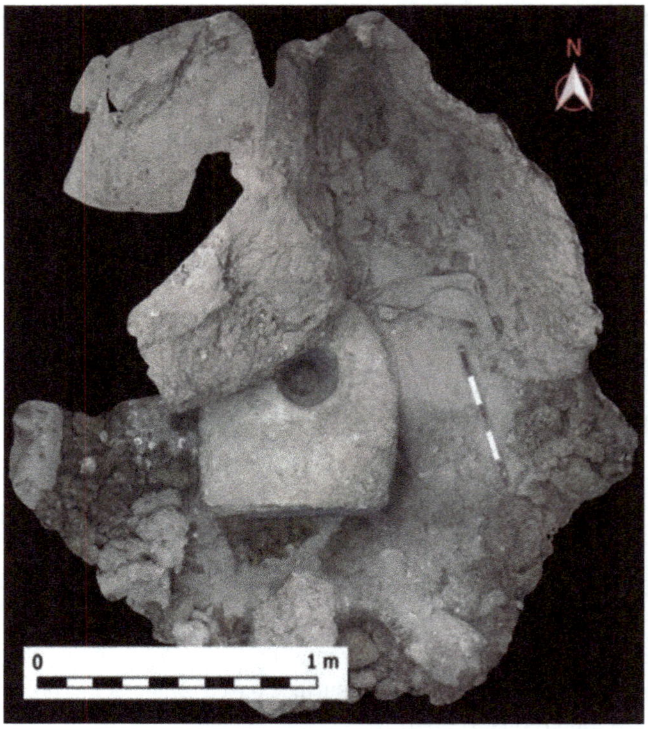

Fig. 5.9 Orthophoto of the anchor (AN.S12.004) found at the bottom of the southern reef of Ras al-Qalaat (M.S. Salama).

5.4.2 The Groups of Anchors

Four groups of anchors, all located scattered at the feet of the southern reef at a depth of 11–12 m below MSL and some 30–80 m from the coast, were surveyed photographically for multi-image photogrammetry (Fig. 5.10). Group 1 is located in Square S11 and includes four small-sized anchors, one of which was two-holed (AN.S11.273), with the rest being one-holed (AN.S11.286–288) (Fig. 5.11). Group 2 is located in Square S11 and includes four single-hole anchors of which two were quite large (AN.S11.292 and AN.S11.293) and the other two modest in size (AN.S11.291 and AN.S11.274) (Fig. 5.12). Group 3 is located in Square T11 and is comprised of seven single-hole anchors (AN.T11.294, 295, 296, 297, 298, 299, and 300) (Fig. 5.12). Group 4 consists of a single-holed anchor along with two stone mill elements (Anchor: AN.T10.146) (Fig. 5.13).

5.4.3 The Isolated Masonry Blocks

A few isolated blocks were recorded. These are mainly located to the south of Ras al-Qalaat. A rectangular block (AN.U9.240) was recorded in 2013 via GPS at a depth of 10.8 m about 100 m off the southern coast of the Anfeh vil-

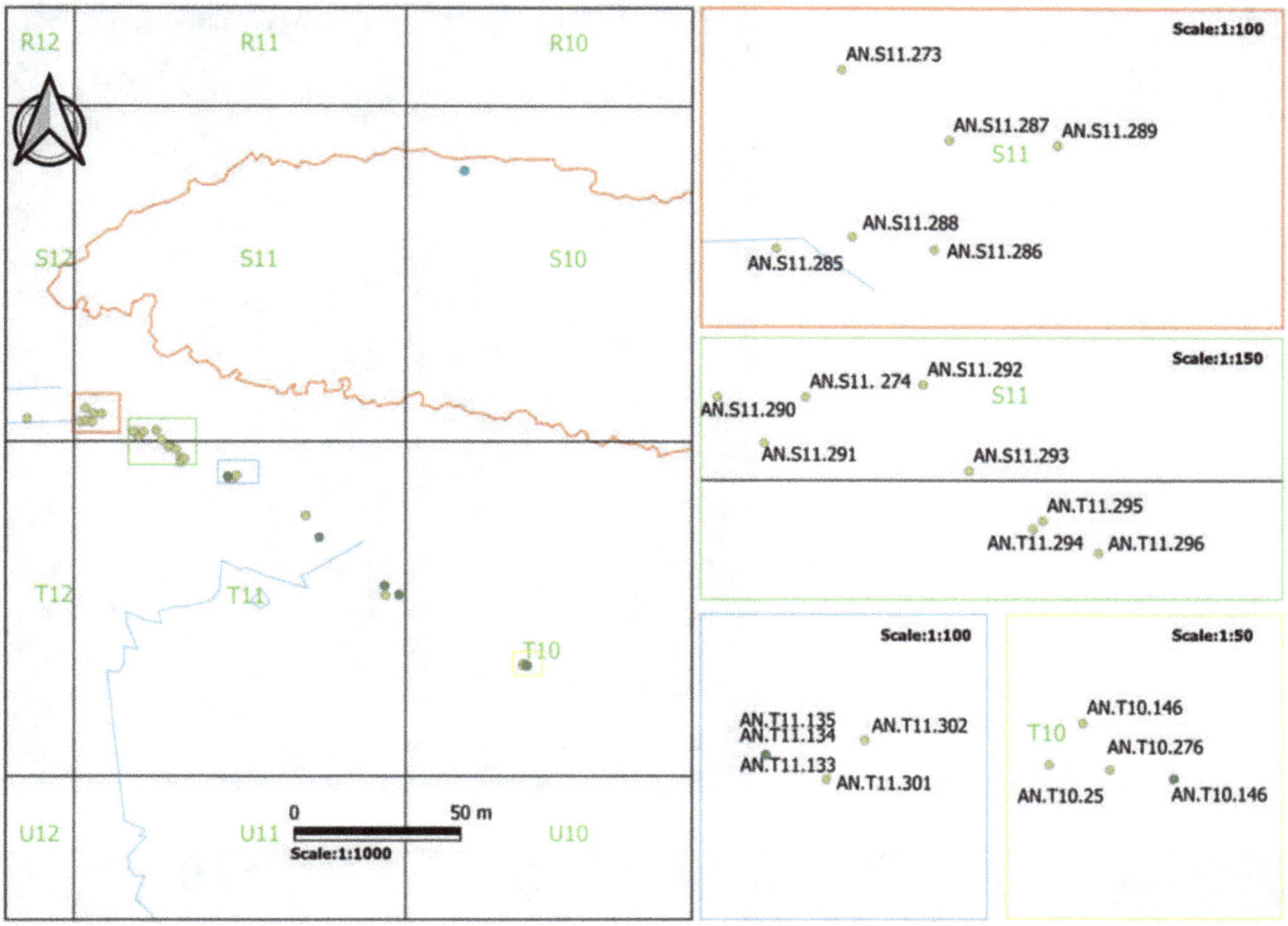

Fig. 5.10 Map of all the artefacts recorded during the surveys of 2013 and 2016 on the southern reef of Ras al-Qalaat (E. Cocca)

lage. It measures 1.98 × 0.45 × 0.30 m. In 2016, this masonry block was found completely covered under some 40 cm of sediments (Fig. 5.14) (Table 5.4). It was only partially uncovered as no excavation dredge was available.

5.4.4 The Masonry Blocks in Groups

5.4.4.1 The North-Eastern Group

This group of masonry blocks is located to the north-east of the Ras al-Qalaat, only a few metres offshore. It consisted of two clusters of blocks located at about 2 m from each other in Square S8. The first cluster is made of three rectangular blocks (AN.S8.185–187) and the second of five blocks (AN. S7.188–192) (Figs. 5.5 and 5.15), and at a maximum depth of 5 m (Table 5.4). The team established a reference grid for this area with four rods and a polygon of approximately 35 m of perimeter rope. The location of these clusters had been recorded with a GPS in 2013.

5.4.4.2 The North-Western Group

This large group of masonry blocks is located in Square R10 on the north-western side of Ras al-Qalaat slightly less than 50 m from the coast (Fig. 5.5). It stretches over an area about 30 m by 20 m, dropping northwards from a depth of 3–13 m. These blocks vary in sizes and shapes; most of them do not seem to be worked, unless they were heavily weathered by the sea conditions (Fig. 5.16).

5.4.4.3 The South-Eastern Group

This group of masonry blocks is located to the south east end of the Ras al-Qalaat, spreading from the foot of the reef to only a few southwards, on a NW-SE axis (Figs. 5.5 and 5.17). It consisted of six spread-out blocks in Square T8 (AN.T8.175–178, 282, 305) and a close-range cluster of eight blocks in Square T9 (AN.T9.306–313). The blocks lay at a depth range of 3–5 m. The location of these clusters had been recorded via GPS in 2013 and georeferenced with the total station through a set of control points in 2017. The study of this group of blocks is ongoing.

Fig. 5.11 Orthophoto of Group 1 of anchors located at the bottom of the southern reef of Ras al-Qalaat (M.S. Salama)

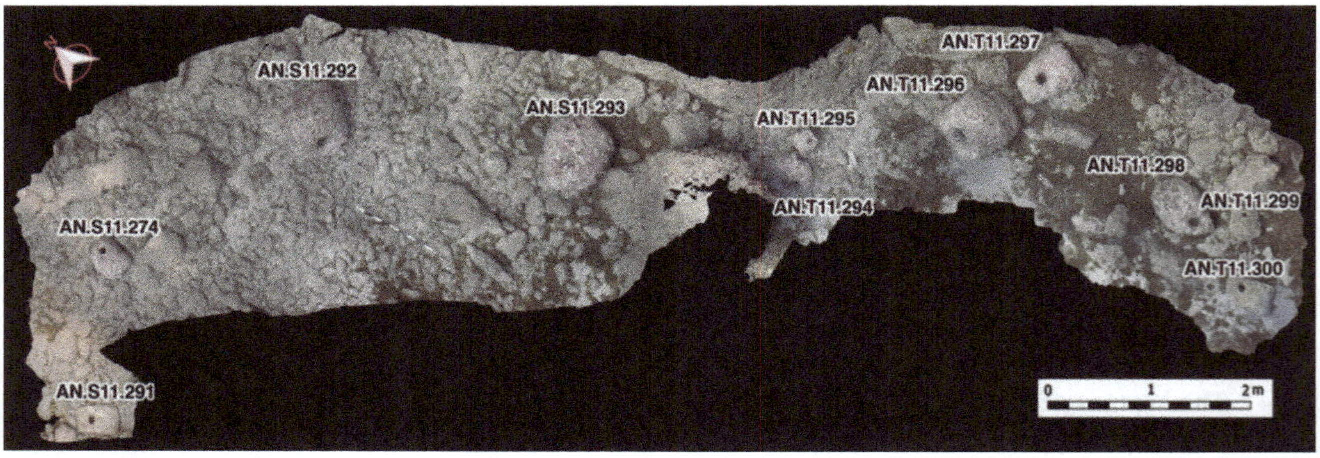

Fig. 5.12 Ortophotos of Groups 2 and 3 of anchors located at the bottom of the southern reef of Ras al-Qalaat (M.S. Salama)

5.5 Accuracy

5.5.1 Accuracy of Georeferencing of the Survey

The archaeological remains located at shallow depths of 3–5 m were recorded using a total station Nikon DTM-322 and a network of coded control points with absolute coordinates. These depths corresponded to the height of the pole holding the prism. The network of control points consisted of tagged square plastic sheets attached to lead weights to sit steadily on the seabed for the duration of the image capture and which were spread conveniently around the targeted areas to be photographed. All the coordinates recorded were georeferenced in the world projection system WGS 84/UTM zone 36N (EPSG: 32636). They were inserted in *Agisoft Photoscan/Metashape* respectively for each of their correspondent control points in order to tie the model into a known coordinate system. This set the data in the right geographic position and with the correct scale, and adjusted distortions in the models.

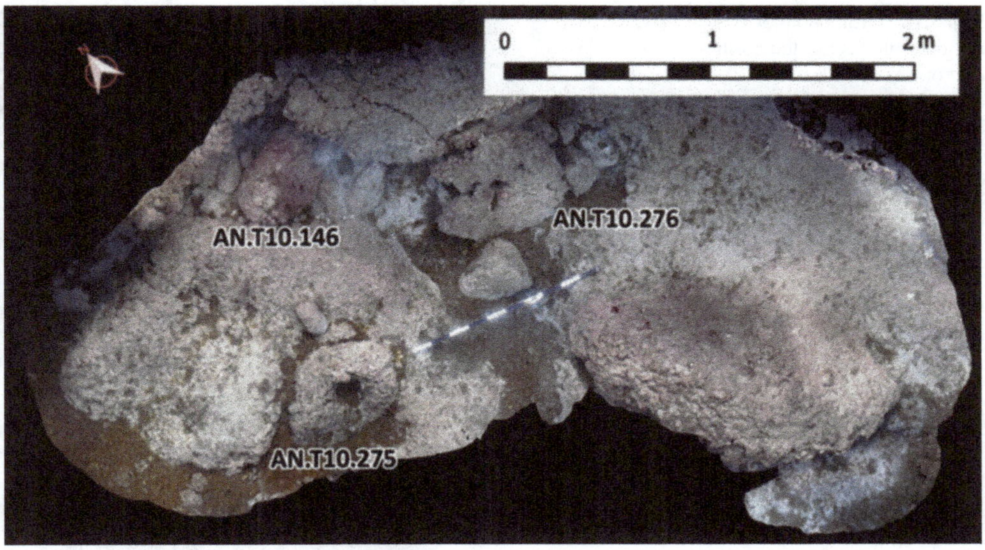

Fig. 5.13 Orthophotos of Group 4 of anchors and mill located at the bottom of the southern reef of Ras al-Qalaat (M.S. Salama)

Fig. 5.14 Orthophoto of masonry block (AN.U9.240) located off of the southern coastline of Anfeh (M.S. Salama).

The master station was positioned on two previously established topographic benchmarks (points 2011 and 2009). Pt.2011 is a geodetic point that had been corrected from the height of 35.87–13.37 m above MSL using the formula by Courbon (2016, 263) 'Mean Sea level=Ellipsoidal height of the GPS—22.35 m.' Knowing the absolute coordinate of Pt.2009, the total station was oriented and then other control points (with absolute coordinate and height corrected) were fixed nearer to the features to be recorded such as the masonry blocks in R10, and T8–T9.

The coordinates of most of the submerged artefacts lying at depths greater than 5 m were recorded from the surface with a handheld GPS (Garmin eTrex 10) using small numbered buoys attached to the objects, during calm weather conditions to reduce error margins as much as possible. The accuracy of this GPS position was 3 m at the surface. The depth and orientation of artefacts was recorded using the wrist compass and dive computer. The wrist compass was an Oceanic SWIV calibrated for the Northern Hemisphere. The accuracy of the depth recorded by the dive computer is slightly variable due to the tidal range but it is considered fit for purpose in this case since it does not exceed a maximum of 0.45 m.

Recording of the masonry blocks in R10 (Fig. 5.16) was undertaken using both methods. The shallow part of the masonry blocks were recorded with control points and the total station while the deeper parts were locked into GIS using GPS readings from the surface of a buoy tied to each rebar in calm weather. Depth of the blocks and rebars were taken using the dive computer under water. The authors are

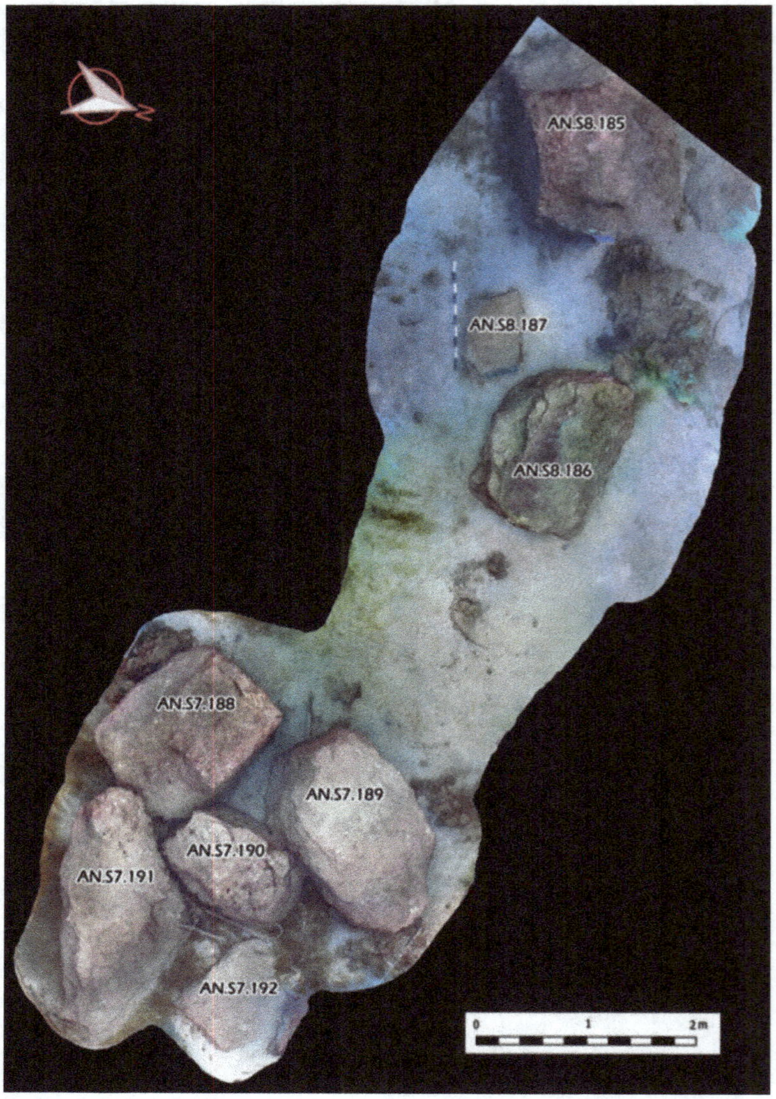

Fig. 5.15 Orthophoto of masonry blocks located north east of Ras al-Qalaat (M.S. Salama)

aware that such a method is not as accurate as the systematic use of control points. Nevertheless it mitigated potential error margins and improved the orientation and accuracy of the resulting models and orthophotos. It is, however, planned in future fieldwork seasons to return to mapped areas (whenever possible) and lock them within a network of control points, georeferenced using a total station.

5.5.2 Accuracy of the Photogrammetric Survey

In order to assess the internal accuracy of the photogrammetric survey, the authors tested results of digital measurements from 3D models against the classic recording method of taking 2D measurements in situ with a tape measure. All models were scaled in reference to the 1 m scale bar which was placed next to the objects prior to taking photographs. Anchors were measured systematically by both the diver and the archaeologist measuring from the 3D model, according to the maximum dimensions available, i.e. the highest numeric value of the length, width, thickness and hole(s) diameter of an anchor.

Table 5.5 compares the measurements of 17 anchors that were taken on the site by divers alongside corresponding measurements extrapolated from the 3D models of these anchors. The comparisons show generally relative negligible differences of 1–3 cm between the two sets of measurements. Two discrepancies of over 3 cm were ascribed to human error rather than the accuracy of the photogrammetric models.

5.6 Challenges

The main set of challenges for underwater photography relates to the operational environment. Parameters such as visibility under water, differential light absorption and colour

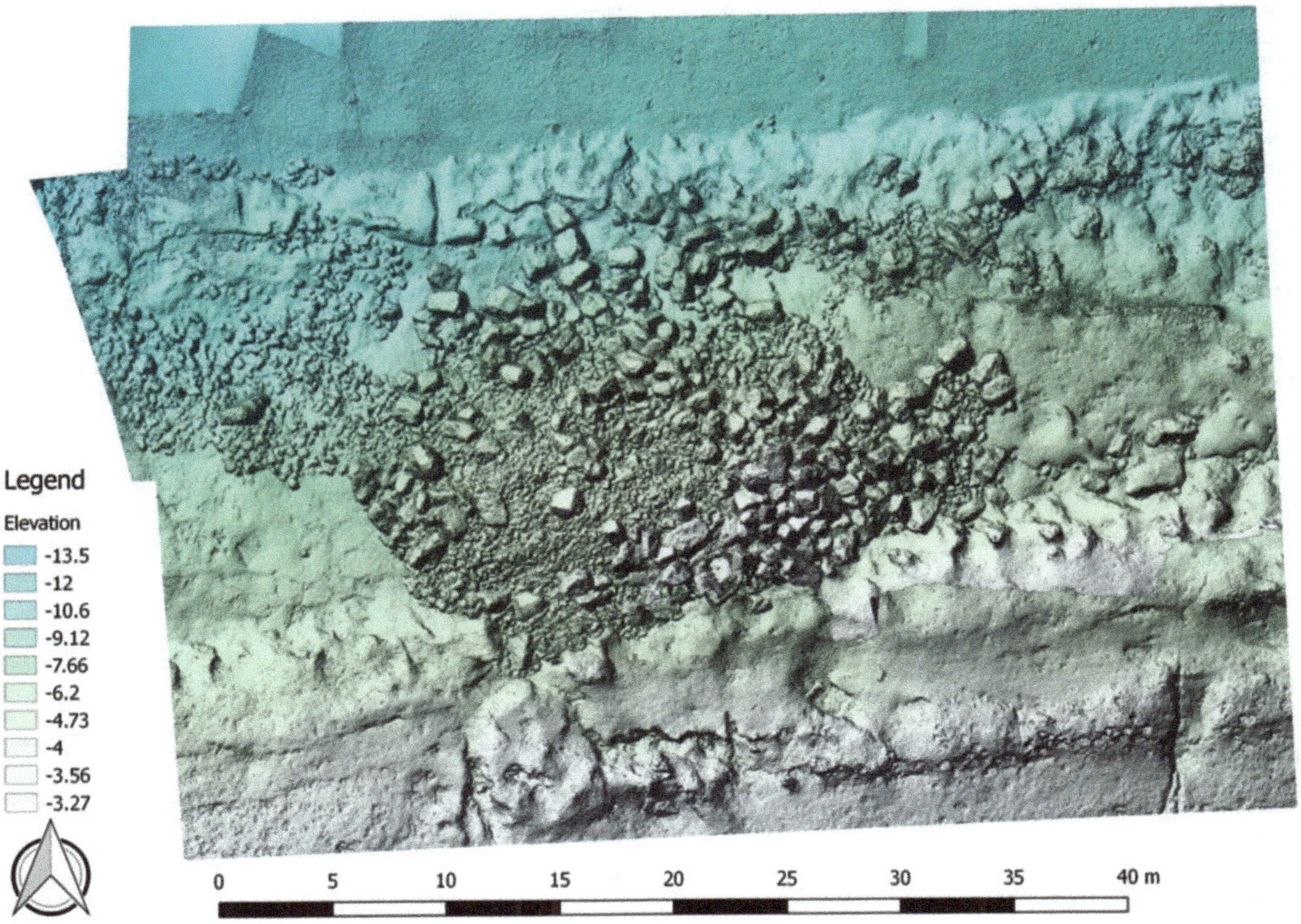

Fig. 5.16 DEM of submerged masonry blocs located to the northwest side of Ras al-Qalaat in square R10 (E. Cocca)

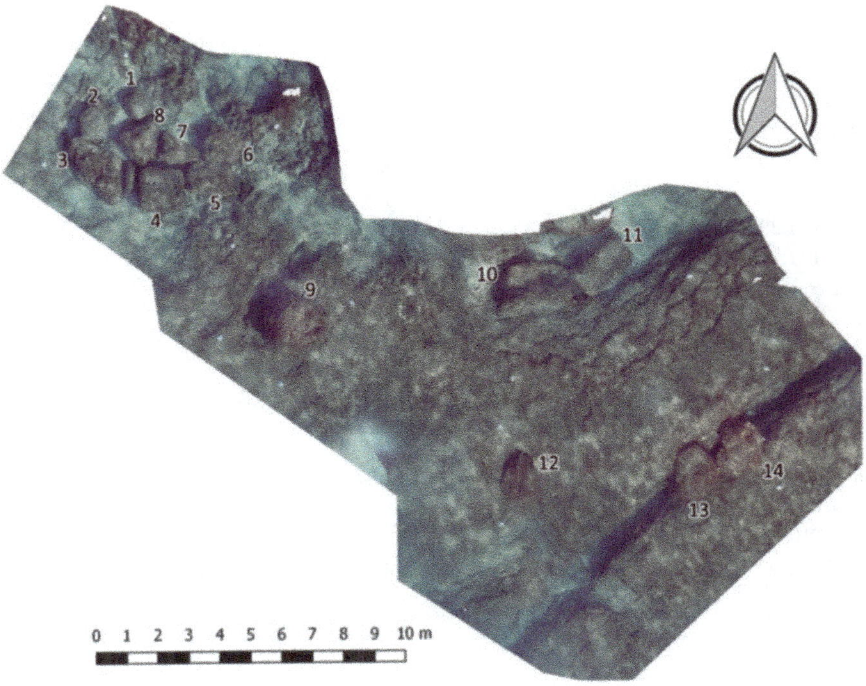

Fig. 5.17 Orthophoto of masonry blocs located to the southeast side of Ras al-Qalaat in squares T8 and T9 (E. Cocca). 1: AN.T9.311, 2: AN.T9.310, 3: AN.T9.309, 4: AN.T9.308, 5: AN.T9.307, 6: AN.T9.306, 7: AN.T9.313, 8: AN.T9.312, 9: AN.T8.305, 10: AN.T8.176, 11: AN. T8.175, 12: AN.T8.282, 13: AN.T8.178, 14: AN.T8.177

Table 5.5 Comparisons of measurements recorded on site by the divers with those taken from the 3D models. N/A is given for missing values, meaning that no comparison could be carried out

Anchor		Length (cm)	Width (cm)	Thickness (cm)	Hole
AN. S12 004		73	65	37.5	24
AN. S12 004	3D	71.8	65.8	36	22
Difference		1.2	−0.8	1.5	2
AN.S11.273		30	18	N/A	N/A
AN.S11.273	3D	35.7	20.8	5.8	4.6–4.1
Difference		−5.7	−2.8	N/A	6.4–7.9
AN.S11.286		46	30	8	9
AN.S11.286	3D	46.7	27.2	7.02	9.4
Difference		−0.7	2.8	0.98	−0.4
AN.S11.287		47	32	4	11
AN.S11.287	3D	46	31.3	4.9	12.3
Difference		0.98	0.7	−0.9	−1.3
AN.S11.288		47	29	6	16
AN.S11.288	3D	47.2	30.1	4.8	16.4
Difference		−0.2	−1.1	1.2	−0.4
AN.S11.274		40	30	N/A	N/A
AN.S11.274	3D	42.5	31.5	11.5	5.3
Difference		−2.5	−1.5	N/A	N/A
AN.S11.291		42	30	9.5	N/A
AN.S11.291	3D	45	30	10.5	8.5
Difference		−3	0	−1	N/A
AN.S11.292		78	57	32	20
AN.S11.292	3D	77.8	58.5	30.8	17.2
Difference		0.2	−1.5	1.2	2.8
AN.S11.293		78	65	34	29x22
AN.S11.293	3D	76	64.6	35	29.1×20.3
Difference		2	0.4	−1	0.1×1.7
AN.T11.294		46	51	15	21
AN.T11.294	3D	49	51.95	14.5	20.8
Difference		−3	−0.95	0.5	0.2
AN.T11.295		24.6	19.3	10	6
AN.T11.295	3D	23.6	20.1	12.5	4
Difference		1	−0.8	−2.5	2
AN.T11.296		69	60	31	25.8
AN.T11.296	3D	67.2	63.3	33	22.9
Difference		1.8	−3.3	−2	2.9
AN.T11.297		63	41.6	22	14.6
AN.T11.297	3D	64.3	41.5	20	14
Difference		−1.3	0.1	2	0.6
AN.T11.298		71.3	48.2	12.7	12.9
AN.T11.298	3D	73.8	47	15.5	12.4
Difference		−2.5	1.2	−2.8	0.5
AN.T11.299		38	30	6.6	5
AN.T11.299	3D	37.9	31	7.2	4.5
Difference		0.1	−1	−0.6	0.5
AN.T11.300		46	28	8.5	9
AN.T11.300	3D	50	30	7.9	8.3
Difference		−4	−2	0.6	0.7
AN.T10.146		60	40	20	10
AN.T10.146	3D	57.6	38.8	19.7	9.9
Difference		2.4	1.2	0.3	0.1

Fig. 5.18 Team member Clara Fuquen examining a masonry element among the algae covering the sea-bottom (Z. Morsy)

loss, capture and dive time at a given depth, and optical distortion due to the lens/water interface, all needed to be mitigated. Optical noise created by moving elements such as fish, floating algae transport and movement, and sediment particles in suspension resulting in substantial backscatter interfered with the process. Anfeh's seabed is covered with fine sediments of clay and sand that become disturbed under the slightest movement. The authors needed to operate with utmost care during the capturing process. Most of the time the photographer needed to keep a near-vertical position with their legs up. Also, the targeted areas of the seabed and especially the masonry blocks were covered to a great extent by a vegetation of *Colpomenia sinuosa*, a type of brown algae (Fig. 5.18), since the diving took place in springtime. These were cleared prior to every survey to enhance visibility and quality of the recording. The process of cleaning large masonry blocks was a tedious one, consuming dive time, and extra care needed to be taken for safety from the presence of bearded fire-worms (*Hermodice carunculata*), which are harmful to people.

Considering that the team was limited in number, and the areas surveyed quite extensive, a considerable amount of recording was accomplished in under a month of fieldwork, through the use of 3D techniques. Only one to two dives dedicated to collecting images for the photogrammetric process was possible each day. Dive teams were formed by two divers with one boat operator on the surface for safety. Depending on the time of day, currents, amounts of sediments and algae, the first available teams of the day would dedicate their dives to cleaning archaeological finds and features; while the divers responsible for photographing had to wait until the debris settled in order to proceed. In certain parts of the seabed, such as at the foot of the southern reef midway across the Ras al-Qalaat, the brown algae formed a floating layer, completely covering the anchors. Although an airlift could have cleared this layer, the lack of this equipment at the time meant that some anchors could not be recorded through underwater photogrammetry.

Weather and sea conditions were not always kind. At times, the divers were faced with strong currents under water that made photographic documentation difficult. This caused some delays since the divers had to pause several times to adjust their position during capture. Also, the underwater visibility deteriorated as the surface surge and underwater swell got stronger, which at times prevented the team from finishing the task at hand. Moreover, the sedimentation rate on the sea-bed of Ras al-Qalaat and adjacent coasts is high. This meant that several finds that had been located during the 2013 underwater survey were either completely or partially buried in sand. As mentioned, the team was only able to partially unearth the masonry block (AN.U9.240), which prevented them from obtaining a complete 3D model of the block. In regard to the post-processing phase of the photogrammetric survey, analyses of the images collected and production of digital models was time-consuming due to staff availability and computer processing requirements and some of the models are still being finalised.

5.7 Discussion and Conclusions

Although the application of photogrammetry has become almost universally adopted in maritime archaeology, this project is the first endeavour of its kind in Lebanon where the research team has demonstrated the impact of 3D photogrammetry on archaeological practice in a marine environment. Despite the limitations and constraints detailed above, the authors found these methods practical for documenting sites where more expensive equipment and methods are not available. The technique also has the advantage, among other diver-based recording methods, to minimize time spent under water compared to manual methods of 2D and/or trilateration. As shown in Table 5.5, measurements from photogrammetry helps in mitigating human errors taken through these traditional real-time measurements while operating in a challenging environment where currents, visibility and depth might impair human judgment and perception of accuracy.

Moreover, underwater photogrammetry is a non-intrusive technique of investigation to extract, integrate, and share archaeological data from 1-to-1 scale 3D models without having to revisit the actual sites (Yamafune et al. 2016, 4). At Anfeh, the authors were able—through the implementation of a time and cost-effective methodology—to capture 3D measurements of submerged artefacts. Having mapped isolated finds as well as their wider underwater context, the 3D photogrammetry offered the possibility of viewing these sites in their entirety that would have otherwise been observed fragmentarily. The authors have effectively gathered a solid baseline of knowledge of Anfeh's seabed, with substantial documentation on which to build further research and interpretation of the site.

Much of the archaeological analysis and interpretation of the archaeological results from the underwater photogrammetry campaigns are still under way. These campaigns form part of the wider research of a program aimed at Anfeh's maritime cultural landscape. However, some preliminary remarks can be made here about the contribution of these surveys to the wider project. As well as allowing a better understanding of the seabed topography and material, the 3D surveys have bridged the physical limitation between land and sea, clearly illustrating the seamlessness connection of the submerged archaeology with the promontory of Ras al-Qalaat. The presence on the south side of Ras al-Qalaat (Fig. 5.10) of a substantial amount of anchors varying in size suggests that this southern reef might have been a popular anchorage location in antiquity, positioned as it is in the lee of the northerly winds. The presence of rock-cut ways of access and stairways on the peninsula's southern face might also indicate that goods were transferred to the shore from boats laying at anchor. The 3D models of anchors have allowed detailed measurement as well as more accurate volumetric and weight estimation. This helps to establish their typology and improves our understanding of their spatial distribution and positioning. Significantly, the estimation of anchor weight can inform the interpretation of ship sizes and types.

The masonry blocks are still under study at the time of writing. However, from a preliminary analysis, it seems that the blocks located in areas S7 and S8 in the north and T8, T9, and U9 south of Ras al-Qalaat, might have been once part of the medieval fortress that stood on the promontory.[8] Some of these have dimensions and shapes that indicate they might have been carved from the quarries on the promontory itself (Abdul Massih 2016). Their presence under water can be explained through two factors: (1) the collapse of built stone architecture on the promontory due to tectonics and weather conditions; or (2) the dismantling and transfer from the promontory to meet the construction needs of the modern village of Anfeh, and/or further afar to the city of Tripoli which is located some 15 km north of Anfeh. As for the large cluster of masonry blocks mapped in Square R10 at the north-western side of the promontory, these might have functioned as a breakwater to shelter the access to the peninsula at some point in time. Research into these blocks is ongoing.

Finally, use of underwater photogrammetry at Anfeh has great potential for public outreach for underwater archaeology in Lebanon. Indeed, one of the future objectives of the research project at Anfeh is to provide virtual access to the underwater world by creating a fly-through video of the anchor area that can be uploaded to the internet. This will help to generate social significance for underwater archaeology, disseminate knowledge, and raise awareness of the importance of Anfeh's underwater cultural heritage.

Acknowledgements The authors would like to extend their gratitude to the Ministry of Culture and the Directorate General of Antiquities of Lebanon; the Honor Frost Foundation for funding the principal author's postdoctoral fellowship; Nadine Haroun-Panayot, head of the Department of Archaeology and Museology at the University of Balamand (UoB), for entrusting the team with this task; the Institute of Environment at UoB for the bathymetry map; the Department of Civil & Environmental Engineering at UoB for lending the team the total station; Crystal Safadi for all things GIS-related and map production; Enzo Cocca for GIS, photogrammetry, database services and his valuable comments on parts of this draft; Ralph Pedersen for his kind advice on and review of earlier drafts; and finally to the other team members without whom this and other endeavours at Anfeh would not have been possible in the past years: Samer Amhaz, Mohammad Azzam, Rupert Brandmeier, Enzo Cocca, Hadi Choueiri, Salvatore Colella, Clara Fuquen, Dylan Hopkinson, Menna-Allah Ibrahim, Mario Kozaily, Ziad Morsy, Lorine Mouawad, Julian Jansen van Rensburg, Crystal Safadi, Serge Soued, and Maii Tarek.

References

Abdul Massih J (2016) Notes sur les carrières maritimes d'Enfeh. In: Panayot-Haroun N (ed) Mission archéologique d'Enfeh: résultats préliminaires des travaux de prospection et de fouille de 2011 à 2015, 16. Direction Générale des Antiquités, Beirut, pp 285–287

Agisoft (2016) Agisoft photoscan user manual: professional edition. http://www.agisoft.com/pdf/photoscan-pro_1_2_en.pdf. Version 1.2. Accessed 27 Oct 2017

Balletti C et al (2015) Underwater photogrammetry and 3D reconstruction of marble cargos shipwreck. Int Arch Photogramm Remote Sens Spat Inf Sci XL-5(W5):7–13. Available at: http://www.int-arch-photogramm-remote-sens-spatial-inf-sci.net/XL-5-W5/7/2015/

Balletti C, Beltrame C, Costa E, Guerra F, Vernier P (2016) 3D reconstruction of marble shipwreck cargoes based on underwater multi-image photogrammetry. Digit Appl Archaeol Cult Herit 3(1):1–8. https://doi.org/10.1016/j.daach.2015.11.003

Bruno F, Lagudi A, Gallo A, Muzzupappa M, Davidde Petriaggi B, Passaro S (2015) 3D documentation of archaeological remains in the underwater park of Baiae. Int Arch Photogramm Remote

[8] For details on this fortress see Chaaya (2016).

Sens Spat Inf Sci XL-5(W5):41–46. https://doi.org/10.5194/isprsarchives-XL-5-W5-7-2015

Capra A, Dubbini M, Bertacchini E, Castagnetti C, Mancini F (2015) 3d reconstruction of an underwater archaeological site: comparison between low cost cameras. Int Arch Photogramm Remote Sens Spat Inf Sci XL-5/W5:67–72. https://doi.org/10.5194/isprsarchives-XL-5-W5-67-2015

Chaaya A (2016) Le château médiéval d'Enfeh. In: Panayot-Haroun N (ed) Mission archéologique d'Enfeh: résultats préliminaires des travaux de prospection et de fouille de 2011 à 2015, Bulletin d'archéologie et d'architecture Libanaises, vol 16. Direction Générale des Antiquités, Beirut, pp 281–283

Costa E, Beltrame C, Guerra F (2015) Potentialities of 3D reconstruction in maritime archaeology. In: Giligny F, Djindjian F, Costa L, Moscati P, Robert S (eds) Concepts, methods and tools: proceedings of the 42nd annual conference on computer applications and quantitative methods in archaeology. Archaeopress, Oxford, pp 549–556

Courbon P (2016) L'étude topographique. In: Panayot-Haroun N (ed) Mission archéologique d'Enfeh: résultats préliminaires des travaux de prospection et de fouille de 2011 à 2015, Bulletin d'archéologie et d'architecture Libanaises, vol 16. Direction Générale des Antiquités, Beirut, pp 262–263

Drap P (2012) Underwater photogrammetry for archaeology. In: Carneiro Da Silva P (ed) Special applications of photogrammetry. IntechOpen, pp 111–136. https://doi.org/10.5772/33999

Drap P, Merad D, Hijazi B, Gaoua L, Saccone MMNM, Chemisky B, Seinturier J, Sourisseau J-C, Gambin T, Castro F (2015) Underwater photogrammetry and object modeling: a case study of Xlendi wreck in Malta. Sensors 15:30351–30384. https://doi.org/10.3390/s151229802

Frost H (1997) Stone anchors: the need for methodical recording. Indian J Hist Sci 32(2):121–126

Fulton C, Viduka A, Hutchison A, Hollick J, Woods A, Sewell D, Manning S (2016) Use of photogrammetry for non-disturbance underwater survey—an analysis of in situ stone anchors. Adv Archaeol Pract 4(1):17–30

Gawlik N (2014) 3D modelling of underwater archaeological artefacts. MA Thesis, Norwegian University of Science and Technology

Green J, Matthews S, Turanli T (2002) Underwater archaeological surveying using photo modeler, virtual mapper: different applications for different problems. Int J Naut Archaeol 31(2):283–292. https://doi.org/10.1006/ijna.2002.1041

Lawrence TE (1988) Crusader castles. Clarendon, Oxford University Press, New York

McCarthy J (2014) Multi-image photogrammetry as a practical tool for cultural heritage survey and community engagement. J Archaeol Sci 43:175–185. https://doi.org/10.1016/j.jas.2014.01.010

McCarthy J, Benjamin J (2014) Multi-image photogrammetry for underwater archaeological site recording: an accessible, diver-based approach. J Marit Archaeol 9(1):95–114

Nordiguian L, Voisin JC (1999) Châteaux et églises du Moyen Age au Liban. Editions Terre du Liban/Trans-Orient, Beyrouth

Panayot-Haroun N (2015) Anfeh unveiled: historical background, ongoing research and future prospects. J East Mediterr Archaeol Herit Stud 3(4):396–415

Panayot-Haroun N (2016) Mission archéologique d'Enfeh. Panayot-Haroun N Mission archéologique d'Enfeh: résultats préliminaires des travaux de prospection et de fouille de 2011 à 2015, 16. Beirut, Direction Générale des Antiquités, 255–294

Polzer M, Casaban J (2012) Photogrammetry: a legacy of innovation reaching back to Yassıada. INA Q 39(1/2):13–17

Semaan L (2016) Surveying the waters of Anfeh: preliminary results. Skyllis 16(1):54–67

Semaan L, Fuquen C, Hopkinson D, Jansen Van Rensburg J, Morsy Z, Safadi C (2016) The underwater visual survey at Anfeh. Panayot-Haroun N Mission archéologique d'Enfeh: résultats préliminaires des travaux de prospection et de fouille de 2011 à 2015, 16. Beirut, Direction Générale des Antiquités, 287–291

Skarlatos D, Demesticha S, Kiparissi S (2012) An 'open' method for 3d modelling and mapping in underwater archaeological sites. Int J Herit Digit Era 1(1):1–24

Skarlatos D, Demesticha S, Neophytou A (2014) The 4th century BC shipwreck at Mazotos, Cyprus: new techniques and new methodologies in the 3D mapping of shipwreck excavations. J Field Archaeol 39(2):134–150. https://doi.org/10.1179/0093469014Z.00000000077

Van Damme T (2015) Computer vision photogrammetry for underwater archaeological site recording in a low-visibility environment. Int Arch Photogramm Remote Sens Spat Inf Sci XL-5/W5:231–238. https://doi.org/10.5194/isprsarchives-XL-5-W5-231-2015

Yamafune K, Torres R, Castro F (2016) Multi-image photogrammetry to record and reconstruct underwater shipwreck sites. J Archaeol Method Theory 24(3):703–725. https://doi.org/10.1007/s10816-016-9283-1

6 Using Digital Visualization of Archival Sources to Enhance Archaeological Interpretation of the 'Life History' of Ships: The Case Study of HMCS/HMAS *Protector*

James Hunter, Emily Jateff, and Anton van den Hengel

Abstract

In 2013, researchers affiliated with the South Australian Maritime Museum and University of Adelaide's Australian Centre for Visual Technologies [ACVT] conducted an archaeological and laser scanning survey of the former Australian warship HMCS/HMAS *Protector*. Between its launch in 1884 and service in the First World War, *Protector* was substantially modified. Once decommissioned, the ship again underwent drastic changes. While several archival photographs exist that depict *Protector* at various stages of its life, they provide only scant understanding of the transformative processes applied to *Protector*'s hull. Researchers at ACVT have developed methods of generating 3D models from archival photographs, and are using *Protector* as a case study. Models have been created that depict the vessel at three specific periods of its service life, which in turn has enabled archaeologists to identify gradual variations to *Protector*'s hull that, in some cases, were so subtle they could not be discerned in existing archival photographs and other historic media.

Keywords

Australia · Heron Island · HMAS *Protector* · Life history · Archival resources · Shipwrecks

J. Hunter (✉) · E. Jateff
Australian National Maritime Museum, Sydney, Australia
e-mail: james.hunter@anmm.gov.au; emily.jateff@anmm.gov.au

A. van den Hengel
University of Adelaide, Adelaide, Australia
e-mail: anton.vandenhengel@adelaide.edu.au

6.1 Introduction

During the latter half of 2013, a team of researchers affiliated with the South Australian Maritime Museum [SAMM] and University of Adelaide's Australian Centre for Visual Technologies [ACVT] conducted a comprehensive archaeological and laser scanning survey of the former Australian warship *Protector* (for additional information about this project, see Hunter and Jateff 2016; Hunter et al. 2016; MacLeod et al. 2014). Originally constructed as a gunboat and commissioned as a light cruiser for South Australia's colonial navy in 1884, *Protector* would later be integrated into the Commonwealth Naval Forces and Royal Australian Navy. It served in both the Yihetuan Movement (Boxer Rebellion) and First World War before being decommissioned from naval service in the 1920s. Subsequently converted into an unpowered lighter, the vessel operated in the waters of Port Phillip Bay, Victoria before being recalled for a brief period of military service during the Second World War. Damaged while en route to participate in the American assault on Papua New Guinea, *Protector* was condemned, sold, and towed to Heron Island on Australia's Great Barrier Reef in 1944, where it was installed as a breakwater.

Although exposed to the ravages of time and tide for over 70 years, *Protector*'s hull has remained largely intact and is still a prominent aspect of Heron Island's seascape (Fig. 6.1). The midships structure, however, has experienced accelerated deterioration and collapse during the past three decades, and corrosion and the elements both continue to exact a toll on the vessel's surviving fabric. The effort to archaeologically document *Protector* was partly borne of the realization that it had not been previously surveyed and its surviving hull was progressively more compromised and in danger of complete structural collapse. In addition, four significant anniversaries in the vessel's history—its launch and delivery to South Australia (1884), involvement in Australia's capture of New Guinea during the First World War (1914), decommissioning from Royal Australian Navy service (1924), and

J. K. McCarthy et al. (eds.), *3D Recording and Interpretation for Maritime Archaeology*, Coastal Research Library 31,
https://doi.org/10.1007/978-3-030-03635-5_6

Fig. 6.1 *Protector*'s surviving hull, as it appeared in 2013 at the time of the archaeological and laser scanning survey (J. Hunter)

installation as a breakwater at Heron Island (1944)—would occur in 2014, and SAMM planned to develop an exhibition to coincide with and commemorate these dates.

Central to the exhibition was a desire to digitally capture *Protector* and exhibit it virtually to a South Australian audience. Faced with this challenge, SAMM sought the advice and expertise of ACVT, and a resulting collaborative research initiative secured project funding from the Australian Research Council's Linkage Grant Program. The primary project goals were to laser scan *Protector*'s hull where exposed above the waterline and develop the means to produce 3D digital models of the vessel from archival images. *Protector* was relatively well-photographed over the course of its military and civilian careers, and the resulting image archive proved ideal for the project. Consequently, it was chosen by ACVT as the pilot study in a potentially groundbreaking initiative to digitally model historic photographs. Models of *Protector* rendered from archival images have revealed subtle changes to its hull, superstructure, armament and fittings over time, which in turn have informed the overall interpretation of its evolution as a military and civilian watercraft.

6.2 Iconography and Maritime Archaeology

The use of iconographic images—including historic photographs—as an interpretive dataset in maritime archaeology has existed for nearly as long as the discipline itself. Indeed, prior to the advent of maritime archaeology as an area of formal study, iconographic investigations were often the only means by which ancient watercraft were analysed and interpreted, and served to 'initiate the whole of what eventually became known as "maritime archaeology"' (Flatman 2004, 1276, 2014, 4668). Following the emergence of shipwreck-focussed archaeological investigations in the 1960s, the use of iconography as a source for ancient and historic ship studies declined, as did its prominence as an area of academic inquiry. Its existence never ceased entirely, however, and maritime iconography continued to inform archaeological studies—particularly those with a ship reconstruction focus (see Crumlin-Pedersen 2000; McGrail 1998; Steffy 1994; Villain-Gandossi 1979). The use of iconography in maritime archaeology persists to this day, has expanded to cover broader thematic topics such as maritime cultural landscapes (see Flatman 2004), and continues to benefit from a variety of image types, including paintings, sketches, ship plans and schematics. Historic photographs have increasingly become a more common dataset in maritime archaeology as the discipline's scope has expanded to encompass ships and shipwrecks from the late-nineteenth and early-to-mid twentieth centuries.

Maritime archaeologists have made qualitative statements about historic vessels through quantitative analysis of iconographic data. For example, Winter and Burningham (2001) employed the use of multiple iconographic images to identify and differentiate variants of early seventeenth-century Dutch watercraft. The authors originally intended to utilize contemporary Dutch maritime art to develop a reconstruction of *Duyfken*, a small *jacht* for which no existing ship plans are known to exist (Winter and Burningham 2001, 57). Univariate and multivariate morphometric statistical analyses of marine paintings, engravings and pen-and-ink drawings revealed the existence of four distinct classifications of three-masted, square-rigged Dutch sailing vessels during the early seventeenth century. These vessel types—*ships*, *jachts*, large *jachts* and small *jachts*—were not previously identified in historical or archaeological literature, opening up new potential research avenues in Dutch ship design. In terms of utility, the authors contend morphometric analysis of iconographic data provided 'objective confirmation of the validity of the design of the *Duyfken* reconstruction…and should be seen as a necessary part of all ship design reconstruction' (Winter and Burningham 2001, 72–73).

Although there are no published studies in which 3D models of archival iconography have been specifically created and used as an interpretive tool in maritime archaeology, there have been efforts to integrate historical imagery with existing 3D models of heritage sites. A team led by Mathieu Aubry (see Aubry 2015; Aubry et al. 2013) has recently developed an algorithm that reliably aligns 3D models of historic structures with corresponding 'arbitrary' 2D depictions, including 'drawings, paintings and historical photographs.' The algorithm builds upon prior research in computer vision that aligns images using 'local features', 'contours' and 'discriminative learning' to search for 'discriminative visual elements' within rendered 3D models of a given structure (Aubry et al. 2013, 1–3; for discussions of prior research in this area, see Baatz et al. 2012; Baboud et al. 2011; Dalal and Triggs 2005; Doersch et al. 2012; Felzenszwalb et al. 2010; Hartley and Zisserman 2004; Huttenlocher and Ullman 1987; Lowe 1987, 2004; Rapp 2008; Russell et al. 2011; Shrivastava et al. 2011; Sivic and Zisserman 2003; Snavely et al. 2006). Using a technique similar to object detection, it then matches these visual elements with matching elements in 2D images. The application is able to make these alignments 'despite large variations in rendering style (e.g. watercolor, sketch, historical photograph)…and structural changes [within] the scene' (Aubry et al. 2013, 1). Significantly, it also produced better results than a variety of baseline methods, including human detection.

While ground-breaking, Aubry's technique relies upon 3D models of existing objects/structures of interest, and it is therefore of limited or no use in instances where a 3D model cannot be generated because the structure has either changed significantly or is no longer present. This is particularly true of most shipwreck and vessel abandonment sites. Surviving hull structure is often limited to the area beneath the waterline, which is rarely—if ever—documented in iconographic sources. In addition, surviving elements of wrecked and abandoned vessels above the waterline are usually so fundamentally altered by natural and cultural transformative processes that they offer few 'discriminative visual elements' with which to compare them to archival imagery. One way to combat this problem is to generate 3D models of vessels as they originally appeared, and then use these models in turn to search for variations in their hull, superstructure, armament and fittings over time.

6.3 A Means for Interpretation: 3D Modelling of Archival Images

Producing 3D models from photographs is one of the fundamental challenges of image analysis, and has long been studied within the field of computer vision. Although a range of methods have been devised (see Szeliski 2010 for a compilation of available techniques), the primary example applicable to archival photographs is what is known as the 'Structure-from-Motion' (SfM) approach (see Hartley and Zisserman 2004 for technical details). SfM is based on a simplified model of image formation, which reflects the geometry of an idealized pin-hole camera. The fundamental challenge with interpreting images from a pin-hole camera, or indeed any traditional camera, is that an infinite number of scenes may give rise to any particular image. This is because cameras effectively remove depth information from a captured scene, and produce images that are flat, 2D representations (see Hartley and Zisserman 2004 for technical details). The images they create do not record the depths or sizes of individual objects within the scene, or the elements they are made up of. In theory, it would not be possible to differentiate between a close-up image of a toy car, and an image of an identical, but larger (i.e., real) car that is further away. Humans overcome this problem using prior knowledge of the size of general objects, but this is knowledge that the SfM method does not possess.

One means for resolving the problem is the use of image pairs instead of single images. The SfM approach exploits a mathematical model of the mechanism by which multiple images of the same scene are created. It takes into account that light travels in a straight line, and explains the manner in which the relative positions of objects in an image change as the camera's position changes. The model has a set of parameters that must be adjusted to match a particular scene and the camera(s) that took the image(s). Key to generating an accurate 3D reconstruction of a scene captured by a simplified camera is to find SfM parameters that best reflect the original scene, and the position(s) and internal characteristics of the camera(s) used.

Unfortunately, archival photographs were not typically taken with 3D modelling in mind, and the cameras that captured the images are long gone. Consequently, in the vast majority of cases it is impossible to take a suitable set of scene measurements to analyse camera positions, or inspect the camera geometry, so as to guide the SfM process. All that is available are the images themselves. An interesting aspect of the projective geometry that underpins SfM is that, by making a suitable set of assumptions or adding external information, it is often possible to estimate the missing parameters. Assumptions such as the water or ground in an image being horizontal, walls being vertical, and objects being symmetric can be used to imply constraints in the optimization problem that SfM solves. By optimising the parameters of the SfM model so that it best reflects these assumptions, in addition to the image-based measurements, a mathematical model of the image-creation process may be recovered. Armed with such a model, it is possible to extract components of the model that identify the 3D characteristics of various objects within the scene.

One significant drawback to the SfM approach is that it only delivers a set of 3D position estimates for various points within the image. This set of position estimates is referred to as a point cloud, and represents a relatively poor model of 3D shape. Point clouds cannot be replicated with a 3D printer, nor rendered as anything but a set of dots. Consequently, a major avenue of inquiry in computer vision is the development of methods for converting point clouds into polygonal models such as may be used to 3D print facsimile objects. Offshoots of this research include visualization of archival imagery in augmented reality systems, and analysis of subtle design changes to historic structures over time. Critical to the successful achievement of these and other goals is the development of a means for converting point clouds into a set of polygons, or a 'polygonal mesh', that accurately reflects the original surface(s) of object(s) depicted in an historic image.

At a minimum, a set of polygons can be constructed that span the points within the point cloud. This method generates a surface, but not one that accurately reflects the surface of the actual object. Other methods already exist in which better polygonal meshes may be generated from a point cloud and set of images, and new techniques are currently in development. All, however, optimize a given polygonal mesh's parameters so that it best reflects the set of measurements derived from an archival image. VideoTrace, a software program used to generate realistic 3D models of objects from digital video, embodies one such approach and was used by ACVT to develop digital models of *Protector* from archival imagery (Fig. 6.2).

The VideoTrace 3D modelling process proceeds as follows: The user takes a video, and loads it into the software, which then carries out a standard structure-from-motion analysis of it. The result of this process is a spare 3D point cloud, and an estimate of the internal parameters and position of the camera for each frame of the video. What distinguishes VideoTrace from other image-based methods for generating low polygon count 3D models is the nature of the interaction, and the level of control that this gives the user over the final polygonal 3D model. VideoTrace uses the results of the structure-from-motion process to inform the interpretation of user interactions, which allows the user to focus on adding high-level, often semantic, information. This is particularly critical for archaeological applications, as it allows the user to exploit their expert knowledge to compensate for the fact that the data required to drive more traditional image-based modelling is unavailable.

6.4 The Challenge of Digitally Modelling Archival Imagery

The process outlined above applies a mathematical SfM model to a set of measurements acquired from available images. The model's parameters are then adjusted until they accurately align with each image's measurements—with the ultimate goal that the resulting model will reveal the original conditions under which the image was formed. A major drawback to SfM is that it relies on a set of assumptions, foremost of which is that all of the available images depict an identical historic scene. In reality, however, the scarcity of historical images of an object of interest make it unlikely that two images of the same object will have been taken at the same location, let alone the same time. SfM modelling attempts to reconstruct the image formation process by analysing differences between two images of the same scene. It does so by performing a mathematical analysis of the dependency between the position of a given object in an image, and the position(s) of the camera(s) that captured it. By analysing the differences between two images taken from different locations, it is possible to extrapolate the relative position(s) of the camera(s), and the relative positions of the depicted objects. This is the same mechanism by which humans see in 3D. Subtle differences in images that we see are caused by the relative positions of our eyes in our head. An analysis of those differences allows our brain to develop an estimate of the 3D shape of the scene we see. Significantly, this approach assumes the only differences in appearance between two images are caused by the respective locations of the cameras (or eyes).

Very few examples exist of multiple images taken of the same scene at exactly the same moment. Early stereoscopy represents one notable exception, and is a valuable data source. Unfortunately, stereoscopic images of historic objects, locations, or events are rare; consequently, those wishing to develop 3D models from archival photographs are faced with doing so from a body of imagery that is largely unsuitable for the purpose. Many historical images may share one or more objects or scenes, but because photography was once far less common than it is today, the total number of available photographs within a data set is typically small. This has a follow-on effect on the overall number of images featuring objects or scenes that were captured on film at the same time. The ability to confirm that two or more archival photographs were generated simultaneously is critical, as it is the best way to guarantee the objects and scenes they depict are no different from one another. Most historical images, even those that feature scenes that appear identical, were often produced at different times.

Fig. 6.2 Screen capture of a polygonal mesh superimposed over an archival photograph of HMAS *Protector* using the VideoTrace software program. (Image: ACVT; base image: South Australian Maritime Museum [John Bird Collection])

Consequently, the objects and scenes within them may appear unchanged, but often exhibit subtle variations.

Another challenge to developing 3D models of archival photographs is that the content that is actually common to all of the images in the set (typically the object of interest) makes up only a small part of the image. In the case of *Protector*, the image archive—while relatively large—depicts the vessel from different points of view in a variety of locations and configurations. A particular problem from a SfM standpoint is that *Protector* is the only common element among the majority of photographs, but comprises a relatively small percentage of each image. In addition, the shape of the ship changes quite dramatically between photographs, to the extent that certain features (such as armament or elements of superstructure) cannot be assumed to have remained in the same position, or even extant, from image to image. As mentioned, the ability to analyse variations in the appearance of common content between two images is the core component of the SfM approach. In instances where common content comprises a small percentage of images within a data set, instability is created within the mathematical model, which in turn can generate significant error in the estimates it generates.

6.5 A Partial Solution

Many of the problems described above can be counteracted by incorporating more photographs into the data set. However, this only helps if the additional images provide new views of the object. Because early photographs were difficult to produce and develop, great care was often taken to capture a scene from the best viewpoint, or depict an object from a 'canonical', or preferred position. As a consequence, many historic photographs effectively show the same point of view, and don't add new information to a given SfM dataset.

These problems affected efforts to model *Protector*'s photo archive, but were corrected in part by the creation of 'pseudo images' from the vessel's construction plans (Fig. 6.3). In order for the SfM model to identify features in common with the archival photographs, the plans had to be manually processed. All photographic images are an accurate reflection of reality and consequently were processed using a standard, automatic SfM pipeline. By contrast, the hand-drawn nature of the construction plans meant they were a reflection of the real (photographic) images, and as a result the SfM approach was inapplicable. Ultimately, the identification of correspondences between pseudo (hand-drawn)

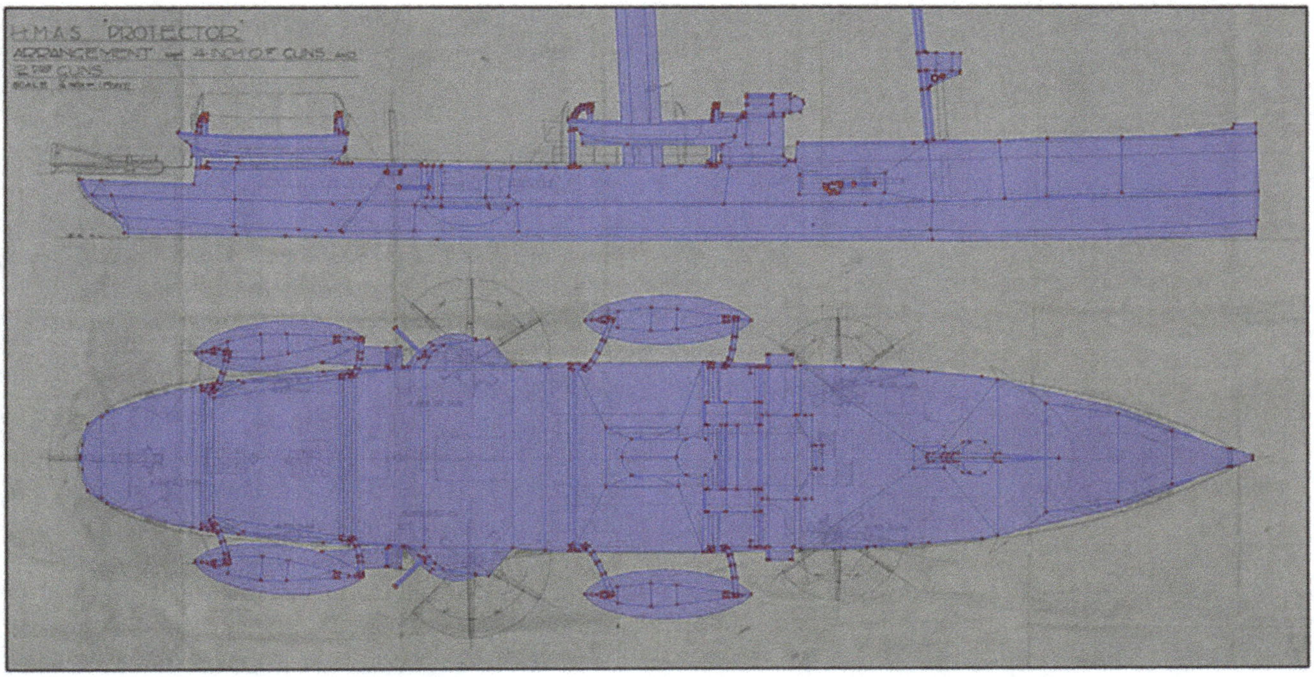

Fig. 6.3 Example of a 'pseudo image' developed from a 1922 plan of *Protector*'s armament arrangement (ACVT; base image: National Archives of Australia [MP551/1:92/14])

images and real (photographic) images had to be carried out manually. These correspondences were significantly outnumbered by correspondences generated automatically between photographic images. While this proved somewhat of a drawback, creation of pseudo images generated alternate views of *Protector* from vastly different perspectives, which in turn contributed valuable geometric information that could be incorporated into the overall models. ACVT staff also consulted with project archaeologists to identify consistent hull and superstructure features within *Protector*'s image archive.

6.6 A Better Solution

Despite the problems outlined above, human beings have very little trouble interpreting historical images. Human interpretation is less accurate than what is required for SfM modelling, but is far more flexible in its approach and applicability. While not necessarily able to generate a model suitable for 3D printing, the human process of image interpretation can be applied to a single photograph, derive information from previously unseen images, and is resistant to various forms of image degradation and deformation. The human mind is able to achieve this feat due to cumulative experience of real-world scenes and objects. This is accomplished through a lifetime of observing the world and interacting with it, which endows humans with the ability to easily recall and exploit their brain's 'information archive' when interpreting an image.

A current issue in SfM is how to incorporate the human mind's mechanism for reading historic photographs into the computing process of image interpretation. ACVT is currently conducting research in this area, and has made gains in generating 3D models from individual archival images (see Li et al. 2015). The neural network-based methods, however, that ACVT has developed require vast volumes of data for testing and manipulation, and these data sets tend to comprise extremely large format images and corresponding sets of laser scans. These data are often difficult to acquire, and the image content they contain frequently limits the methodologies that may be employed during the modelling process. In many cases, the models generated from these data sets tend to be in the form of height maps rather than full 3D models—which makes them wholly inappropriate for many of the modelling tasks that are of interest to historians and archaeologists. ACVT's research to generalize this approach continues, as do its efforts to generate accurate, full 3D models of historical objects and scenes.

6.7 Applying 3D Archival Imagery to Interpret *Protector*'s 'Life History'

Utilising available image assets, ACVT generated composite digital models that depicted *Protector*'s external hull and superstructure during three general phases of its career (Fig. 6.4). The first showed the vessel in its 'as-built' configuration in 1884 (as a South Australian colonial gunboat),

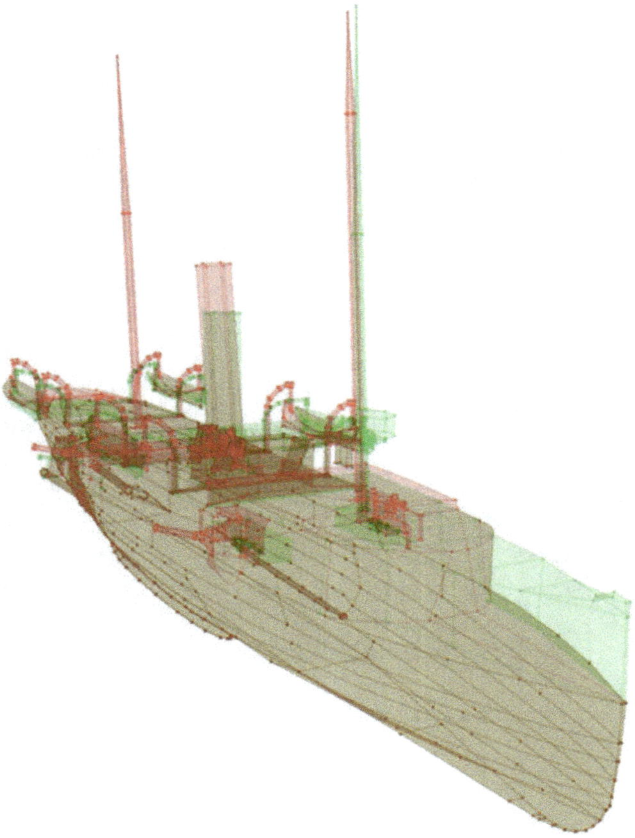

Fig. 6.4 Isometric view of superimposed digital models of *Protector*'s external hull as it appeared in its colonial (red) and national (green) navy configurations. Brown indicates where both models overlap (ACVT/authors)

while the second and third highlighted its appearance as a Royal Australian Navy auxiliary warship (1911–1924) and civilian lighter (1924–1944), respectively. To fully understand and appreciate the transformative processes that were applied to *Protector*'s hull over the course of its life as a functional watercraft—a period totalling 60 years—a brief discussion of its historical background is necessary.

Her Majesty's Colonial Ship (HMCS) *Protector* was originally purchased by the South Australian colonial government in 1882 in response to several 'Russian Scares' that plagued the Australian colonies during the 1860s, 1870s and 1880s (Jeisman 2012, 37–38; Nicholls 1988, 80–84). It was constructed according to a 'flat-iron' gunboat design first introduced during the Crimean War (1853–1856) and exhibited a unique silhouette that included a forward section with incredibly low freeboard, a prominent centrally positioned funnel, and armament that exceeded what was considered normal for a warship of its size (Gillett 1982, 62; Jeisman 2012, 44, 49–55; Jones 1986, 62–63). For the next 40 years, *Protector* was an active asset of the South Australian colonial navy, Commonwealth Naval Forces, and Royal Australian Navy. It participated in two major conflicts: suppression of the Yihetuan Movement in China in 1900, and the First World War (1914–1918). Upon integration within the Royal Australian Navy in 1911, the vessel was renamed His Majesty's Australian Ship (HMAS) *Protector* and its armament altered. Its First World War service included acting as a tender to the Australian submarines *AE1* and *AE2*, guarding the port of Rabaul in New Guinea, and serving as a patrol vessel in Australian coastal waters (Jeisman 2012, 289). During its final years of naval service, *Protector* was renamed HMAS *Cerberus* and operated as a tender at the Flinders Naval Depot in Victoria.

In 1924, *Protector* reverted to its original name and was decommissioned from the Royal Australian Navy. It was subsequently purchased by civilian interests and stripped of armament, engines and machinery, which were sold at auction. The remaining hulk was also sold, converted into a lighter, and in 1931 ended up in the possession of Melbourne-based Victorian Lighterage Pty Ltd. Renamed *Sidney*, the vessel stored and transported bulk commodities, including fuel oil and wool, in Port Phillip Bay for several years (Gillett 1982, 68; Jeisman 2012, 291; Pennock 2001, 95). During the Second World War, *Sidney* was requisitioned by the US Army and served the Small Ships Section of the US Army Services of Supply under the vessel designation *S-226*. Loaded with various army stores, *Sidney* was in transit to New Guinea in September 1943 when it collided with another vessel at the Queensland port of Gladstone. Four months later, the US Army was preparing to strip and scuttle the vessel when Cristian Poulson, proprietor of a tourist resort at nearby Heron Island on the Great Barrier Reef, offered £10 for the hulk (Gillett 1982, 68; Jeisman 2012, 291; Pennock 2001, 95–96). In April 1944, *Sidney* was towed to the western side of Heron Island and installed as a breakwater.

Between integration into the Commonwealth Naval Forces in 1901 and the outbreak of the First World War in 1914, *Protector*'s hull underwent a series of significant modifications. One of the most substantial alterations was to the vessel's bow, which in 1912 was built up to the level of existing superstructure amidships. This created additional working and living space, increased *Protector*'s freeboard in the forward section by approximately 4 metres, and almost certainly improved the ship's overall seaworthiness. The stern bulwarks, by contrast, were lowered slightly during this period—possibly in preparation for the installation of minesweeping gear. A series of planned alterations to *Protector* include an early twentieth-century schematic detailing a proposed arrangement for minesweeping gear and associated equipment. These plans feature an annotation stating the stern bulwark would be 'cut away' to facilitate deployment of a water kite, used for minesweeping, and its towing array

(National Archives of Australia [NAA], MP551/1, 92/20; United States Navy 1917, 5). Although the height of the stern bulwarks was reduced, and two sections were removed entirely, the minesweeping apparatus does not appear to have ever been installed.

While changes to *Protector*'s bow and stern are evident from a cursory examination of archival photographs and plans, superimposition of ACVT models depicting the vessel in its colonial and national navy configurations reveals far more subtle hull and superstructure variations (Fig. 6.5). For example, the height and rake (the incline from the perpendicular towards the stern) of *Protector*'s funnel was reduced when it became a Royal Australian Navy fleet asset. The rake of the foremast was also reduced, and an observation platform installed on the front of the foremast, while the aft mast was removed entirely. Slight alterations were also made to *Protector*'s bridge structure, including an increase to its overall height and outboard expansion of the port and starboard wings. Even the four sets of davits that accommodated the vessel's Montagu whalers and other ship's boats were slightly altered and shifted outboard. Finally, while the increase in the height of *Protector*'s bow is obvious in archi-

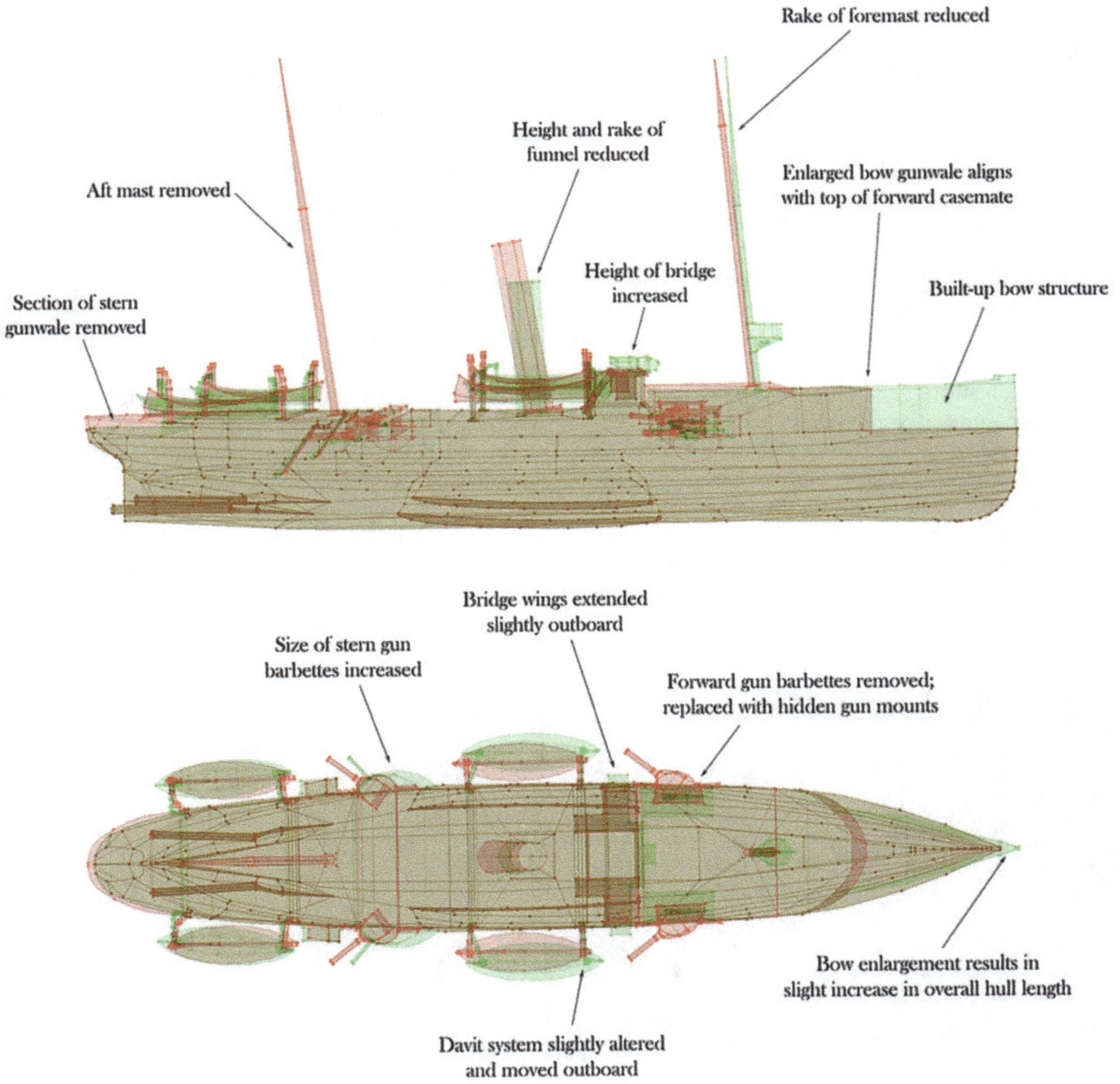

Fig. 6.5 Superimposed digital models of *Protector* derived from archival sources, showing the vessel in its colonial and national navy configurations (ACVT/authors)

val sources, a barely perceptible increase in its length is also evident—but only when the superimposed models are viewed in plan (Fig. 6.5, bottom).

Alterations to *Protector*'s armament over the course of its naval career are relatively well documented, but only in terms of the types of shipboard artillery with which it was outfitted. For example, the vessel's original complement of fixed weaponry included a bow-mounted 8-inch Armstrong rifled breech-loading gun, five 6-inch Woolwich-Armstrong rifled breech-loading guns, four 3-pounder Hotchkiss quick-firing cannon, and five 10-barrel Gatling machine guns. Following *Protector*'s transition to the Royal Australian Navy, the Gatling guns were removed and its primary armament replaced with two 4-inch, two 12-pounder, and four 3-pounder guns (Gillett 1982, 62; Jeisman 2012, 44, 49–55, 289; Jones 1986, 62–3). Far less is known about the manner in which these weapons were mounted and deployed, and whether existing infrastructure (such as gun mountings and protective armour) was retained, adapted or removed. Some changes, such as the complete removal of *Protector*'s forward gun barbettes (semi-circular armour plating that surrounded the vessel's rotating gun mounts) are fairly obvious in archival photographs and plans. By contrast, subtle variations to the size and form of the ship's stern barbettes are not as easy to discern but become obvious when the colonial and national navy models are superimposed over one another (see Fig. 6.5, bottom).

Superimposition of these models directly aided archaeological interpretation of surviving elements of *Protector*'s weapons systems. Components of the vessel's forward gun mountings still remain in situ on the foredeck, including cast-steel pedestals that originally accommodated the base of each 12-pounder and served as the point on which it pivoted from side to side (Fig. 6.6, left). Because the forward gun positions were adapted from barbettes to 'disappearing' variants concealed behind closed watertight hatches, a logical assumption is that their corresponding mountings were moved and modified. While the barbette was clearly removed and the pedestals may have been altered to accommodate new weaponry, comparison of the digital models reveals the mounting location itself remained unchanged (Fig. 6.6, right).

Following its removal from naval service, and subsequent conversion to the lighter *Sidney*, the vessel was stripped of most of its equipment and machinery. The removal of items of considerable weight, including the engines, boilers, guns, and deck machinery, caused it to rise 'up from its original water line' to such an extent that its armour belt—normally located at and slightly below the waterline—was reportedly 'well up [and] out of the water' (State Library of South Australia [SLSA], RN 100, bd 994.23 R432b). Taking the adjusted waterline of the vessel into consideration, and no doubt hoping to negate its top-heaviness and improve overall stability, *Sidney*'s civilian owners cut the hull down so that the run of the deck was essentially level from the bow to the break of the poop. From the break, the deck rose slightly to the height of the original poop and continued aft to the end of the stern. The ship's conversion to a quasi-flush-decked configuration necessitated disassembly of the augmented bow structure, including bulwarks, decks and internal architecture back to its pre-1912 height. In addition, the superstructure amidships, including the funnel, bridge and stern bulwarks, were completely removed.

This wholesale reduction of the hull and removal of superstructure is obvious in archival photographs, and a comparison of the national navy model with that of *Sidney* further graphically illustrates the significant degree to which the vessel was altered in 1924 (Fig. 6.7). In addition to creating a flush deck, *Sidney*'s civilian owners also sought to improve the run of the hull. As a warship, *Protector* was out-

Fig. 6.6 Left: *Protector*'s surviving hull, showing the locations of its forward 12-pounder gun mountings; right: superimposed digital models of the vessel's colonial and national navy configurations reveal the positions of the forward gun mountings were not altered. (*left*, James Hunter; *right*, ACVT)

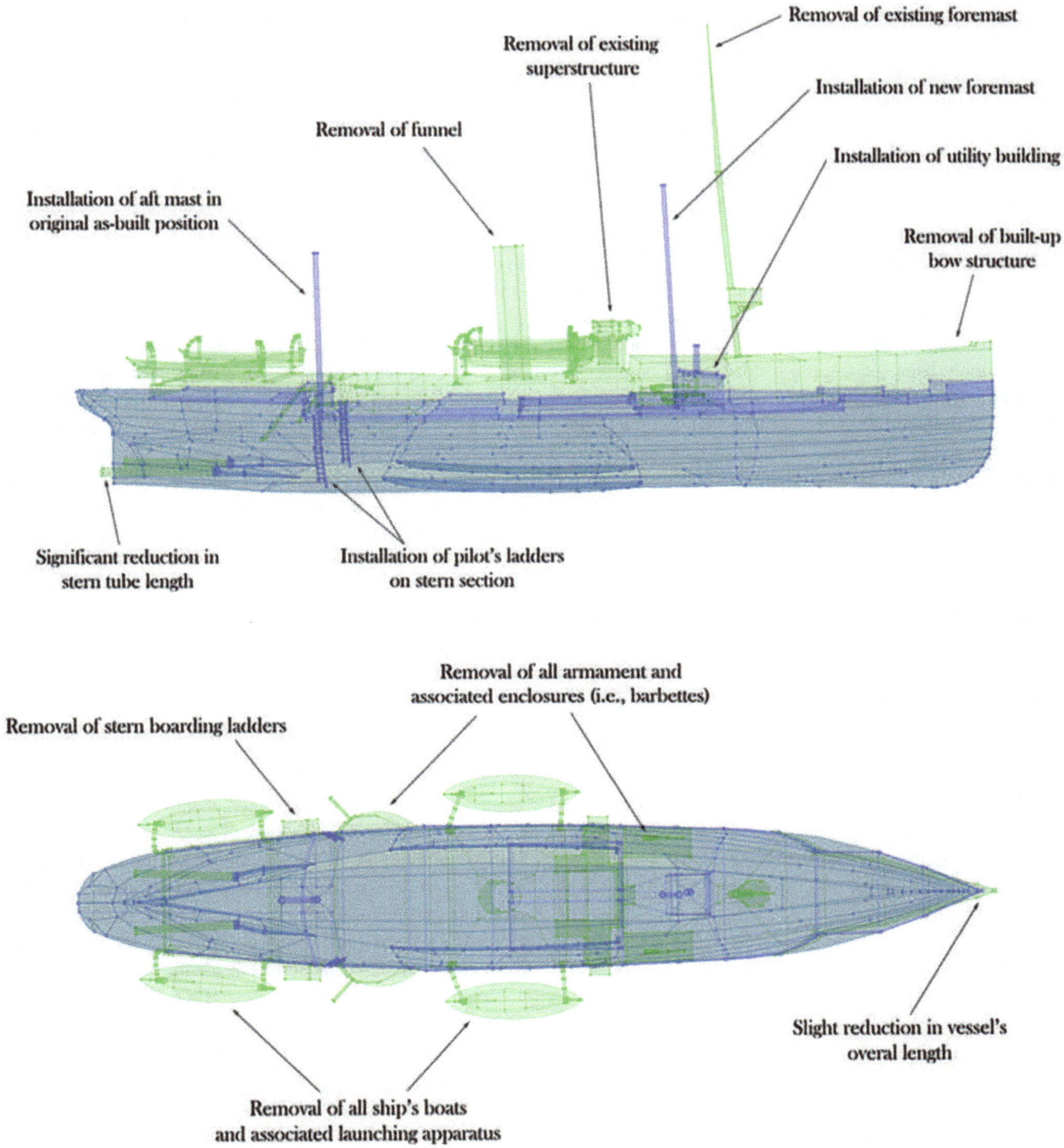

Fig. 6.7 Superimposed digital models of HMAS *Protector* and the lighter *Sidney* derived from archival sources, showing the vessel in its national navy (green) and civilian (blue) configurations (ACVT/authors)

fitted with a number of features—such as barbettes—that disrupted its otherwise clean silhouette. During its conversion to a lighter, efforts were made to modify and correct these disruptions, almost certainly as a means to improve the vessel's seaworthiness, handling qualities, or both. Such modifications included the reduction and/or removal of protruding structural elements, and the addition of steel hull plating to cover what remained and create a clean run fore and aft. When viewed in plan, the superimposed digital models show these changes very clearly as well. Also of note is the complete absence of the vessel's auxiliary boats and their launching apparatus, and reduction of the hull—via removal of bow structure—back to its original overall length (Fig. 6.7, bottom).

More subtle alterations to *Protector* that occurred during its conversion to *Sidney* include positioning of the vessel's masts. *Protector*'s foremast was one of its most prominent features during its naval career but was removed when the vessel was converted to a lighter and replaced by a much shorter guyed mast that formed the basis of one of two boom derricks (simple derricks such as those aboard *Sidney* would have been used to move cargo into and out of the vessel's hold). Curiously, the models reveal the forward guyed mast was not stepped into the same location as the original foremast, but instead positioned slightly aft (see Fig. 6.7). This was likely done to place the derrick immediately adjacent to *Sidney*'s forward hatch, where at least half of its cargo would have routinely been loaded and offloaded. By contrast, the aft guyed mast was stepped in the same position as the mizzenmast used aboard *Protector* in is original colonial navy configuration (Fig. 6.8). This appears to have been more a matter of circumstance than design, as archival photographs and archaeological investigation of the surviving hull reveal this mast was located immediately behind the section of hull transformed into *Sidney*'s aft cargo hatch.

The superimposed models also reveal alterations to the manner in which personnel boarded and departed the vessel both prior to, and after, its conversion to a lighter. As a colonial warship, *Protector* was outfitted with one boarding ladder on either side of the hull immediately astern of its aft gun barbettes. This configuration remained in place following the vessel's transition to the Royal Australian Navy, with very minor adjustments to the angle and length of the ladders. When *Protector* was sold and converted into *Sidney*, the boarding ladders were removed and replaced with Jacob's ladders (see Fig. 6.7, top). Although far simpler in terms of their overall design, and easier to deploy and retract, the Jacob's ladders were positioned in approximately the same location as the boarding ladders, which suggests this section of the hull remained a preferred point of entrance and egress.

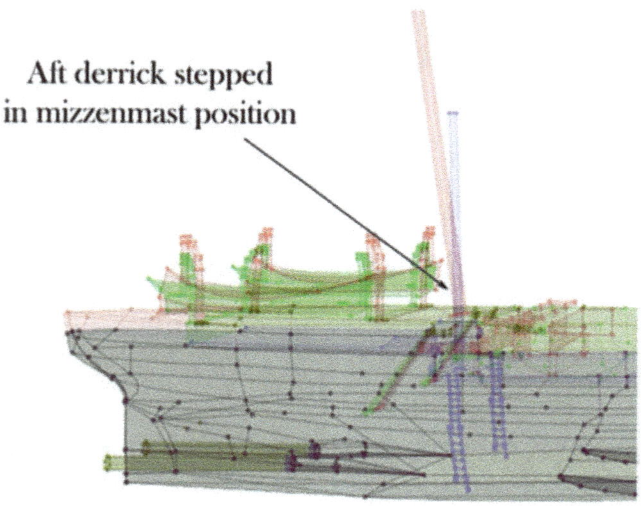

Fig. 6.8 Superimposed digital models of the stern section of *Protector/Sidney* derived from archival sources, showing the position of the vessel's mizzenmast and aft derrick in its respective colonial navy (red) and civilian (blue) configurations. Grey represents where all models overlap (ACVT/authors)

In the wake of the decision to convert *Protector* into a towed lighter, the vessel's propulsion system was no longer necessary and occupied valuable space that was targeted for conversion into one or more cargo storage areas. As a consequence, its engines—and their associated propellers and propeller shafts—were removed. However, the stern tubes (cylindrical iron housings through which the propeller shafts penetrated the hull to connect the propellers with the engines) were partially built into the ship's architecture and could not be completely removed without making significant alterations to *Protector*'s lower hull. In an effort to correct the problem each stern tube was cut off at the point where it entered the hull. Additionally, the propeller shaft struts that supported the aft ends of the stern tubes were completely cut away. The removal of these unnecessary structural members would have eliminated drag, as well as a potential source of fouling.

Archaeologically, the stern tube removal process is represented by extant remnants of *Protector*'s upper propeller shaft struts, each of which exhibits clear evidence of having been cut away at the point where it intersects the hull (see Hunter and Jateff 2016, 432, Fig. 7c). While the exact manner by which their removal was carried out is unclear, the gouges and other marks that remain on each shaft stay stump are prominent and easily discernible. Remnants of the lower propeller shaft struts, if they exist, were buried beneath the seabed and could not be examined. Removal of the majority of *Protector*'s stern tubes may have improved the hull's hydrodynamics and handling qualities, but the structures that remained were essentially open conduits to the sea. To counteract this problem, measures were taken to close the aperture of each stern tube. This is evidenced by the presence of a circular steel plate welded just inside the port side tube that completely seals its opening (see Hunter and Jateff 2016, 432, Fig. 7d). The starboard stern tube presumably was modified in a similar fashion, but was buried beneath the seabed and could not be examined during the survey.

While the majority of archival photographs depict *Protector/Sidney* afloat, and effectively obscure the appearance of the hull below the waterline, a very small handful of images show the vessel completely—or almost completely—out of water. These images are significant because they provide much-needed depictions of the vessel's hull below the waterline, which in turn have augmented the overall interpretive value of the ACVT models. They also supplied useful data that informed archaeological analysis of the stern tube removal process. Three photographs were taken in 1884 while *Protector* was still on the stocks at the shipyard of Sir William Armstrong & Co. in Newcastle-on-Tyne, and two include different views of the ship's stern tubes and propellers. Three others taken in 1943 depict *Sidney* aground at Facing Island in Queensland shortly after an initial ill-fated attempt by Cristian Poulson to tow it to Heron Island. Again, two photographs show the vessel's stern tubes, and clearly reveal a sizeable portion of their overall length was cut away. All of these images were integrated within the ACVT modelling dataset, and reveal approximately two-thirds of each stern tube's total length was removed (see Fig. 6.7, top)—a figure that would have been difficult to discern simply by looking at the photographs themselves, or even the archaeological signatures of their removal.

6.8 Discussion and Conclusions

As demonstrated by the *Protector* case study, creation of 3D digital models from archival sources has utility in the interpretation of the 'life history' of watercraft. Further, the technique can be a useful analytical tool in the archaeological investigation of wrecked and abandoned vessels. This is especially true in instances where alterations to hull and/or superstructure are no longer evident in the archaeological record, or so subtle they are practically unrecognisable in surviving archival photographs and plans. *Protector*'s known photographic archive covers the complete span of the vessel's life from its 1884 launch until the present day. Indeed, the ship's surviving hull continues to be documented yearly (if not monthly or weekly) by visitors to Heron Island, and the volume of images depicting it has increased exponentially with the advent of digital photography and smart phones. Many of the more significant alterations to *Protector*'s hull and superstructure during its operational career are immediately evident in historic images, but the presence of many others—such as the variation in rake of the ship's foremast—were only identified from a comparison of digital models generated as a result of this research.

While useful, the technique also has clear limitations that may inhibit or prevent its use in certain circumstances. Perhaps one of the biggest drawbacks is the need for a large photographic archive that depicts the vessel from as many different perspectives as possible. In the case of some watercraft, the available collection of historic images is vast and varied; however, the majority of historic vessels have few, if any, existing photographs from which to derive reliable 3D models. Because it is based entirely on photographs, the technique also precludes the creation of digital models of vessels built and/or operated prior to the advent of photography. Finally, comparative models that show a vessel's change over time can only be generated if the photographic archive includes several different images that share multiple hull and/or superstructure features in common. Because they form consistent reference points, these features are absolutely necessary to the success of the SfM method. Without the right imagery, application of the technique is difficult, if not impossible.

The development of *Protector* 'pseudo images' from plans and schematics was the primary means by which ACVT compensated for image shortfalls, and may serve as a launching point from which to explore the feasibility of adapting the technique to other iconographic formats (such as marine paintings, sketches and engravings). The integration of other iconographic data would serve as an ideal complement to photographic material, particularly in instances where a vessel's photographic archive is limited. Similarly, it would provide a method for modelling watercraft for which no photographic archives exist, and vastly expand the temporal span of the technique's applicability.

Although it has greatly enhanced our understanding of one ship's particular history, perhaps the biggest outcome from this study is that it has expanded the potential scope and utility of iconographic data to the archaeological interpretation of historic shipwrecks and abandoned vessels generally. Archival photographs have been a mainstay in the investigation and analysis of (primarily) iron and steel vessels for some time, but have in most cases merely served to complement or verify information revealed in the archaeological record. 3D modelling of archival imagery has taken this valuable dataset a step further, and demonstrated a new means by which changes to vessels over time may be recognized, analysed and interpreted. It has also provided an opportunity for maritime archaeologists to move beyond broad qualitative statements about vessel modification and adaptation, and quantitatively interrogate how, when and in what manner specific alterations occurred. This has potentially significant ramifications for ship studies specifically, and maritime archaeology more broadly. The same technique(s) that were applied to *Protector*'s photographic archive could be used to interpret a variety of other site types—such as elements of land-based maritime and other infrastructure—for which historic photographs exist. This can only enhance our understanding of the past, and better serve the goals of our discipline as a whole.

Acknowledgements The initiative to document, digitally model and interpret *Protector* would not have been possible without the assistance, advice and input of several individuals and institutions. Many thanks are due to our colleagues in the survey, research, and digital modelling teams: Daniel Pooley and John Bastian (Australian Centre for Visual Technologies), Kevin Jones, Lindl Lawton and Adam Paterson (South Australian Maritime Museum), Ian MacLeod (Western Australian Museum) and Ed Slaughter. Staff at Heron Island Research Station were excellent hosts, and must be commended for sharing information about *Protector*'s surrounding environment and providing logistical support during the survey. Paddy Waterson (Queensland Department of Environment and Heritage) kindly provided data from previous visual inspections of *Protector*, and assisted with the acquisition of archaeological permits. This project was supported by an Australian Research Council Linkage Award (LP130101064), the Commonwealth of Australia's Your Community Heritage Program, and the Silentworld Foundation.

References

Aubry M (2015) Representing 3D models for alignment and recognition. École des Ponts ParisTech, Paris

Aubry M, Russel B, Sivic J (2013) Painting-to-3D model alignment via discriminative visual elements. HAL online open-access archive (hal-00863615v1)

Baatz G, Saurer O, Köser K, Pollefys M (2012) Large scale visual geo-localization of images in mountainous terrain. In: Fitzgibbon A, Lazebnik S, Perona P, Sato Y, Schmid C (eds) Proceedings of the 12th European conference on computer vision, Florence

Baboud L, Cadik M, Eisemann E, Seidel HP (2011) Automatic photo-to-terrain alignment for the annotation of mountain pictures. In: Proceedings of the 2011 IEEE conference on computer vision and pattern recognition, Washington DC

Crumlin-Pedersen O (2000) To be or not to be a cog: the Bremen cog in perspective. Int J Naut Archaeol 29(1):230–246. https://doi.org/10.1006/ijna.2000.0321

Dalal N, Triggs B (2005) Histograms of oriented gradients for human detection. In: Schmid C, Soatto S, Tomasi C (eds) Proceedings of the 2005 IEEE conference on computer vision and pattern recognition, San Diego

Doersch C, Singh S, Gupta A, Sivic J, Efros AA (2012) What makes Paris look like Paris? ACM Trans Graph 31(3):101

Felzenszwalb PF, Girshick RB, McAllester D, Ramanan D (2010) Object detection with discriminatively trained part based models. IEEE Trans Pattern Anal Mach Intell 32(9):1627–1645

Flatman J (2004) The iconographic evidence for maritime activities in the Middle Ages. Curr Sci 86(9):1276–1282

Flatman J (2014) Maritime iconography. In: Smith C (ed) Encyclopaedia of global archaeology. Springer, New York, pp 4666–4671

Gillett R (1982) Australia's colonial navies. Naval Historical Society of Australia, Garden Island

Hartley R, Zisserman A (2004) Multiple view geometry in computer vision. Cambridge University Press, Cambridge

Hunter J, Jateff E (2016) From battleship to breakwater: post-military adaptive reuse of the Australian warship Protector. Int J Naut Archaeol 45(2):423–440. https://doi.org/10.1111/1095-9270.12177

Hunter J, Jateff E, Herath N, van den Hengel A (2016) Protector revealed: an initiative to archaeologically document, interpret and showcase an historic Australian warship with laser scanning technology. J Cult Herit Conserv 37:25–40

Huttenlocher DP, Ullman S (1987) Object recognition using alignment. In: Proceedings of the first international conference on computer vision, London

Jeisman S (2012) Colonial gunboat: the story of HMCS Protector and the South Australian naval brigade. Adelaide

Jones C (1986) Australian colonial navies. The Australian War Memorial, Canberra

Li B, Shen C, Dai Y, van den Hengel A, He M (2015) Depth and surface normal estimation from monocular images using regression on deep features and hierarchical CRFs. In: Proceedings of the IEEE conference on computer vision and pattern recognition, pp 1119–1127. https://doi.org/10.1109/CVPR.2015.7298715

Lowe D (1987) The viewpoint consistency constraint. Int J Comput Vis 1(1):57–72

Lowe D (2004) Distinctive image features from scale-invariant keypoints. Int J Comput Vis 60(2):91–110

MacLeod I, Jateff E, Hunter J (2014) Corrosion on a wrecked colonial vessel: HMCS protector, 1882–1944. Corros Mater 39(3):50–54

McGrail S (1998) Ancient boats in Northwest Europe: the archaeology of water transport to AD 1500, Longman archaeology series. Routledge, London

National Archives of Australia, Canberra. HMAS protector: proposed arrangement of minesweeping gear, etc. MP551/1, 92/20

Nicholls B (1988) The colonial volunteers: the defence forces of the Australian colonies, 1836–1901. Allen & Unwin, Sydney

Pennock R (2001) A warship for South Australia. Robin Pennock, Blackwood

Rapp JB (2008) A geometrical analysis of multiple viewpoint perspective in the work of Giovanni Battista Piranesi: an application of geometric restitution of perspective. J Archit 13(6):701–736. https://doi.org/10.1080/13602360802573868

Russell BC, Sivic J, Ponce J, Dessales H (2011) Automatic alignment of paintings and photographs depicting a 3D scene. In: Proceedings of the third international IEEE workshop on 3D representation for recognition, Barcelona

Shrivastava A, Malisiewicz T, Gupta A, Efros AA (2011) Data-driven visual similarity for cross-domain image matching. In: ACM transactions on graphics/proceedings of ACM SIGGRAPH Asia 2011 30(6)

Sivic J, Zisserman A (2003) Video Google: a text retrieval approach to object matching in videos. In: Proceedings of the ninth IEEE international conference on computer vision, Nice. https://doi.org/10.1109/ICCV.2003.1238663

Snavely N, Seitz SM, Szeliski R (2006) Photo tourism: exploring photo collections in 3D. ACM Trans Graph/Proc ACM SIGGRAPH 25(3):835–846

State Library of South Australia, Adelaide. H.B. Treacy to Lt. H.M. Cooper. Research notes [RN] No. 100, bd 994.23 R432b

Steffy JR (1994) Wooden ship building and the interpretation of shipwrecks. Texas A&M University Press, College Station

Szeliski R (2010) Computer vision: algorithms and applications. Springer, London

United States Navy (1917) Mine sweeping manual. United States Navy, Washington, DC

Villain-Gandossi C (1979) Le navire medieval à travers les miniatures des manuscrits français. In: McGrail S (ed) Archaeology of medieval ships and harbours in Northern Europe, BAR international series 66. BAR Series, Oxford, pp 195–225

Winter W, Burningham N (2001) Distinguishing different types of early 17th-century Dutch jacht and ship through multivariate morphometric analysis of contemporary maritime art. Int J Naut Archaeol 30(1):57–73. https://doi.org/10.1016/S1057-2414(01)80007-8

7 The Conservation and Management of Historic Vessels and the Utilization of 3D Data for Information Modelling

Dan Atkinson, Damien Campbell-Bell, and Michael Lobb

Abstract

The increased use of laser scanning and photogrammetry has given rise to new opportunities in disseminating information about historic maritime assets and are of great use in conservation management initiatives. This chapter discusses the current state of 3D survey of historic vessels and how this has been applied more recently for historic vessel conservation management. Key questions such as how this data is utilized, and what it is that the capture of such data is trying to achieve for the conservation and management of historic ships and vessels will be explored. In addition, this chapter will introduce information modelling, most commonly seen as Building Information Modelling (BIM) as an approach for furthering the effective management of historic ships and vessels, as well as other historic marine and maritime assets. It will demonstrate that the majority of attempts at utilizing BIM in the heritage sector have been limited to buildings, and that the full potential of this technique has not been realized. Through the use of the 'VIM' project at HMS *Victory* the chapter will then explore how information modelling can be applied to a highly complex historic ship.

Keywords

Historic ship · Conservation management · Information modelling · BIM · HMS *Victory*

D. Atkinson (✉)
Wessex Archaeology Ltd, Edinburgh, UK
e-mail: d.atkinson@wessexarch.co.uk

D. Campbell-Bell
Wessex Archaeology Ltd, Salisbury, UK
e-mail: d.campbell-bell@wessexarch.co.uk

M. Lobb
HMS Victory, National Museum of the Royal Navy, Portsmouth, UK
e-mail: michael.lobb@NMRN.org.uk

7.1 Introduction

The increased application of 3D datasets in the heritage sector in recent years has significantly improved the ability to optimize how we visualize and disseminate information about underwater and intertidal archaeological sites (Sanders 2012). It is also the case that the utilization of 3D data has increasingly formed an element of the survey and recording strategies for extant historic ships and vessels, either as stand-alone projects, or as part of conservation management initiatives.

Conservation Management for historic vessels follows the more general heritage conservation principles set out in the 'Burra Charter' (Australia ICOMOS 1999) first adopted in 1979 and updated in 2013; the seminal work undertaken by Semple Kerr (2013); and the 'Barcelona Charter' (Heidbrink 2003) which outlines approaches to best practice for the conservation and restoration of operational historic watercraft in Europe. In the United Kingdom, national heritage bodies have also published guidance on conservation management, including Historic Environment Scotland (2000); Historic England (formerly English Heritage) (English Heritage 2008); and national funding bodies such as the Heritage Lottery Fund (2012). In essence, these publications provide the benchmark for effective heritage conservation planning and management, and this formed the basis for dealing with the same considerations in relation to historic vessels in the United Kingdom, delivered through guidance developed by National Historic Ships (NHS 2010) in the UK. Connected to extant historic ships are those that represent examples of largely complete ships that have been recovered from the seabed. A good example is the *Vasa* ship (1628) in Sweden where the real need for effective conservation management planning has been clearly stated (Malmberg 2003).

The traditional approach to historic vessel survey is based very much in the 2D world and previous publications outline these traditional approaches that could be employed by the

J. K. McCarthy et al. (eds.), *3D Recording and Interpretation for Maritime Archaeology*, Coastal Research Library 31,
https://doi.org/10.1007/978-3-030-03635-5_7

investigator (Anderson 1994; Kentley et al. 2007; Lipke et al. 1993). The outputs comprise the traditional expression of the line drawing, in addition to plans, elevations and sections of a vessel. While more generally, advances in technology and surveying techniques and the resultant outputs in recent years have demonstrated the effectiveness of 3D survey, it is only in the last 10 years or so that 3D survey has begun to be utilized more fully for historic vessels; the details for which are more commonly found within unpublished papers and client reports. Indeed, until the last 5 years or so, the 3D outputs created for historic vessel survey have generally followed the aims of the traditional 2D outputs, these being the expression of hull shape, line drawings, and associated elevations, plans and sections. Developments in integrated modelling and data management software, particularly in a BIM context, has demonstrated the importance of accurate survey data for the effective management of complex projects (for which historic vessels certainly form a part). Whilst the primary focus for software developers has been the Architecture, Engineering and Construction (AEC) sector, the benefits such software, and the processes which the AEC sector has developed to underpin its use, are now being explored more fully within the heritage sector.

For such a new concept, there is a surprizing amount of literature about heritage BIM, including a recent edited volume devoted to it (Aracyici et al. 2017); however, there is little that fully addresses BIM as software and process. The majority of the academic literature on BIM and heritage focuses on single aspects of BIM, be that storage of heritage information (Counsell and Edwards 2014; Edwards 2017; Simeone et al. 2014; Yajing and Cong 2011; Zhang et al. 2016); 3D representation of historic buildings (Brumana et al. 2013; Fai and Sydor 2013); or automated modelling (Dore and Murphy 2012a, b, 2013). Whilst these approaches no doubt have things to offer, they do not exploit the full benefits of BIM as a project and asset management tool (Breeden 2015, 2016; Campbell-Bell 2015, 2016 for an introduction to how BIM could be fully utilized in the heritage sector). It is these aspects of BIM that have the most to offer complex conservation management projects; and this is certainly the case with historic ships, given the inherent challenges that large and extremely complex artefacts pose.

This chapter discusses the current state of play with 3D survey of historic vessels and how this has been utilized more recently for historic vessel conservation management, focusing on experiences in the UK. Key questions will be explored such as: how this data is utilized; what is it that the capture of such data is trying to achieve for the conservation and management of historic ships and vessels; and what mechanisms developed for other sectors can be utilized to best enhance the effective conservation management of historic vessels—especially given their inherent complexity and heritage value? The chapter will then focus on the application and integration of 3D data into information modelling (such as BIM) as a useful management tool, and on the concept of BIM and its application to historic ship conservation management in the UK. We explore how the opportunities presented in 3D realization can be employed to best effect in relation to the conservation management of historic ships and vessels—ultimately using the iconic HMS *Victory* as a case study to demonstrate mechanisms that can be employed for effective conservation management, and to highlight some of the limitations of existing systems.

Finally, the chapter will address the way forward, and the challenges and opportunities that lie ahead—particularly in relation to the practicalities of utilizing 3D data for information modelling in the effective conservation and management of historic ships and vessels more specifically, and indeed the wider maritime heritage resource more generally.

7.2 Historic Vessel Conservation Management Practice

The principles set out in the UK National Historic Ships guidance *Conserving Historic Vessels* (NHS 2010) outline the considerations for approaches to effective historic vessel conservation through 'gateways.' These gateways influence the priorities that face a vessel owner when establishing the way in which a vessel is potentially conserved and maintained into the future; essentially through *preservation*, *restoration*, *adaptation*, and *reconstruction*. The basis of the 'gateway' approach is established and realized through the preparation of Conservation Management Plans or Conservation Statements, which aim to highlight and understand the key significances of a vessel, and present ways in which the opportunities and threats facing the asset can be effectively managed into the future. Indeed, there have been a number of such plans and statements undertaken for a variety of historic ships and boats in recent years; select examples include SS *Great Britain* (1999)*; Cutty Sark* (2005)*; Lively Hope* (2007)*; City of Adelaide* (2012)*:* HMS *Caroline* (2014); HMS *Victory* (2015), and the *Scottish Fisheries Museum Fleet* (2017), some of which are discussed further in this chapter. Many of the earlier surveys, however, were produced at a time when 3D survey for historic vessels, and the understanding of its potential, was in its relative infancy, and due to a number of limiting factors—particularly in terms of data quality, processing, and the issues relating to time constraints and cost—survey initiatives were either not included (in the case of SS *Great Britain*), or where undertaken (for *Cutty Sark* and *City of Adelaide* for example), were not fully utilized for effective conservation management purposes.

It is also worthy of mention that whilst it is not the intention of this chapter to discuss the conservation management of ships recovered from the seabed (such as *Vasa* and *Mary Rose*), and submerged archaeological sites in general, the interesting parallel alongside the conservation management of historic vessels is the use of conservation management plans and statements for wreck sites. While the approach is not new, national heritage bodies such as Historic England in the UK have recently commissioned the preparation of a suite of conservation statements in helping to understand the key management requirements for the protection, enhancement, and wider accessibility of designated wreck sites in English territorial waters (examples include Dunkley 2008; May et al. 2017). Allied to this is the increasing synergy between the management objectives outlined in the reports and the use of 3D survey imagery for sites, where investigators visiting sites are encouraged to use 3D outputs to provide information relating to ongoing monitoring and management. This is an important aspect of shipwreck conservation management, and while beyond the scope of this chapter, represents a clear avenue for future research and publication.

7.3 3D Survey for Historic Vessels

3D survey techniques, particularly laser scanning, and the increased use of photogrammetry, for the survey of extant historic vessels in recent years has resulted in a very broad based, and albeit fairly simplistic application. Early surveys, such as that employed on the clipper ship *Cutty Sark* (1869) in 2005, utilized hardware and software that has been improved in recent years—particularly in terms of factors such as data quality and processing, and the relative reduction in the time and cost associated with a project. The nature of the application aside, the objectives apparent in the utilization of the resultant data were primarily based within the realms of engineering, simple 2D outputs, or for virtual reality applications. Data acquisition for archaeological purposes during most historic vessel survey resulted in the provision of an accurate archaeological record, which in most cases was used for outputs such as the creation of basic wireframes, elevations, plans, and sections, or the creation of more traditional line drawings. This was perhaps without any real understanding of how the 3D data could and should be utilized for archaeological and heritage conservation purposes.

Perhaps one of the earliest surveys that prompted the question of archaeological enquiry and the use of the data for archaeological purposes was the laser scan survey of the ex-whaling vessel *Charles W Morgan* (1841), based in Mystic on the eastern seaboard of the United States (Classic Boat 2010). In this case, the resultant data was utilized for a number of key purposes, including gathering detailed information about the hull and the individual components during each phase of the restoration, and the production of detailed plans of the ship, for which no historic archive was available.

The further use of laser scanning within an historic vessel context occurred around the same time as *Charles W Morgan*, in 2008, with the survey of the emigrant clipper ship *City of Adelaide* (1864) (also named *Carrick* during service as a Royal Navy Reserve Headquarters in Glasgow) whilst located at the Scottish Maritime Museum in Irvine, Scotland (Fig. 7.1). In this case the ship was under threat and the museum had limited options for the sustainable future of the vessel. As such, an application was sought with the national and local heritage curators to deconstruct the ship—in line with the National Historic Ships guidance *Deconstructing Historic Ships* (NHS 2007). The status of the ship as a protected 'Listed Building' meant that certain conditions were required to be met by the museum, prior to the final decision on the future of the ship. This included a full archaeological record of the ship hull and constituent parts, and a contextual record of the ship and the site to help aid engineering requirements (Figs. 7.2 and 7.3). In light of these requirements, a laser scan survey was commissioned which succeeded in providing an accurate record of the ship and the site upon which the vessel was situated (Atkinson et al. 2009). The ultimate use of the data however, was to help provide engineers with an accurate rendition of the hull of the ship in preparation for the design and fabrication of a bespoke cradle on which the ship was to be removed from the site in advance of transportation back to Port Adelaide in South Australia (Fig. 7.4).

Similar to the *City of Adelaide* survey, engineering requirements also formed part of a laser scan survey for the hull of the *Research*, an early twentieth-century first class Zulu sailing herring drifter, based at the Scottish Fisheries Museum in Anstruther, Fife. In addition to providing an archaeological record of the vessel, the data was used to help provide a control on which the hull could be monitored for movement as part of a static exhibit within the museum (Atkinson et al. 2010) (Figs. 7.5, 7.6, and 7.7).

A further example of the laser scanning of a large hulked clipper ship is that of *El Ambassador* (1868) located in Chile. In 2013, the *Instituto de Arqeologia Nautica y Subacuatica* surveyed the surviving hull of *El Ambassador* and succeeded in gaining detailed data about the nature of the remains, and the conditions and environment in which the ship survives. Like previous surveys however, the use and application of the data was limited to producing digital plans. It was a landmark survey representing the first laser scan survey of an extant historic vessel undertaken in Chile (Pollet and Pujante 2013).

These surveys were clearly successful in capturing the detail of the ship, and could be used for a range of purposes. This included providing control for the preparation of ship

Fig. 7.1 The emigrant clipper ship *City of Adelaide* on the slip at Irvine on the east coast of Scotland prior to survey and removal to Port Adelaide in South Australia (Headland Archaeology Ltd)

Fig. 7.2 Orthographic representation of the resultant point cloud acquired form the laser scan survey (Headland Archaeology Ltd.)

plans used in restoration works, in the case of the *Charles W Morgan*, providing an archaeological record and engineering control, for the design of a bespoke cradle, in the case of the *City of Adelaide*, and for ongoing structural monitoring in the case of the *Research*. The exploration of the conservation management potential of the data was minimal, however—mainly due to the difficulties in using the data for effective archaeological enquiry and the financial constraints in integrating 3D survey techniques such as laser scanning into project designs at the time. Furthermore, the primary aim was to aid in the understanding of the priorities in stabilizing the shape and condition of vessels in the early stages of conservation, not to form part of an ongoing conservation management programme throughout the lifecycle of a

Fig. 7.3 External and internal long sections of the *City of Adelaide* as acquired from the survey data (Headland Archaeology Ltd)

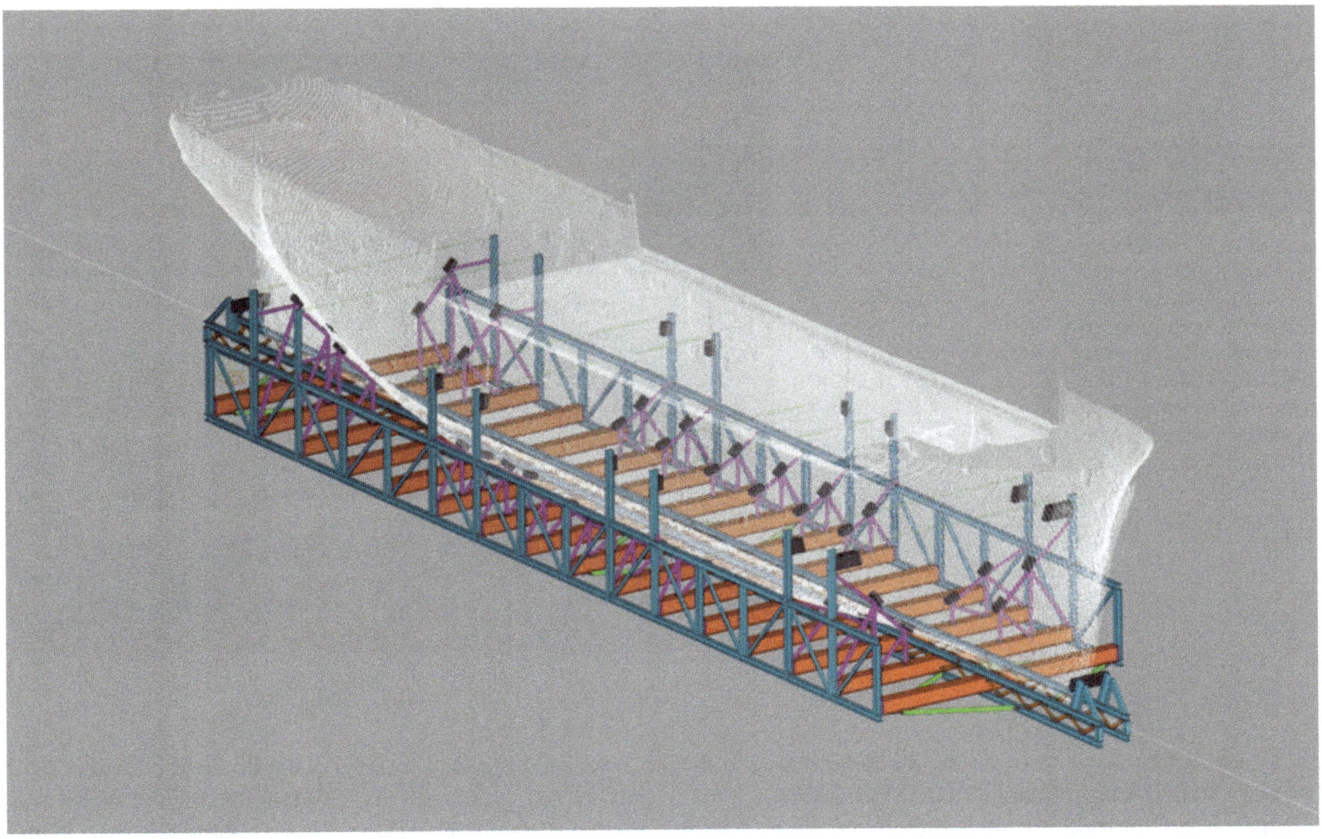

Fig. 7.4 Examples of the resultant engineering modelling of the *City of Adelaide* developed during the design of the bespoke cradle used during transportation to Port Adelaide in South Australia (Clipper Ship 'City of Adelaide' Ltd)

conservation project. The same can also be said for projects where the principal aim was to provide data for virtual reality platforms and displays to augment educational dissemination; recent examples include the laser scan survey and digital modelling of the Qatar Museums watercraft collection in Doha (Cooper et al. 2018); the RRS *Discovery* (1901) located in Dundee, Scotland (Digital Surveys 2014); *Edwin Fox* (1853) based at Picton in New Zealand (3-D Scans

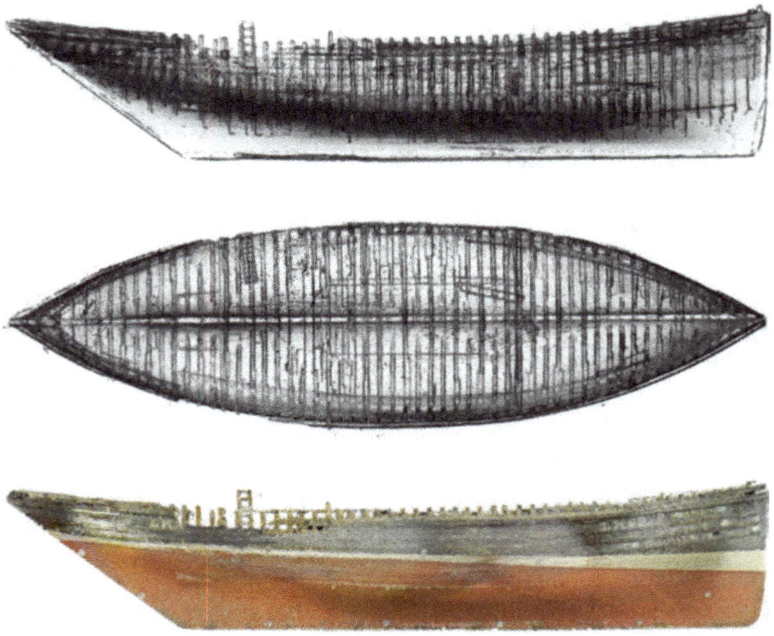

Fig. 7.5 Plans and elevations derived from the laser scan data of the Zulu Research showing the main characteristics of the internal and external hull as surviving (Headland Archaeology Ltd)

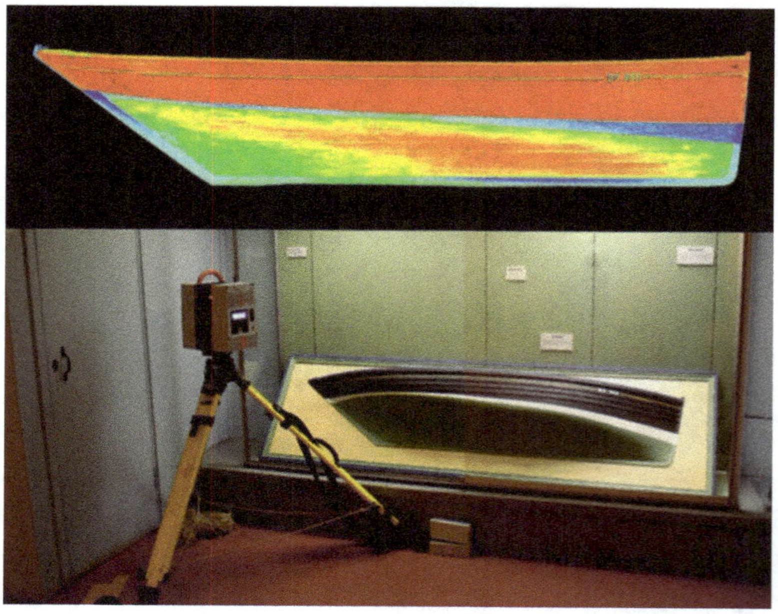

Fig. 7.6 Laser scan point cloud from the survey of the builders' half model of a Zulu herring sailing drifter for a vessel built at the same boatyard (and period) as the *Research*. The model was scanned to provide a control from which to overlay the laser scan data from the *Research* to help understand the movement in the hull of the vessel to inform future conservation management and engineering considerations. (Headland Archaeology Ltd)

2017); and more recent work on the SS *Robin* based at the Royal Victoria Docks in London (SIAD Ltd. 2017).

More recent historic vessel conservation projects have included the preparation of Conservation Management Plans (NMRN 2014, 2015), a key part of which includes the use of 3D survey applications. The most notable is the laser scan survey undertaken for HMS *Victory* (1765) in 2013, which will be discussed in more detail within the following case study. Other examples include the laser scan survey in 2014 of HMS *Caroline* (1914) a British First World War light cruiser and last extant survivor of the Battle of Jutland (1916), and the laser scan survey in 2017 of the Second World War era British submarine HMS *Alliance* (1945) (Fig. 7.8). In the case of HMS *Caroline*, a laser scan survey was undertaken with the principal aim of producing an accurate representation of the ship, its constituent parts and its internal spaces. This provided a basis for the production of accurate plans and elevations to assist with the recording phases associated with the ship's biography—essentially identifying parts of the ship relating to various phases of the vessel's life, from construction

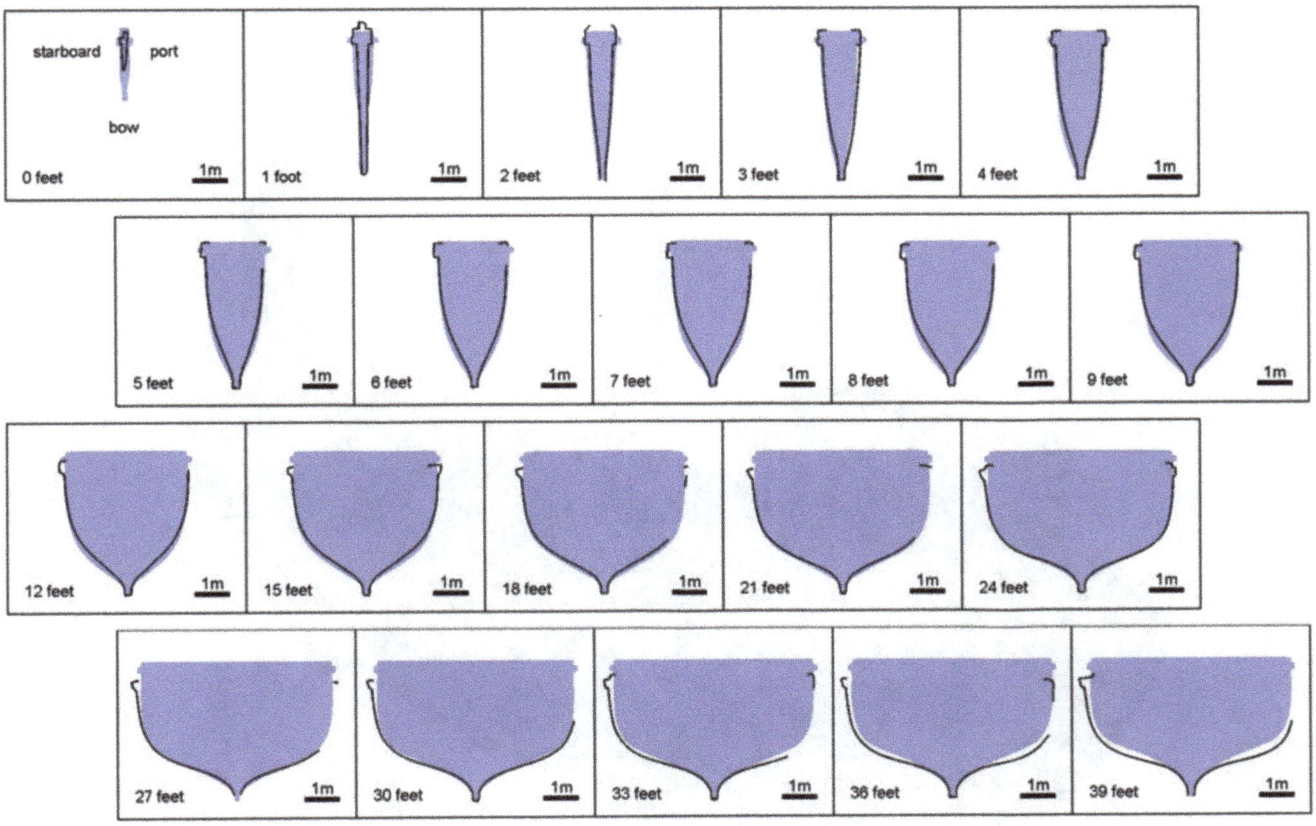

Fig. 7.7 Sections along the vessel (bow top left, towards the mid-section, bottom right) denoting the data derived from the scan of the half model (in purple), and the corresponding section of the hull derived from the scan of the *Research* (black line). This clearly indicates 'sagging' and 'pinching' of the hull along the length of the vessel, providing important insight into future support strategies for the vessel in the museum gallery. (Headland Archaeology Ltd)

to decommissioning, following a long period of service as a Royal Navy Reserve Headquarters located in Belfast (Figs. 7.9 and 7.10).

Aside from laser scanning, 3D data has also been acquired using photogrammetry, and this is becoming an increasingly popular method of carrying out rapid and cost-effective surveys across the heritage sector (demonstrated in the various chapters throughout this volume). A good example of the use of photogrammetry in relation to conservation management for historic vessels was the production of a series of Conservation Management Plans and Statements associated with the fleet of vernacular Scottish fishing vessels based in the collection at the Scottish Fisheries Museum at Anstruther in Fife, Scotland (Wessex Archaeology 2017). Similar to the laser scan surveys noted above, the principal outputs were relatively simplistic, and in most cases provided data to enable the production of plans and elevations, and the ability to understand detail of the vessel components and construction characteristics (Fig. 7.11). In this case however, it is important to note the usefulness of capturing 3D data for smaller vessels and boats that, due to their vernacular nature, have no drawn record from which to undertake informative recording, assessment, and interpretation. Further effective use of photogrammetry within historic ship conservation is the use of the technique to provide an accurate, quick and cost-effective means of recording areas of a vessel that are undergoing active conservation work. During the restoration works on HMS *Caroline* (carried out between 2014 and 2016), the technique was used to great effect in the recording of historic deck planking from the starboard waist area, and parts of the original floor in the former drill hall, introduced to the ship in the early 1920s when in use as a Royal Navy Reserve Headquarters. Despite working alongside contractors, it was possible to record and interpret the characteristics of the historic decking and the relict features on the deck plating below (Fig. 7.12). The resultant interpreted plans were then provided as part of proactive 'live' mitigation to help assist the ship's managers with the Heritage Impact Assessment (HIA) process, and the means to understand how the significance of the historic ship fabric is retained.

There is no question that the capture of 3D data during the survey of historic vessels has proved useful, particularly in terms of providing an accurate archaeological record, assisting with the preparation of plans, elevations and section of vessels; particularly where documentary sources are limited or do not exist. The use of the data to assist with engineering

Fig. 7.8 Orthographic representation of the Second World War era submarine HMS *Alliance* following laser scan survey (Wessex Archaeology Ltd)

considerations has also proved very useful in many cases. In order to begin to realize the true power of the data however, opportunities to embed the data in more meaningful and long-term conservation strategies are now being realized—particularly with the development of information modelling applications.

7.4 The Concept of Building Information Modelling (BIM)

There have been a range of approaches to aggregating information and spatial data since the advent of computing technology, the most widely used likely being GIS. In recent years however, Building Information Modelling (BIM) has seen widespread adoption in the AEC industry and increased interest amongst heritage professionals. BIM is a term which has been used and understood in many ways since it was conceived, but perhaps the most useful way to consider it is in terms of the UK Government driven push towards greater collaboration and data sharing within the construction industry. The aim is to reduce capital expenditure (CAPEX) by 20%, and to offer significant reductions in operational expenditure (OPEX) (HM Government 2011). Within this frame of reference, BIM can be seen as a collaborative process of project design, management and implementation, guided by common specifications and data standards. This combination of process and information management, along with the fully 3D nature of BIM is what distinguishes it from other computing approaches.

At its heart BIM is about more efficient working. Through early engagement of stakeholders, collaboration and careful documentation of the project timelines, responsibilities and deliverables, BIM aims to ensure that time is not wasted through reworking, mistakes, confusion and data loss. Whilst some people see it as such, BIM is not about software alone, it is a way of working within set guidelines. It is also important to note that the 'Building' in BIM, is a verb, not a noun.

Fig. 7.9 HMS *Caroline* in Alexandra Dock in Belfast in 2018, following the programme of restoration which aimed to re-introduce as far as possible the 1916 'Battle of Jutland' appearance and configuration (Wessex Archaeology Ltd)

Whilst BIM was developed for constructing buildings it is increasingly being implemented in other areas, such as infrastructure. Whilst there have been calls to proliferate the process name, where associated with use in different sectors, this distances these use cases from the core elements of BIM and offers little in terms explanatory value. We could talk of Heritage BIM, or Ship Information Modelling, but prefer to follow the standard adopted by the many AEC industry special interest groups and use BIM for historic vessels.

By mandating its use on major government-funded projects and establishing the BIM Task Group to oversee its development (www.bimtaskgroup.org), the UK is one of those at the forefront of global BIM implementation. Specifications such as the CIC BIM Protocol, BS 1192:2007, BS 1192-4, PAS 1192-2 3, 5 and 6, and the upcoming PAS 1192-7, define standard ways of working which should be adhered to when using BIM on a project.

In addition to using these specifications, project documentation such as Employer Information Requirements (EIR) and BIM Execution Plans (BEP) are key, respectively, for defining project requirements at the tendering stage and for tenderers demonstrating how they will meet those requirements. After a contract has been awarded, a Post-Contract BEP is created collaboratively to plan the project implementation.

Software is still very important to the implementation of BIM. There are many BIM authoring tools available, such as *Graphisoft ArchiCAD* or *Autodesk Revit* which allow the integration of 3D geometry and non-geometric information, facilitating the implementation of BIM in a way which will lead to the desired benefits. Again, it is important to stress that the use of these standard software packages do not make a project BIM compliant, nor is their use necessarily required.

These software packages allow the creation of 3D models with information rich objects, meaning that all the information about a construction project, including time scales, costs, responsibilities and attributes of each individual item can be included. By including this information, the model serves as a single source of truth for the project and can be used for the automatic generation of documents, such as schedules and 2D drawings. This model can also serve as a valuable asset in the future maintenance of the structure. The use of IFC as a standard data format allows this information to be interoperable, i.e. it can be shared between different software packages and maintain integrity, allowing different practitioners to work on the model.

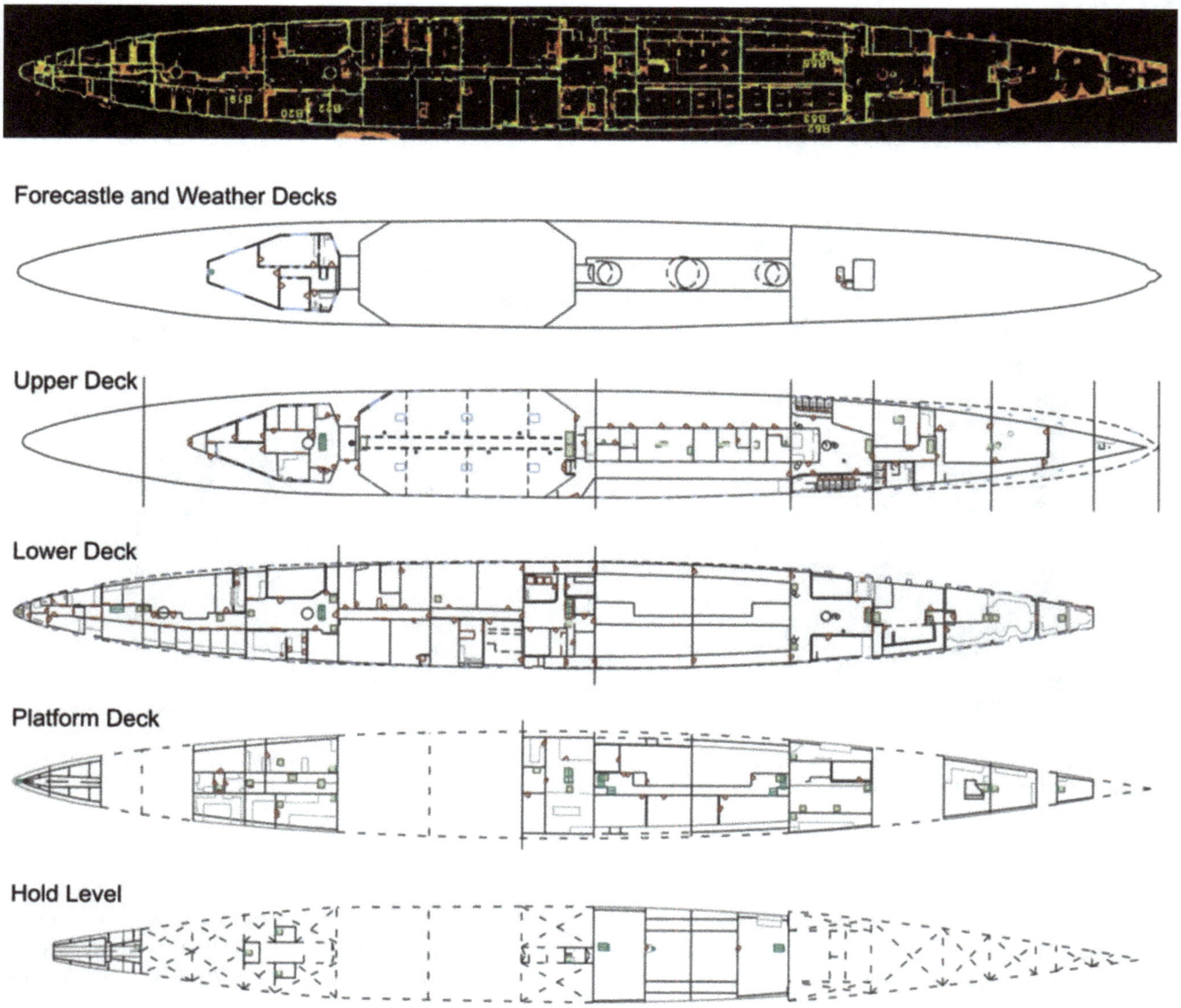

Fig. 7.10 2D output of the lower deck of *HMS Caroline* derived from the laser scan data (top) as indicated in a traditional plan (bottom). This allowed the most current representation of the compartments and spaces along the deck to assist with the detailed recording and interpretation phase during the compilation of the CMP (Wessex Archaeology Ltd)

Three key principles underlying the use of these models are:

1. Level of Detail (LOD)—the amount of geometric detail included in the model, ranging from a generic place holder shape of the right dimensions, to a fully modelled example of the actual object;
2. Level of Information (LOI)—the amount of information attached to an object, ranging from basic information such as what the object represents, and its dimensions, through to full specifications, and links to maintenance schedules and operating manuals for the object; and
3. Level of Definition (confusingly also abbreviated LOD)—a combination of Level of Detail and Level of Information.

Much has been achieved in driving towards BIM Level 2, but true BIM Level 3 (Fig. 7.13) remains out of reach at this time, due to technical, legal and procedural limitations. BIM's implementation within heritage however, has been patchy, and rarely meets its full potential, yet it is clear that this potential exists, particularly in the management of complex heritage assets such as historic vessels. As often very complex structures, both in terms of construction and history, there is a vast amount of information associated with these vessels. This information is important to understand and to manage the vessels, but it is often widely spread, inconsistently structured, and disassociated. Although in a different context, it was precisely these difficulties that BIM was conceived of to address, and so, as with operational management of a building through BIM, by combining all

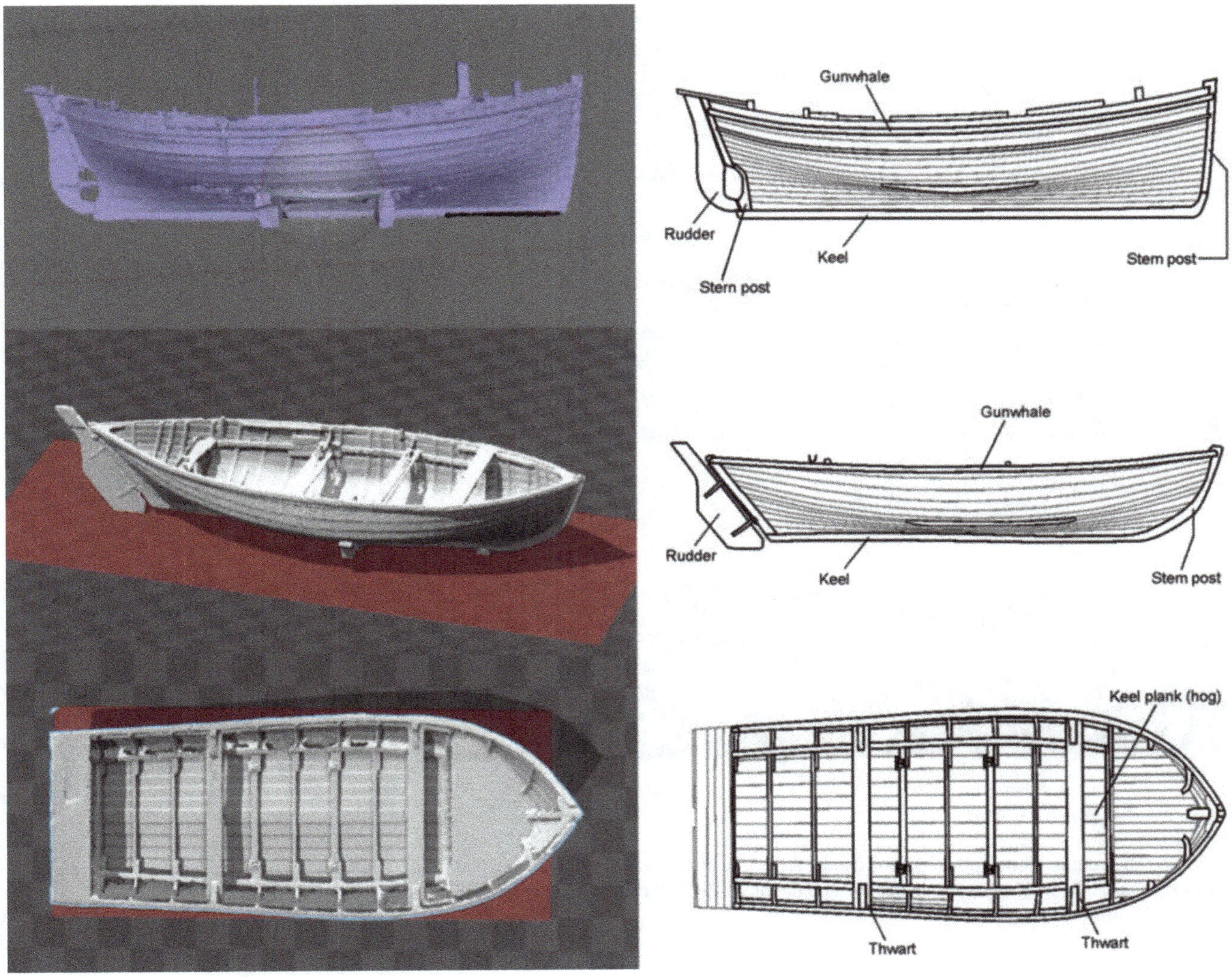

Fig. 7.11 Vernacular fishing vessels from the Scottish Fisheries Museum fleet. Photogrammetric renditions of the Baldie class *White Wing* (top); the Scaffie *Maggie* (middle); and the Montrose Salmon Coble *Jubilee* (bottom). Examples of the resultant construction detail drawings for each vessel are noted to the right of each (not to scale) (Wessex Archaeology Ltd)

the pertinent information into one place the conservation management needs of historic vessels can be much better met.

7.5 Use of BIM in the Heritage Sector

BIM development across the world has been firmly focused on new construction; heritage has been mostly ignored outside academic circles. This is despite calls, in the UK, for '*all contractors*' to engage in the BIM process, and the importance of existing assets highlighted in *Digital Built Britain* (HM Government 2015). Over 90% of buildings which will be used in the UK over the next 25 years currently exist; many of these are, or will eventually become, historic buildings. The result of this lack of attention amongst policy makers and the AEC industry is that there are no robust specifications or standards for BIM in the heritage sector.

Given the current lack of standards, it is unsurprizing that BIM, as defined above has yet to see widespread use for heritage purposes. BIM has been applied in a number of projects, however, these all focus on historic buildings. Just as BIM is now being applied to infrastructure in the AEC industry, it can be applied to the operation of commercial archaeology projects and other elements of the built environment, such as ships, as is demonstrated below with HMS *Victory*. The lack of data standards and consequent limitations of software, as well as the broad array of heritage assets makes BIM's utilization somewhat problematic at this time, but examples such as the Manchester Town Hall project show the benefits it can

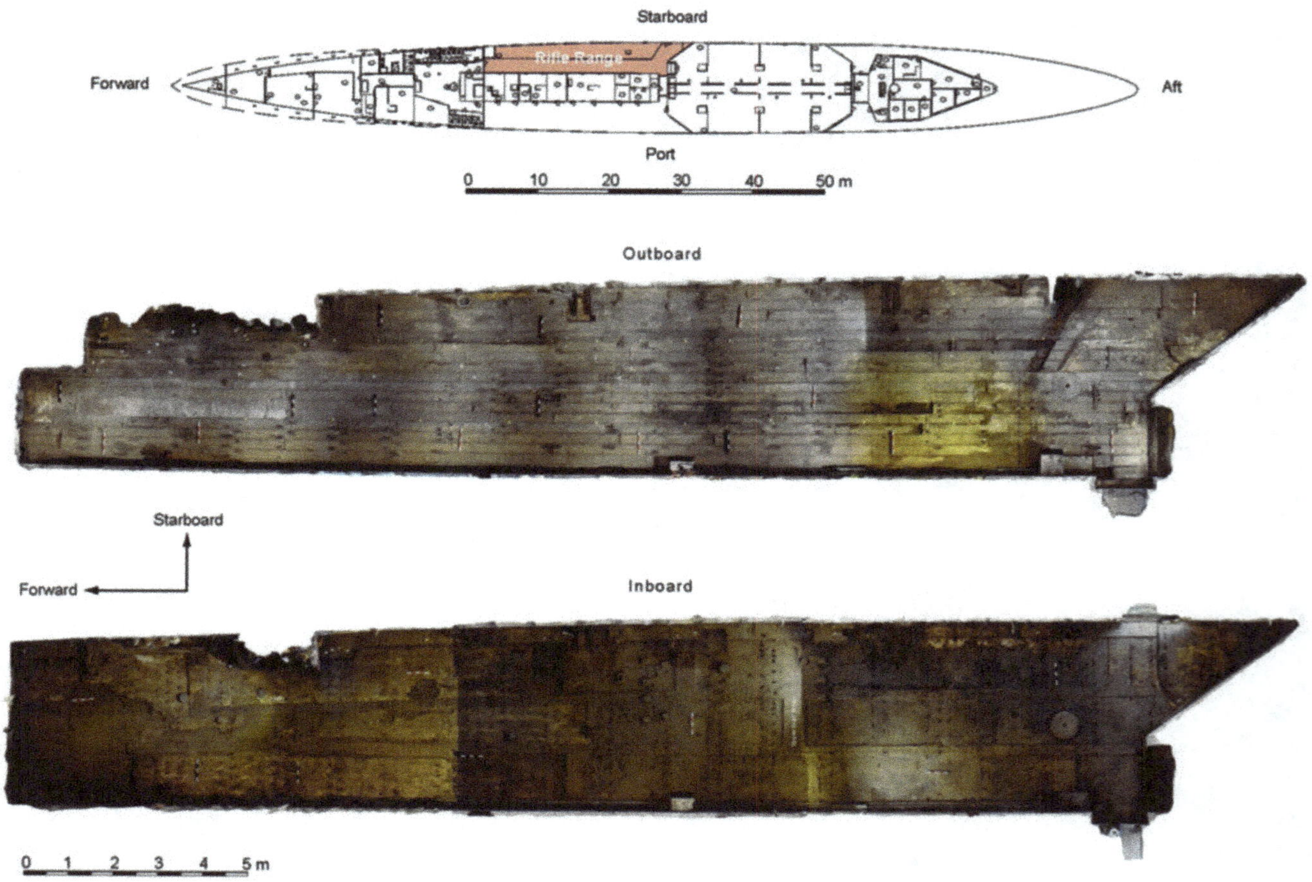

Fig. 7.12 The outputs from a photogrammetric record of the historic deck structure and underlying deck plating from the starboard waste area of HMS *Caroline* (indicated in the red shaded area on the plan—top) (Wessex Archaeology Ltd)

bring (HM Government 2013). Heritage BIM projects have tended to fall into three broad categories:

1. BIM as Process—Renovation of historic buildings, using BIM as part of the project management, but not incorporating historic information which may be of relevance to the project;
2. BIM Model as Archive—Creating a BIM model of a building in order to serve as an archive for information about it. These may be used in the management of the asset, but there is no use of BIM as a project management process; and
3. BIM Model for Renovation—Creating a BIM model of an historic building with an emphasis on geometry over information, often for renovation or repair.

The first of these approaches gains the benefits of BIM as a project management process, but misses the fact that these buildings have particular planning needs that can impact work. This risk is therefore not controlled in the same way as others in the project. This also limits the model's future use in Facilities Management (FM) (or indeed conservation management), it is however how construction professionals generally make use of BIM in a heritage context.

The second approach misses the core of BIM as a driver for efficiency in projects and looks straight to the technology as the latest method of storing heritage information. This view of BIM is widespread within the heritage industry and acts very much like an advanced GIS (see for example Counsell and Edwards 2014; Simeone et al. 2014; Yajing and Cong 2011; Zhang et al. 2016; Edwards 2017). Whilst such an approach could undoubtedly have value for the management and research of heritage assets, for many needs the data included is too much, and renders the model too large to practically use (Zhang et al. 2016).

The final approach is also very common within heritage circles, treating BIM modelling as a way to better understand historic buildings and either ignoring, or passing over the issue of historic information within the model (see for example Brumana et al. 2013; Fai and Sydor 2013). This can be seen as an extension of the use of 3D survey data discussed above. A similar underlying focus can be seen in the use of BIM modelling for virtual reconstruction of excavated buildings (Garagnani et al. 2016). The problem with this approach

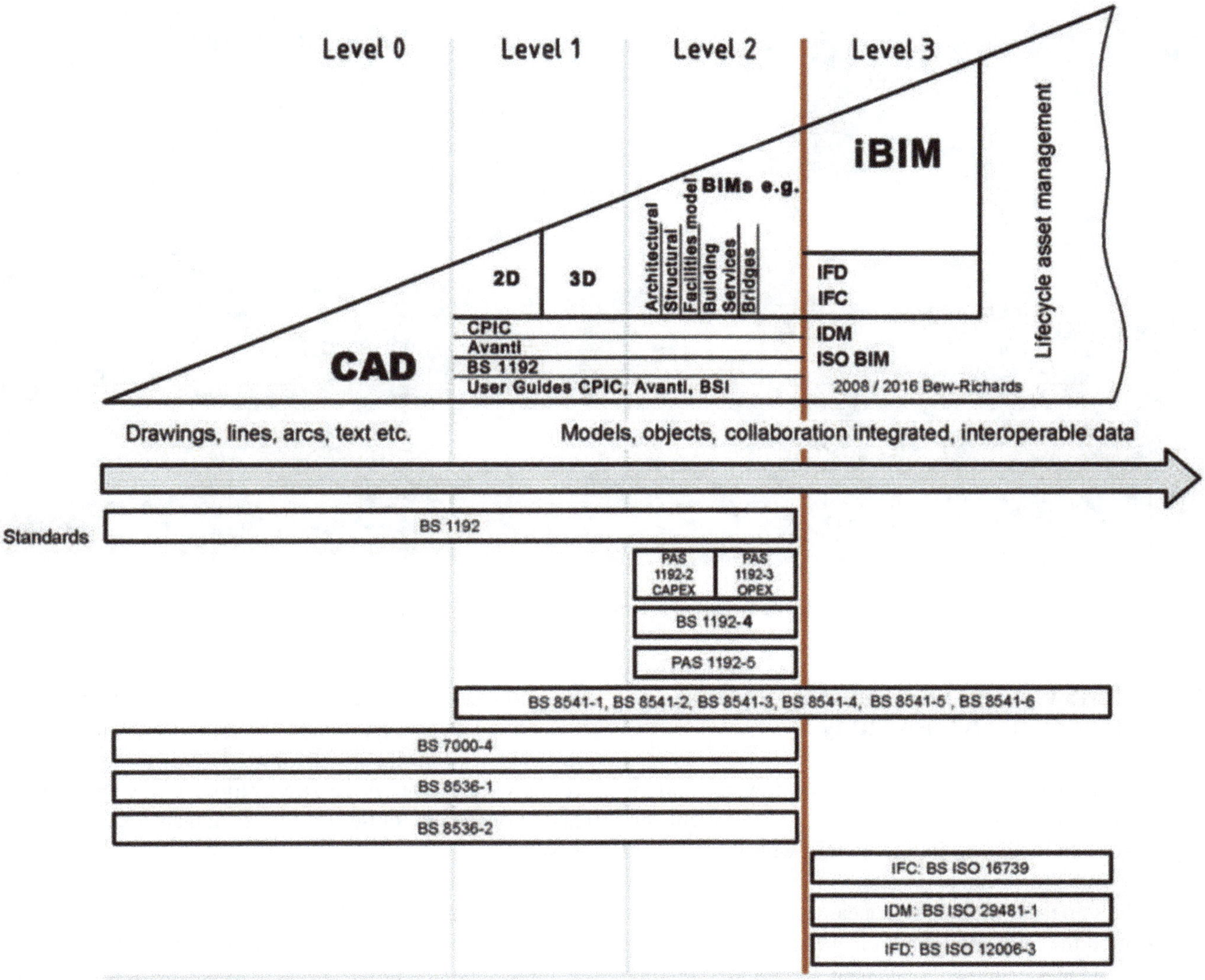

Fig. 7.13 The UK BIM maturity model, detailing the basic requirements for each level of BIM implementation. Level 2 has been clearly defined with British and International Standards. Level 3 still requires technical, procedural and legal progress, and as such it is not currently fully defined by standards and is unachievable. Some projects do exceed the requirements of Level 2 however (bim-level2.org- Derived from Bew and Richards 2008)

is twofold; it is the information which truly makes BIM valuable, and as the software packages were made for designing new buildings, representing the irregularity of historic assets is very difficult. A model will never be able to represent the geometry of a building as well as the laser scan data it is usually based upon, making attempts at exact reproduction redundant. Attempts to use the geometry of BIM models of historic buildings for designing renovations have therefore proven unsuccessful (Bryan 2015).

Whilst a certain Level of Detail is important, it is Level of Information which is key for heritage application of BIM. For the heritage industry to gain the full benefits of BIM it must also incorporate the process aspects; this will allow the efficient and joined up management of conservation management projects and for data requirements to be clearly understood throughout. Doing so will bring the benefits seen in other industries to heritage, and avoid some of the missed opportunities seen in early uses of 3D survey data. Before this can be fully realized however, the lack of data standards, interoperability and specifications must be resolved. Until then, terrestrial and maritime built heritage (archaeological works are somewhat less constrained) cannot explore the full value of BIM.

With respect to the use of 3D survey, in this case laser scanning, to help obtain accurate and up to date information about the condition of the hull and associated components, and for informing ongoing and future conservation management objectives, we now turn to HMS *Victory*. This is an example of a conservation project that aims to utilize integrated 3D data and heritage information, through information modelling, for effective ongoing conservation management.

Fig. 7.14 HMS *Victory* in No. 2 Dock at the Historic Dockyard in Portsmouth (National Museum of the Royal Navy)

7.6 HMS *Victory* (1765) and Information Modelling: A Case Study

The survival of HMS *Victory*, and the continued efforts to restore and conserve the ship throughout the twentieth century, are due in totality to its association with the Battle of Trafalgar and the death of Nelson (Fig. 7.14). Efforts to restore the ship to its 'Trafalgar' appearance were the driving force behind the initial restoration in 1922–1928, as well as the major repairs undertaken between 1955 and 1964, and the continuation of repair and restoration works through to the bicentenary of the Battle of Trafalgar in 2005. More recently, however, there has been a recognition that *Victory* is equally important as an object in and of itself. This is in no small part due to the attrition of the numbers of wooden ships of the line during the nineteenth and twentieth centuries. Increasingly HMS *Victory* represents an archaeological repository which can inform our knowledge of the history and technology used in naval architecture of the eighteenth and nineteenth centuries, while the extensive works carried out in its restoration and conservation during the twentieth century provide a narrative for the early conservation of historic vessels as well as providing essential information on the efficacy of various approaches used in the conservation of historic ships. HMS *Victory* is an extremely complex structure, with over 16,000 individual timber components used in its construction, few of which date to its initial build. Various phases of rebuilding, refitting and restoration over the ship's 250-year life span have resulted in a mix of materials of different age, type and significance. In addition to the archaeological evidence contained within the ship's fabric, a large amount of documentary evidence exists relating to the restoration and conservation of the ship through the twentieth century. From the 1920s onwards drawings, plans and reports were created to chart alterations to the ship for the restoration to as close to the Trafalgar appearance as possible.

The next 15–20 years will see a large programme of works designed to stabilize the structure of the ship and ensure its long-term survival. Core to this is the preservation of significant parts of the structure to prevent further loss of archaeologically significant material. This approach requires the ability to understand the impact of works on the fabric of the ship, and to understand the significance of each component so that proposed works can be designed in such a way as to minimize impact on significant material. BIM serves as a potential management tool to assist in this analysis, by relating the information generated from both archaeological and documentary research with a 3D model of the ship. The use of BIM on HMS *Victory* therefore represents a fuller utilization than those discussed above.

7.7 Development of the VIM

In 2012 ownership of HMS *Victory* passed from the Ministry of Defence to the National Museum of the Royal Navy (NMRN), and work was undertaken to understand the current condition of the ship and underpin the conservation and planned maintenance of the vessel, established through the preparation of a comprehensive Conservation Management Plan (NMRN 2015). As part of this work a measured survey

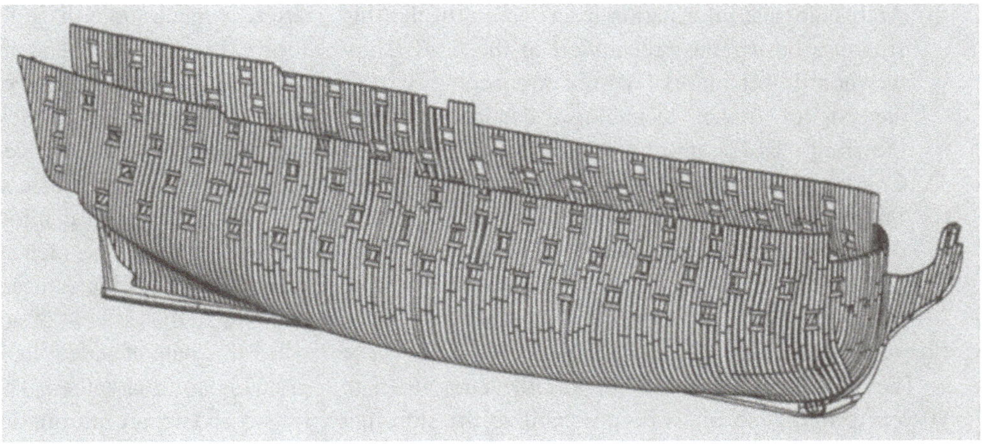

Fig. 7.15 Frames of HMS *Victory*

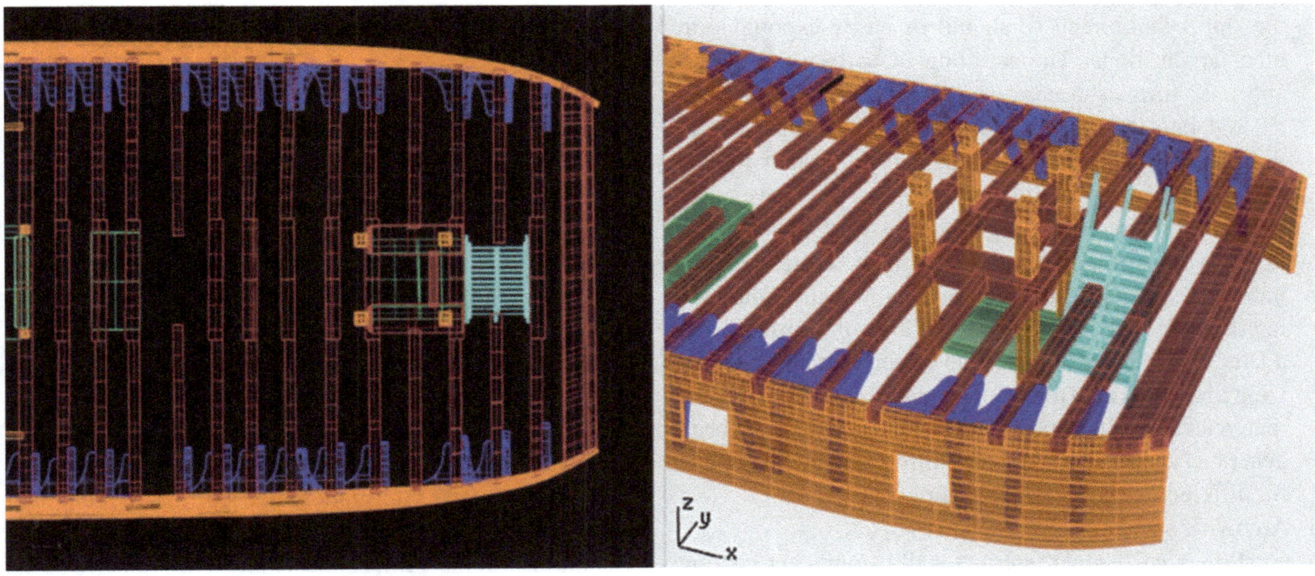

Fig. 7.16 Plan and isometric view of the different components comprising the upper gundeck

was undertaken using terrestrial laser scanning to create a detailed 3D point cloud of the ship (Fenton Holloway Ltd 2014). The point cloud was used to create a solid surface model of each significant component within the ship's structure, resulting in a complex 3D model with over 15,000 individual elements (Fig. 7.15). This model was used to facilitate the structural analysis of the ship, generate accurate plans and sections and to form the basis of a 3D model for the management of information relating to each component within the ships structure. This management model, which has been dubbed the VIM (or *Victory* Information Model), seeks to apply the concepts of BIM to the ship. The aim of the VIM is first and foremost to support and inform the ongoing conservation works of the ship by providing the ability to understand the significance of each individual component within the structure to ensure that planned alterations and interventions do not impact on the most significant elements (Fig. 7.16). In planning the creation of the VIM, there were several key considerations:

1. The nature of information recorded:

   A wealth of information exists on HMS *Victory*, but not all this information is relevant to the ship's conservation, and while the model could conceivably be used to collate and organize a lot of this information, the priority of the project is collating information that will facilitate the decision-making process in advance of HIA. This information can loosely be divided into three types:

   (a) Physical information contained in the 3D model itself, such as dimensions, shape and weight, along with additional material information such as timber species, presence of insect action and pest management (McCormack 2016), fruiting bodies, structural failure;

(b) Archaeological information from surveys of the ship structure during the preparation of the CMP, shipwright's timber marks (Wessex Archaeology 2014), analysis of timber technology, dendrochronology (Nayling 2014), and paint analysis (Crick-Smith Conservation 2014); and

(c) Subjective values such as various types of significance (evidential, historical, aesthetic and communal) as identified in the CMP, and other research (Leggett 2016)

2. Longevity of data:

   The VIM is designed to support the current 15 to 20-year programme of works to stabilize the ship, but aims to go beyond this into the long-term management of the vessel. In addition, the data contained in the VIM is potentially useful in other forms of research relating to the ship's fabric, and so should be easily exportable to other applications. The selection of software packages and data formats needs to ensure that all information entered into the database is futureproofed against software change or redundancy. Equally important for the long-term management of an asset which may have many different individuals and organizations working on it over its lifecycle is standardization of data structures and terminologies. Establishing this early on will allow the easy interrogation of data and less reworking of information at a later date. This also raises the importance of standardizing and checking (for completeness, accuracy and current relevance) data before entry, as it will have been generated over many years by different individuals and for different purposes (Historic England 2017).

3. Archiving standards:

   HMS *Victory* is an internationally significant historic ship. Data generated as part of the ongoing works to the ship must be compatible with the guidelines developed by National Historic Ships, as well as heritage bodies such as Archaeological Archives Forum, ARCHES, CIfA and ADS.

   In order to support both 2 and 3, a key requirement of any software is that it uses, or can easily export to, non-proprietary or open file formats. This is advantageous for a number of reasons: (1) Straightforward archiving of data; (2) Import/export to other software to minimize replication of work; and (3) Futureproofing by accommodating transfer of data to newly developed software.

Several significant issues were encountered when trialling BIM software. Firstly, Level of Detail (LOD) of the existing 3D data is not readily compatible with existing off-the-shelf BIM packages. Most BIM software is created to facilitate the design process, and simplifies and manages 3D data by using standard forms and set templates for elements of buildings. Entering 'as-built' information for historic assets is much more difficult, with non-standard geometric forms (i.e. ships timbers) not being easily realized in the software. Secondly, the Level of Information (LOI) required was outside the remit of most BIM programs. Whilst material, dimensions, appearance and even cost are common factors, information such as condition, presence of rot, etc. as well as archaeological information such as timber marks, dendrochronology, evidence for reuse, value and significance are not attributes used as standard in BIM. This can complicate the process of adding this information. Thirdly, when this data is added, its interoperability with other programs is not guaranteed. This makes it difficult to directly import and export information to other software through the standard format of IFC.

	BIM	VIM
Level of detail	Simple 3D geometry	Complex 3D geometry
Structure	Buildings/ infrastructure	Ship
Process stage	Design and operate	As-built
Future structural change	Minimal	Major
Level of information	Moderate, standardized	High, specialized

To satisfy the conditions required of the VIM, an alternative to an off-the-shelf BIM package had to be developed which could deliver the key concepts of BIM in the context of an historic vessel. This involved the combination of two key elements: (1) A 3D modelling software that allows the import of existing 3D data; and (2) A database program managing the key information needed for the ongoing works.

In the case of the VIM, Rhinoceros 3D was selected as the modelling software. This software was selected due to its ability to handle the complexity of the 3D data, its ability to both import and export to a range of open and non-proprietary file formats, and the availability of an SDK allowing the potential for a custom-built plugin to be developed. Microsoft Access was chosen as the database management program. To access the information held in the database from the Rhinoceros 3D model the object is selected and the Hyperlink option within Rhinoceros 3D is used to link to an html version of the database entry. Whilst this does not follow a standard BIM software approach, it was deemed the best way to implement BIM in this context.

7.8 Future Development of the VIM

Creation of the VIM is an ongoing process, and is by no means a finished product, indeed as the VIM is designed to support the next 15–20 years of work on the ship, it needs to be a fluid and responsive database. A large amount of time is

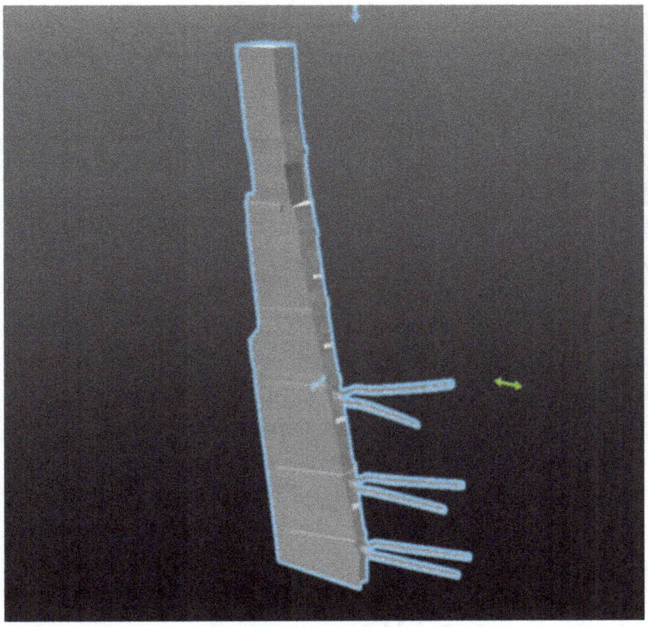

Fig. 7.17 Model of the rudder from HMS *Victory* in .stl format opened in simple viewer

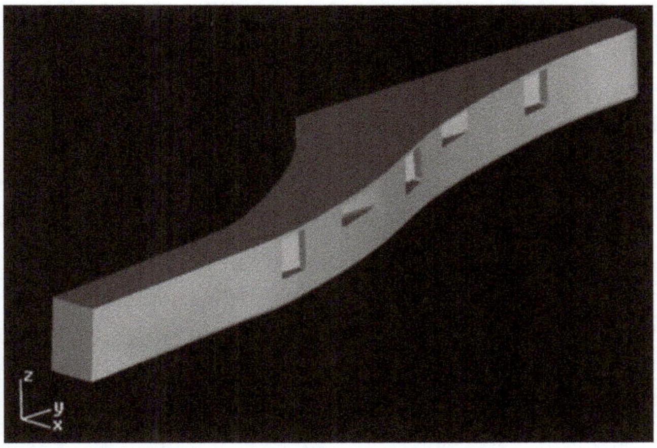

Fig. 7.18 Model of a beam arm from the lower gundeck showing LOD of individual elements within the VIM

needed to input a range of data, with over 100 years of reports and surveys, along with additional information from photographs and plans, many of which are now themselves historically significant items. Establishing data formats, standards and terminologies for digitizing this information, in response to the lack of accepted industry standards, will ensure that work is not replicated. In addition, an archaeological watching brief on the ongoing works on the ship is exposing previously obscured material which helps to establish the age and condition of the ship both now and in the past. This information will both inform the future management of HMS *Victory* and further contribute to an understanding of its history.

In the short term, the aim is to develop a plugin that better links the 3D dataset to the accompanying database, as well as allowing the filtering and display of elements by each of the database fields. One of the key reasons for using open or non-proprietary file formats is to ensure that the database and model are flexible enough to be transferred into new software packages as they develop (Fig. 7.17). BIM is a rapidly developing industry with an ever-increasing range of applications, and working within already well established open file formats, although not optimum, future-proofs the dataset for new developments.

7.9 Lessons from the VIM

The development of the VIM provides a single source of truth for the HMS *Victory* project, uniting a detailed model of the physical object with information necessary for conservation and management. The VIM straddles the categories of BIM usage in Heritage described above. It is to be used for archiving information, conservation (as opposed to renovation) and to a degree process management. The first of these is limited in scope, for the reasons described above, but it is nonetheless a living archive of important information about HMS *Victory*. As a process management tool, it is used to understand what work must be done and when, though not in the same detail as some of the 4D uses in the AEC sector. Its development was also not as tightly controlled with documentation, being driven more by an opportunistic use of existing data/knowledge rather than existing standards or future goals. The key learning experience of the VIM project has been to understand the purpose of an information model. Ultimately, this would allow for a better scope to be defined, especially in terms of what Level of Information and what type of information is required in any future project (Fig. 7.18). This in turn would lead to process changes in how data is gathered and incorporated in to the model, reducing costs and producing an easier to use model.

7.10 Discussion

Past historic vessel conservation has striven to understand the key significances that require to be retained, whether it is physical historic fabric, or maintaining elements or spaces within a vessel that magnify significance in other areas—essentially the different values that make the asset important. The opportunities to better manage the retention of significance and understand how conservation work throughout the lifetime of a project is best understood can be helped significantly through the acquisition of 3D data and effective information modelling. There is a real motivation and willingness of the wider marine and maritime heritage sector to engage with and embrace 3D survey initiatives; and where the key integration of 3D outputs in projects represents a central goal. This is the case not only with large organizations such as the National Museum of the Royal Navy in the UK (examples include HMS *Victory*, HMS *Caroline*, HMS *Alliance*), but also smaller organizations such as the Scottish Fisheries Museum in Scotland (Fifie Sailing Drifter *Reaper*, the Baldie *White Wing*, and other small craft within the collection).

In relation to HMS *Victory*, the challenges inherent in the development of the VIM are clear to see, but more generally, there is also a real opportunity to push the adoption of information modelling in historic vessel conservation, as something that is seen as beneficial, rather than potentially seen as something that is costly and time consuming.

Further to this, the non-selective nature of 3D data collection techniques such as laser scanning and photogrammetry means that there is a wealth of data, which have already been collected, that may be used in the development of information models for historic maritime assets. This depends on accuracy and completeness, and some of the older surveys may fall short of what is needed, but importantly there is no need to start afresh, all of the previous work is still valuable.

The importance of early engagement in the development of maritime applications of information modelling cannot be overstated. In order to maximize efficiency and long-term value, real progress needs to be made on standards, both here and in heritage in general. The heritage industry has a long history of standardization to aid interpretation and research, with information modelling it has the added value of saving a lot of time and expense for individual projects. The best approach to conservation management needs to be further explored too. Whilst the VIM shows one way of taking conservation management forward into information modelling, it has developed under a very specific circumstance, with technical restrictions as they exist at this time. Going forward, with enough pressure, these technical restrictions may change, offering better ways to handle the uniquely complex assets we deal with.

A certain amount of education of the AEC industry and software developers may be required, but it need not be solely for our benefit. The assets we deal with have a long history. One day the new builds that are being managed with BIM will have a history to contend with too; considering the issue of history now could save a lot of trouble later. The immense Level of Information required for the management of an historic vessel will of course outstrip a new building for some time, but one day they too may be considered important heritage assets with associated restrictions on what can and cannot be done to them. It would be better to record information when it is current, and allow for its integration into BIM than to have to retrofit that information at a later date. This requires a long-term view certainly, but that is the purpose of BIM.

There is a need to be pragmatic about what is represented in an information model of an historic vessel. If the represented geometry is too complex it will serve to increase file sizes, making models more difficult to use, and greatly increase the cost and difficulty in creating them in the first place. The Level of detail does need to be sufficient for the needs of the CMP however. In the case of HMS *Victory* for example, it would be no good to represent the hull with a handful of objects since it has such a complicated interwoven history of conservation and repair. We may find however, that in the future dealing with complex geometry is made easier by the application of automated object generation from shape grammars, as demonstrated by Dore and Murphy (2012a, b, 2013) in neo-classical buildings, especially with the ongoing improvements in AI recognition. Recent work using procedural modelling to create parametric models of ships by Suarez et al. (this volume), for example, brings the 3D modelling process for historic assets more in line with that usually seen in BIM projects, and reduces the time and expertise required to create iterative models. Their focus is on reconstruction from partial information, rather than the creation of a digital twin of a fully recorded vessel, but the procedurally generated model could be used as a starting point for the geometry. This geometry could then be edited to adequately represent the vessel. The complexity of this task will likely scale with the required LOD, with more complete representations requiring more edits, as real vessels do not necessarily follow all of the rules, and can have very complex histories of modification. It may be that for the most complex models it is actually faster to start from scratch than to check and modify the many individual elements, but this procedural modelling approach would need to be tested upon a complex extant vessel.

7.11 Conclusions

Whilst it is still too early to gauge the long-term success of the VIM project, the easy access to information that it provides, as with BIM projects, should lead to significant benefits. These will include improving project outcomes through better management (whether on existing assets, new assets or through expanded opportunities for education), better use of resources, and ultimately, expanding our knowledge of historic vessels and their effective conservation management. Combining the principles of information modelling with a CMP, effectively embedding that CMP within a visual interface that makes access to information easier and more contextual than ever, remains the key challenge. 3D survey was once seen as too expensive to be used in heritage, but it is now common place in the industry. Information modelling is a costly process to kick start, both for an industry and individual organizations, but its success within the AEC sector, and the benefits highlighted in this chapter show that it is worth the investment. In addition to extant historic ships, it is also important to stress the potential application for information modelling in the wider marine and maritime archaeological sphere. A key area of application concerns the ongoing management of wreck sites (either on the seabed or subsequently recovered), whether it is in the domain of research projects, or through the management perspectives and requirements of the national curator, particularly in relation to designated wreck sites. There is a clear window of opportunity, and the researcher, heritage practitioner, and curator alike has the opportunity to engage with the development of a solid platform on which to effectively integrate 3D applications through information modelling with the conservation management of the marine and maritime historic environment well into the future.

References

3D Scans (2017) http://www.3ders.org/articles/20170704-3d-scanning-helps-preserve-one-of-the-worlds-oldest-ships-in-new-zealand.html. Accessed 15 Aug 2017

Anderson RK (1994) Guidelines for recording historic ships. National Park Service, US Department of the Interior, Washington, DC. Historic American Engineering Record (reprinted 1995, new edition 2004)

Aracyici Y, Counsell J, Mahdjoubi L, Nagy GA, Hawas S, Dweidar K (2017) Heritage building information modelling. Routledge, Abingdon

Atkinson DE, Dalland M, Van Wessel J (2009) City of Adelaide laser scan survey. Headland Archaeology. Unpublished client report

Atkinson DE, Dalland M, Van Wessel J (2010) Zulu sailing drifter 'research' laser scan survey. Headland Archaeology. Unpublished client report

Australia ICOMOS (1999) The Australia charter for the conservation of places of cultural significance (the Burra charter). Australia ICOMOS, Canberra

Bew M, Richards M (2008) BIM maturity diagram, building SMART Construct IT Autumn Members Meeting, Brighton

Breeden C (2015) Archaeology, BIM and large infrastructure (Part 1). BIM Today (February 2016):100–101

Breeden C (2016) Archaeology, BIM and large infrastructure (Part 2). BIM Today (February 2016):82–83

Brumana R, Oreni D, Raimondi A, Georgopoulos A, Bregianni A (2013) From survey to HBIM for documentation, dissemination and management of built heritage: the case study of St. Maria in Scaria d'Intelvi. In: 2013 digital heritage international congress (DigitalHeritage), Marseille, France, 28 October–1 November 2013, pp 497–504. https://doi.org/10.1109/DigitalHeritage.2013.6743789

Bryan P (2015) BIM & heritage—are they a good match. TSA BIM Conference, 11 November 2015

Campbell-Bell D (2015) Archaeology and BIM: an introduction. BIM Today (August 2015):150–151

Campbell-Bell D (2016) Built heritage and BIM. BIM Today (May 2016):52–53

Classic Boat (2010) Restoration of the Charles W Morgan. Classic Boat Mag:264 (June 2010). https://www.classicboat.co.uk/articles/restoration/restoration-the-charles-w-morgan-to-sail-again/2/. Accessed 10Aug 2017

Cooper JP, Wetherelt A, Zazzaro C, Eyre M (2018) From Boatyard to museum: 3D laser scanning and digital modelling of the Qatar Museums watercraft collection, Doha, Qatar. Int J Naut Archaeol 47(2):419–442. https://doi.org/10.1111/1095-9270.12298

Counsell J, Edwards J (2014) A sense of the past. Build Conserv J (December 2014/January 2015):28–29

Crick-Smith Conservation (2014) HMS Victory: paint analysis. Unpublished client report

Digital Surveys (2014) http://www.digitalsurveys.co.uk/case-study/rrs-discovery. Accessed 30 Sept 2017

Dore C, Murphy M (2012a) Integration of HIBIM and 3D GIS for digital heritage modelling digital documentation. In: Digital documentation international conference, Edinburgh, 22–23 October 2012

Dore C, Murphy M (2012b) Integration of historic building information modelling and 3D GIS for recording and managing cultural heritage sites. In: 18th international conference on virtual systems and multimedia: virtual systems in the information society, Milan, 2–5 September 2012, pp 369–376

Dore C, Murphy M (2013) Semi-automatic modelling of building facades with shape grammars using historic building information modelling. Int Arch Photogramm Remote Sens Spat Inf Sci XL-5/W1:57–64. https://doi.org/10.5194/isprsarchives-XL-5-W1-57-2013

Dunkley M (2008) *Rooswijk*, Goodwin Sands, off Kent. Conservation statement & management plan. Unpublished report, English Heritage, Swindon

Edwards J (2017) It's BIM—but not as we know it. In: Aracyici Y, Counsell J, Mahdjoubi L, Nagy GA, Hawas S, Dweidar K (eds) Heritage building information modelling. Routledge, Abingdon, pp 6–14

English Heritage (2008) Conservation principles, policies and guidance for the sustainable management of the historic environment. English Heritage, Swindon

Fai S, Sydor M (2013) Building information modelling and the documentation of architectural heritage: between the 'typical' and the 'specific. In: 2013 digital heritage international congress (DigitalHeritage), Marseille, France, 28 October–1 November 2013. https://doi.org/10.1109/DigitalHeritage.2013.6743828

Fenton Holloway Ltd. (2014) Modelling and structural analysis of HMS Victory. Unpublished client report

Garagnani S, Gaucci A, Gruška B (2016) From the archaeological record to ArchaeoBIM: the case study of the Etruscan temple of Uni in Marzabotto. Virtual Archaeol Rev 7(15):77–86. https://doi.org/10.4995/var.2016.5846

Heidbrink I (ed) (2003) The "Barcelona charter": European charter for the conservation and restoration of traditional ships in operation. European Maritime Heritage, Bremen

Heritage Lottery Fund (2012) Conservation plan guidance. https://www.hlf.org.uk/conservation-plan-guidance. Accessed 22 Aug 2018

Historic England (2017) BIM for heritage: developing a historic building information model. Historic England, Swindon

Historic Scotland (2000) Conservation plans: a guide to the preparation of conservation plans. Heritage policy. Historic Scotland, Edinburgh

HM Government (2011) Government construction strategy. HM Government, London

HM Government (2013) Construction 2025 industrial strategy: government industry in partnership. HM Government, London

HM Government (2015) Digital built Britain: level 3, building information modelling—strategic plan. HM Government, London

Kentley E, Stephens S, Heighton M (2007) Understanding historic vessels. Vol 1, recording historic vessels. National Historic Ships, London

Kerr JS (2013) The conservation plan: a guide to the preparation of conservation plans for places of European cultural significance. National Trust of Australia, New South Wales. https://australia.icomos.org/publications/the-conservation-plan/. Accessed 22 Aug 2018

Leggett D (2016) Restoring *Victory*: naval heritage, identity, and memory in interwar Britain. Twentieth Century Br Hist, 28(1), 1 March 2017, 57–82

Lipke P, Spectre P, Fuller BAG (1993) Boats: a manual for their documentation. Museum Small Craft Association, Nashville

Malmberg L (2003) Management planning for historic ships. In: Brebbia CA, Gambin T (eds) Maritime heritage, Transactions on the built environment 65. WIT Press, Southampton, 8 pp

May R, Badcock A, Panter I, Parham D (2017) Dunwich bank wreck site: conservation statement and management plan, Historic England Project 7835. Historic England, Swindon

McCormack D (2016) Preservation of HMS *Victory*: pest management techniques. Historic Ships (Dec):7–8

National Historic Ships (2007) Understanding historic vessels, vol 1. Conserving historic vessels. National Historic Ships. Park Lane Press

National Historic Ships (2010) Understanding historic vessels, vol 3. Conserving historic vessels. National Historic Ships. Park Lane Press

National Museum of the Royal Navy (2014) HMS Caroline conservation management plan, vol 1. National Museum of the Royal Navy, Portsmouth

National Museum of the Royal Navy (2015) HMS Victory conservation management plan, vol 1. National Museum of the Royal Navy, Portsmouth

Nayling N (2014) Tree-ring analysis of selected timbers from HMS *Victory*. Unpublished client report

Pollet C, Pujante P (2013) Patrimonio marítimo de Magallanes, proyevto de levantamiento laserométrico del clíper Ambassador, Monunento histórico. Fondo Nacional de Desarrollo Cultural y las Artes

Sanders DH (2012) Virtual reconstruction of maritime sites and artifacts. In: Catsambis A, Ford B, Hamilton DL (eds) The Oxford handbook of maritime archaeology. Oxford University Press, Oxford, pp 305–325. https://doi.org/10.1093/oxfordhb/9780199336005.001.0001

SIAD (2017) https://www.siadltd.com/our-services-interactive-visual-isat. Accessed 12 Oct 2017

Simeone D, Cursi S, Toldo I, Carrara G (2014) BIM and knowledge management for building heritage. In: Proceedings of the 34th annual conference of the Association for Computer Aided Design in Architecture ACADIA 14: Design Agency, pp 681–690

Wessex Archaeology (2014) Survey of timber marks and fabric analysis on HMS *Victory*: technical report. Unpublished client report

Wessex Archaeology (2017) Scottish Fisheries Museum, Anstruther, Scotland: Museum Fleet conservation management plan. Unpublished client report

Yajing D, Cong W (2011) Research on the building information model of the stone building for heritages conservation with the outer south gate of the Ta Keo Temple as an example. In: 2011 international conference on electric technology and civil engineering (ICETCE), Lushan, China, 22–24 April 2011, pp 1488–1491. https://doi.org/10.1109/ICETCE.2011.5776479

Zhang Y, Zhu Z, Li C, Chang L (2016) Integration application system of Chinese wooden architecture heritages based on BIM. In: 2016 International Conference on Logistics, Informatics and Service Sciences (LISS), Sydney, NSW, Australia, 24–27 July 2016. https://doi.org/10.1109/LISS.2016.7854507

8 A Procedural Approach to Computer-Aided Modelling in Nautical Archaeology

Matthew Suarez, Frederic Parke, and Filipe Castro

Abstract

This chapter analyses the functionality and applicability of procedural computer-based modelling techniques in the field of nautical archaeology. To demonstrate this approach, an interactive procedural model of the lower hull timbers of a sixteenth-century European merchant ship was developed through a process of prototype implementation, and an evaluation of the usefulness and effectiveness of the prototypes developed was carried out using the timbers from the hull remains of the Belinho 1 shipwreck, found in Portugal in 2014. The 3D model was created using *Houdini*, a procedural node-based 3D software package. A basic collection of the main timber components of a ship's lower hull was defined, and functional rules were created for each timber, based on real-world ship design and construction processes. Then the rules were incorporated into the logic of the procedural modelling algorithm, and the resulting model was changed by using the Belinho 1 shipwreck scantlings. The results, which will be discussed, were satisfactory, except for the planking, which is a very complex part of the shipbuilding process and deserves future attention.

Keywords

Procedural modelling · Ship and boat archaeology · Shipwreck · Research models

M. Suarez (✉) · F. Parke
Department of Visualization, Texas A&M University, College Station, TX, USA
e-mail: jlcasaban@tamu.edu; frederic-i-parke@tamu.edu

F. Castro
Department of Anthropology, Texas A&M University, College Station, TX, USA
e-mail: fvcastro@tamu.edu

8.1 Introduction

The collaboration between the J. Richard Steffy Ship Reconstruction Laboratory (ShipLAB) and the Department of Visualization (VizLab) started a decade ago and since then the two institutions have discussed and developed several projects pertaining to the modelling of archaeological ship remains. Alexander Hazlett (2007) sought to synthesize the data from various sources on the subject of the Portuguese *nau* to create a timber-by-timber model, following a manuscript which enumerated the timbers necessary to build a merchantman, or *nau da India*. From the model, developed in *Rhinoceros 3D*, Hazlett created, annotated, and illustrated construction diagrams of this ship, and hypothesized a construction sequence (Fig. 8.1). Audrey Wells (2008) created a detailed reconstruction of the Portuguese vessel *Nossa Senhora dos Mártires*, lost near Lisbon in 1606, following a hypothetical reconstruction (Castro 2003, 2005, 2009; Castro et al. 2010). This reconstruction could be viewed using a real-time immersive visualization system (Fig. 8.2). Justus Cook (2012) created a parametric computer model of an Iberian *nau* or *nao* hull, and varied the main dimensions according to a set of measurements from historical manuscripts, in order to better understand the changes in the hull shapes (Fig. 8.3).

The goal of this project was to reduce the large investment of time and expertise that is currently required to create 3D reconstruction models for nautical archaeological research using typical modelling methods. The strategy is an approach that leverages computer-based modelling, both parametric- and rule-based. To demonstrate this, a procedural model of the lower hull of a sixteenth-century European merchant ship was developed through an iterative process of prototype implementation. The resulting model was flexible and versatile, and could be iterated through parametric controls, greatly reducing the traditional change and revision time.

The results of this project provided evidence of the time-saving effectiveness of a procedural approach to create 3D

J. K. McCarthy et al. (eds.), *3D Recording and Interpretation for Maritime Archaeology*, Coastal Research Library 31,
https://doi.org/10.1007/978-3-030-03635-5_8

Fig. 8.1 Alexander Hazlett's model

models as research tools. Once procedural models are created, they provide both accessible and powerful means for researchers to create and test multiple interpretations of the archaeological data. Although the development of a procedural model requires skilled computer operators and a significant investment in design, construction, and trouble-shooting, these problems are outweighed by the flexibility offered by the models.

8.2 Computer-Aided Modelling in Archaeology

Computer-aided modelling offers archaeological researchers quick 3D images of their reconstructions of the past, and a perspective through which they can analyse archaeological data. Such models allow visualizations of the data collected from archaeological sites, including shipwrecks. Models offer researchers an opportunity to view what is left of the remains without the constraints of low visibility, poor lighting, and the partial observations inherent to the excavation process: archaeologists see their sites through layers and trenches. As it is highly destructive to expose an entire site all at the same time, marine archaeologists tend to dig and record partial areas, which are quickly covered—or recovered—as soon as they have been cleaned, tagged, and recorded.

These partial views—of partially preserved shipwreck remains—make it difficult to identify or understand any particular shipwreck. Cargo, construction features, and materials employed are important clues, but a 3D view of a shipwreck site, which can be scaled—zoomed in and out— provides a good tool for the interpretation and reconstruction of a shipwreck.

In any given time period, ship shapes and hull structures varied from region to region. This variation means that a ship's hull shape can give an indication to its provenience. Once an archaeologist has a basic idea of what kind of ship she is excavating, the mental reconstruction process begins and influences the recording process. Ship reconstruction is an iterative and interpretive process, continually evolving as evidence is uncovered. To assist in visualizing the data collected, researchers will often create models; hand-drawn, physically built, or computer based. The benefit of creating models is that they will often unveil or expose patterns in the data that were previously unclear or even unseen. Models can also expose oversights and misinterpretations of the data. Iterating upon these models can also be used to test hypothesis and explore alternatives. There are a number of benefits to creating computer-based 3D models. For instance, model precision is maintained by the computer and therefore models are subject to fewer opportunities for human error and fatigue. The scalability of 3D computer models allows them to be constructed at full scale. Moreover, 3D modelling also allows for the automation of redundant tasks. Another advantage is that computer processing power allows larger quantities of data to be included in a model.

The main drawback of typical 3D modelling techniques is the large overhead of expertise and knowledge in 3D modelling and 3D modelling software required to create a good 3D model. A second problem is that once a model is created, revisions and iterations can be time consuming. These aspects make one-off production computer models less than ideal for the ship reconstruction iterative process, and that is a deterrent for researchers wanting to implement 3D research models.

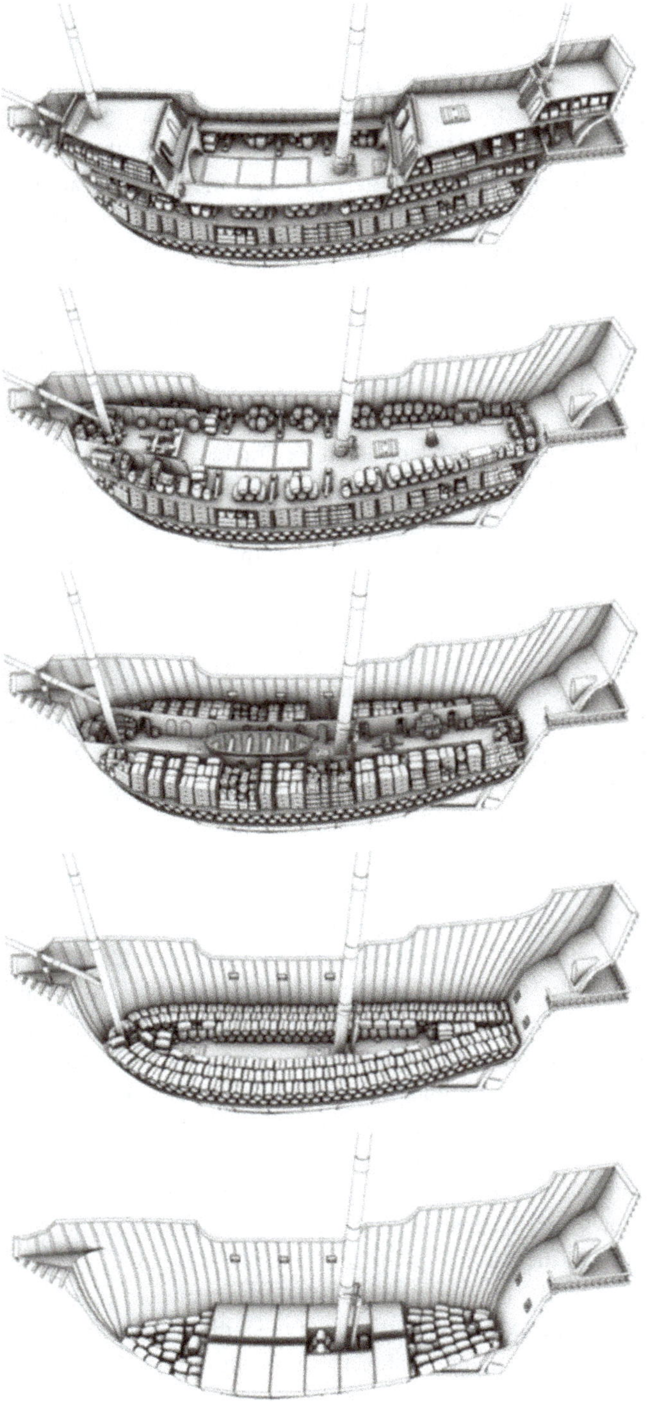

Fig. 8.2 Audrey Wells' model

8.3 Computer-Based Modelling in Archaeology

During the 1960s and 1970s computers were mainly used in archaeological research for statistical applications such as classification and seriation, archaeological techniques which predate computers. One of the first textbooks published on the topic of computers in archaeology, *Mathematics and Computers in Archaeology* (Doran and Hodson 1975), focuses on the application of data classification and quantification in archaeological research. The use of computers at the time was limited, due to their cost and low accessibility. Most computers were only available at universities or other large institutions.

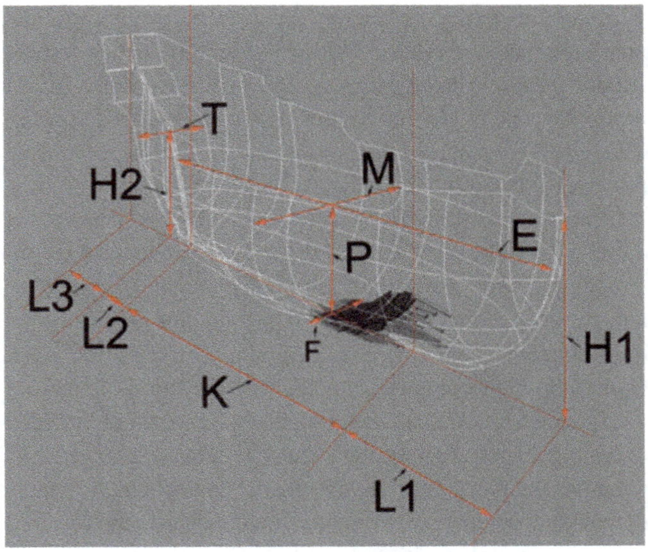

Fig. 8.3 Aspect of Justus Cook's model

The advent of the microprocessor in 1971 and the invention of the first microcomputer in 1975 reduced costs and facilitated access, and by the late 1970s computers were integrated into most areas of archaeological work. According to a 1986 survey of computer usage in British archaeology, computers at the time were mainly focused on theoretical tasks (Richards 1986, 2). These were tasks, not based on theory, but rather on using computers to automate tasks like managing and processing large amounts of data (Lock 2003, 1). The rapid development of graphics, computer-based visualization, and computer software in the mid-1980s and through the 1990s brought about the modern integration of computers into archaeology.

Computer applications within archaeology at this point were now multimedia; integrating the use of text, images, models, animation, and sound. Computers also allowed opportunities to cross-link all this information into different contextual situations. The cross-linking of information encouraged a new data-driven exploratory method of archaeology which we see today (Lock 2003, 211).

8.4 Computer Models

A model is defined as 'a simplification of something more complex to enable understanding' (Lock 2003, 6). In *Models in Archaeology* Andrew Fleming and David Clarke (1973) described models as 'ideal representations of observations which are heuristic, visualizing, comparative, organizational, and explanatory devices.' With this definition, the authors support multiple interpretations and open the possibility of more than one model for any one situation, because they are 'not "true" but a part of the hypothesis generation and testing procedure' (Fleming and Clarke 1973, 316–317).

Prior to the mid-1970s, an archaeological model was either a 2D orthographic set of drawings or a physical reconstruction. These were the traditional methods of exploring and recording ship dimensions. Up until this point, digital developments had been essentially methodological. Computers and computer models provided tools that were considered a theoretical, meaning that they were not intended to explore or inspire interpretations but rather to measure and document existing data (Evans and Daly 2006, 11). In this context, computers and computer models were not yet used to inspire or explore alternative ideas. Once this capacity was developed, computers became tools capable of influencing the creation of theory, adding to the traditional use of 2D drawings or physical models (Haegler et al. 2009, 1). One of the opposing arguments to this view was summarized by Miller and Richards (1995), who claimed that each project was the result of the collaboration between computer scientists and archaeologists, rather than being archaeologically controlled. In most cases the visualization software itself was not accessible to the archaeologists and therefore the computer scientists were perceived as a filter between archaeologists and their data. In other words, archaeologists did not have direct control of the modelling themselves (Miller and Richards 1995, 20).

It is only recently that this 3D software gap of accessibility has been narrowed. Computer-based 3D modelling software is now widely available and widely utilized in most universities and institutions of research. Our research proposes an approach which could further narrow the accessibility gap by leveraging parametric user-interaction and procedural modelling to facilitate the use of computer-based modelling as an exploratory theoretical tool.

The questions raised by Roger Hill (1994; Yamafune et al. 2016), remain more relevant than ever, pertaining to the necessity to understand the relation between the archaeological remains, created by a range of dynamic processes, which he called 'a database in which an imperfect memory of those processes is retained,' and the simplified computer models we use to try to understand and reconstruct the past they represent.

8.5 Procedural Modelling

'Procedural modelling' is a general designation for techniques in computer graphics which create 3D models or textures from a set of rules. L-Systems, fractals, and generative modelling are all included in this family of techniques. For our work, the term procedural modelling refers to the creation of 3D models through rules which are configurable by parameters. In a paper titled 'Procedural Modeling for Digital Cultural Heritage' (Haegler et al. 2009) the authors examine the application of procedural modelling in archaeology. They argue that 'the efficiency and compactness of procedural modelling make it a tool to produce multiple models, which together sample the space of possibilities.' The core of their argument is what they refer to as 'The Problem of Reconstruction Uncertainty'. This is the notion that detailed or realistic visualizations of archaeological research have the potential to falsely lead the viewer to take the 'correctness of every detail for granted.' This can be misleading, they argue, because a reconstruction is an educated guess among several other hypotheses. They further argue that procedural modelling addresses this concern because the variation between different models expresses levels of uncertainty implicitly and they go on to discuss examples of procedural modelling in archaeology, some of which are mentioned in this section, and instances where the notion of 'reconstruction uncertainty' is addressed successfully using procedural modelling.

Procedural modelling has been used to efficiently create a 3D reconstruction of an archaeological site in Mexico (Mueller et al. 2006). This implementation was based on the Computer Generated Architecture shape grammar, or CGA, which is a programming language specified to generate architectural 3D content used in the software *Esri CitiEngine*. Using this shape grammar, a rule set was created which could be used to create 3D models of Puuc-style architecture with a minimal effort. The authors summarize their approach in contrast to traditional 3D modelling: 'Traditional 3D modeling tools often require too much manual work and their application is therefore overly expensive for archaeological projects. In contrast, our procedural modelling approach allows for the testing of several hypotheses by adjusting some of the parameters' (Mueller et al. 2006, 1). The procedural model presented in their paper was created in 3 days. According to the authors, each of the buildings could be created within minutes, using this procedural model (Mueller et al. 2006, 6).

In another seminal study, Chun-Yen Huang and Wen-Kai Tai (2012), present a procedural approach for modelling a detailed Chinese ting, or pavilion. Huang and Tai propose that the use of procedural modelling and a user-friendly Graphic User Interface, or GUI, provide non-professionals with an intuitive means of constructing variants of complex Chinese tings within minutes. They provide evidence for this by way of a user study. They invited twelve users, two 3D artists and ten novice users, to use their modelling tool to accomplish three tasks: (1) model an existing ting from a reference photo; (2) model a ting to match a reference model;

and (3) create a ting with innovation. The researchers documented the time each user spent on each task and how many polygons made up the resulting 3D model. The results of this user study supported their hypothesis that non-professionals could effectively create 3D models using their procedural tools. The average time spent on the three tasks by the twelve users was 9.6 min (576 s), 10.3 min (618 s), and 7.2 min (618 s) respectively. On average, the users rated their experience using the procedural modelling tool to complete the three tasks as a 7.5 out of 10. Huang and Tai's paper provides evidence that a well-designed procedural model can provide researchers with limited 3D modelling experience an efficient means of constructing detailed 3D models (Huang and Tai 2012, 1303).

Another interesting approach, by Marie Saldaña (2015), demonstrates the use of procedural modelling to construct an entire city from GIS data and procedural rules. The project utilizes geographic data and maps to create the terrain. *CityEngine* software was used to describe and generate the different Roman building types and city rules. Once the scenes were generated by *CityEngine*, they were made viewable within the *Unity* game engine. By implementing a procedural approach, the researchers were able to build a comprehensive model of the city of Augustan Rome. The limitations of this approach discussed by Saldaña suggest that the CGA shape grammar does not have vocabulary to describe curved or radial geometry: 'My rules for a theater or stadium, for example, would seem to have been a simple exercise in symmetrical, radial geometry. The procedural grammar, however, was not well-equipped to describe such geometry, which made the writing of this rule a rather tortuous and long-winded process' (Saldaña 2015, 6).

8.6 Methodology

The methodology of this study was to develop a procedural model of the lower hull timbers of a sixteenth-century European merchant ship to demonstrate the effectiveness of a procedural modelling approach to ship reconstruction in nautical archaeology. Our approach was to construct each timber parametrically and maintain a procedural relationship between any given timber and the rest of the ship. Each component can be updated, in real-time, as revisions are made to interdependent components. The intent was to develop an approach which could construct a model with each timber component adjusted automatically; dramatically reducing the iteration time currently required using traditional modelling techniques. The development process for this model was a cycle of prototyping using *Houdini* modelling software.

The goal of this project was to test whether a procedural approach could be used to create a computer-based 3D model of the lower hull of any sixteenth-century European merchant ship, based on a known recipe. Fernando Oliveira's instructions were selected to build an Indiaman, dating to circa 1580. To accomplish this project, we attempted to define: (1) a taxonomy describing each of the ship components considered, and its relationship to the other components; (2) a procedural model for each ship component based on a taxonomy developed from a series of sixteenth- and seventeenth-century technical documents related to shipbuilding; (3) a set of relations connecting the component parts, which generated a procedural model of the main components of a ship's lower hull; and (4) a set of rules to allow an assessment of the usefulness and effectiveness of the models created.

This project used *Side Effects Houdini* Software, a node-based procedural 3D package. The parameter interface for each component was constructed by leveraging Houdini's 'digital asset' file format. This facilitated the design and construction of parametric graphical user interfaces (GUIs). *Houdini* organizes the parts of a model into networks of nodes. Each node defines a part of the parametric data flow, which in turn defines each component. A digital asset is a way of encapsulating a network of nodes, which can then be interacted with, at a high level. Once the network is created and encapsulated within a digital asset, the user interface can be constructed by referencing node parameters inside the asset. And once created, the digital asset file can be loaded into any *Houdini* scene file. The asset can be placed inside of the scene using the software TAB menu or tool palette. We have created a collection of *Houdini* digital assets (HDAs), which can be installed into a scene file and then be used to efficiently create 3D models of a hull.

An important step in the implementation of this methodology was to create Graphical User Interfaces (GUIs), for each of the procedural timbers of the *nau* model. This is basically a library of timbers, which can be changed according to the taste of the user, varying its length, sided and moulded dimensions, scarves, curvature, etc.

The power of this methodology is that the GUI of each component, or HDA, allows the user to set parameters that affect the modelling procedure of each component. Each HDA has its own unique GUI, which is designed to present the user with parametric control of each of the procedural variables in a compartmentalized fashion. For instance, when an HDA encompasses multiple timber components, such as the keel, deck, and transom, each component's respective parameters are organized into labelled tabs. Some HDAs, such as the keel HDA, have context sensitive parameters, which are only activated and displayed when other parameters have certain values. In the case of the keel HDA, depending on the value of the *Rabet Type* parameter, the HDA interface will update with parameters specific to the selected rabbet type (Fig. 8.4).

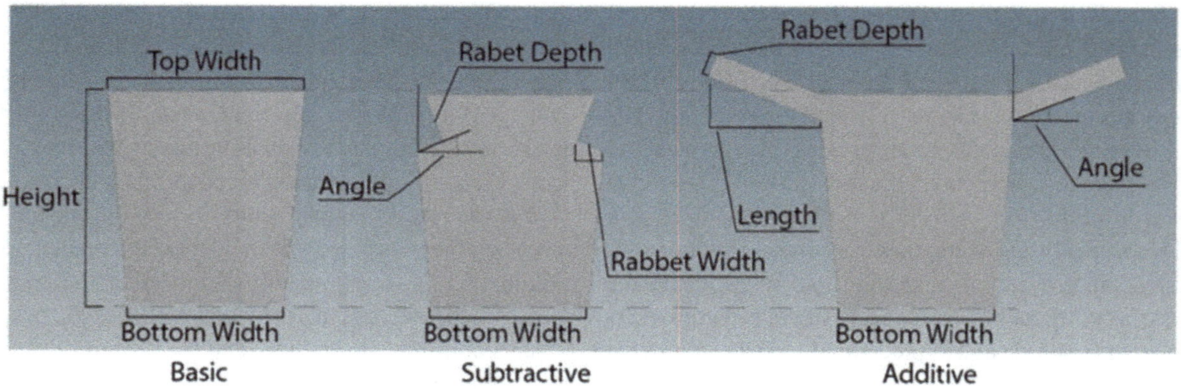

Fig. 8.4 Possible rabbet configurations built in the program

Dynamically populated parameters is another interface mechanism which we implemented in the planking GUI. This GUI proved to be too complex and diverse to be tackled in the preliminary study presented here, but its development is possible and relatively easy, only requiring extensive time to define and test the required parameters and the relations between them. Dynamically populated refers to the fact that the parameters are created and linked to their corresponding variables via a script, which is run to initialize the HDA. For example, upon initialization the planking HDA will automatically create sets of parameters for each frame used. The planking HDA interface also has clear and populate buttons to force updates when the number of input frames is changed.

The GUI for each HDA was designed with the intention of ease of use. The intention behind compartmentalizing parameters is to reduce visual clutter and avoid overwhelming the user with large quantities of parameters on screen at once. Context sensitive parameters also provide a way to limit the number of parameters displayed, only displaying parameters which are relevant to the current state of the model. Dynamically populated parameters provided the ability to design open ended interfaces, which adapt to user needs as the model changes and more parameters are needed.

8.7 Approach

The first step was to set the scope for the project by deciding on a basic collection of main timber components that go into the construction of a ship's hull. This scope defines the breadth of the timber taxonomy. The components considered were divided into three groups: longitudinal timbers, transversal timbers, and planking. Within the longitudinal timbers we have established the following sub-groups:

- Main ship spine—keel, stem, sternpost, and keelson;
- Longitudinal reinforcements—wales, stringers and breast hooks;
- Within the traversal timbers I have established the following sub-groups:
- Frames—floor timbers, futtocks;
- Stern panel timbers—fashion pieces, transoms; and
- Deck timbers—deck beams, knees, clamps, waterways, and carlings.

The deck timbers group contains both longitudinal and transversal timbers, and these were grouped together by their common purpose, the construction of a diaphragm in the shape of a deck (Fig. 8.5).

We established a grammar of spatial relations between these timbers. Functional rules were also created for each timber, based on real-world ship design and construction processes. These rules were incorporated into the logic of the procedural algorithm. We built the ability of varying specific dimensions of each component into this model, while adhering to the rules of the grammar. It often took several attempts to determine a way to implement procedures which mimicked traditional ship construction processes. For every component there was careful attention to achieving a balance between customizability and automation. Automation was reserved for enforcing rules within the grammar and parametric control, and was provided for variables which were traditionally 'eye-balled' by the ship designer.

To define the basic shape of the main components of our model, we used a recipe proposed by Fernando Oliveira for the construction of a merchantman in 1580. Oliveira was a Portuguese priest and intellectual who wrote a shipbuilding treatise describing the main dimensions and shape of a Portuguese *nau* (Oliveira 1991). His text and drawings provided us with a departing model of a generic hull, which served as a base upon which to reason, test, develop new hypothesis, and try to understand the Belinho 1 timbers. The first step was to adopt a system of coordinates that defined the space where our base model was developed: Z-forward, Y-up, right handed coordinate system. This means the keel is created along the positive Z axis. In this coordinate system,

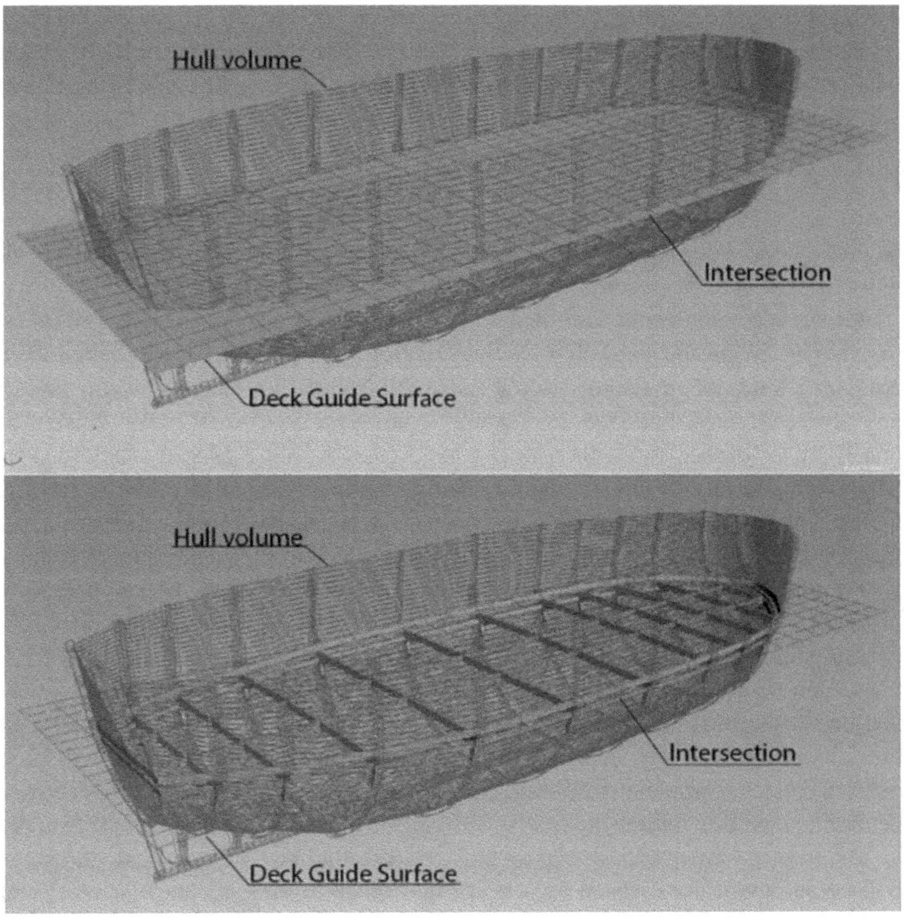

Fig. 8.5 A deck structure, encompassing all the timbers that have a natural symbiotic relation, such as clamps, beams, waterways, deck knees, and carlings

longitudinal reinforcements run along the Z axis and transversal timbers lie in defined XY planes.

8.8 Hull Components Description

The ship's longitudinal axis is composed of three main parts: the sternpost, the keel, and the stem. We divided the keel into five sub-groups of parameters; length, cross-section, sternpost, stern knee, and skeg. The most important parameter, the keel's length, is the length in metres of the horizontal portion of the keel, from sternpost to stem. The cross-section is a 2D shape which is swept along the keel and stem. For this model we provided three cross-section options: basic—a trapezoidal shape with variable top and bottom widths; subtractive—the same as the basic trapezoidal shape but with the subtraction of triangular or rectangular grooves known as *rabbets* (with a variable length, depth, and angles); and additive—with protruding rabbets with variable length, thickness, and angle (Fig. 8.4).

The sternpost is described using three measurements; length, depth, and angle. The stern knee is a curved support timber at the base of the sternpost. It is described by three measurements: height, length, and thickness. The skeg, the protruding timber at the base of the sternpost whose function was to protect the ship's rudder in the event of beaching or hitting a reef, is described using three measurements; back edge Y, skeg height, and length.

The first step in the construction of the keel was defining the shape of its cross-section by creating a trapezoid based on the *Top Width*, *Bottom Width,* and *Height* parameters. The *Rabbet Type* parameter allowed the creation of several geometries for this feature. Next, the main horizontal portion of the keel, extending from the base of the sternpost to the base of the stem, was created by sweeping the cross-section shape along a line of N length, defined by the *Keel Length* parameter. The stem was created according to Oliveira's treatise, using a simple arc. In this case the keel's cross-section was swept along this curve to create the stem.

The stern knee was created using a rectangular polygon placed where the sternpost and keel meet. This polygon has the same width as the sternpost and its bottom edge was placed at the crease where the sternpost meets the keel. The height of the polygon was determined by a *Height* parameter. The polygon can rotate its base edge to the same angle as the sternpost. The bottom edge of the polygon was therefore extruded in the positive Z direction according to the *Length* parameter. The 'L' shaped geometry created was then extruded inward according to the *thickness* parameter. The inside edge was bevelled to create a smooth curve. The skeg was created by extruding the keel's cross-section shape, without any rabbets, in the negative Z direction by its *Length* parameter. The *skeg height* parameter defined how much taller than the keel cross-section height the skeg was supposed to be. The *Back Edge Y* parameter controls the height of the top corner of the skeg. This can be used to create a tapering skeg as seen on some ships.

All the subcomponents described above were assembled together to create the final ship axial structure (Fig. 8.6). *Houdini* allows assignment of arbitrary data as attributes to any geometry and we utilized this feature to pass data from one HDA, or timber, to another. Keel length, cross-section height, cross-section width, and stern angle measurement are assigned as attributes to the keel geometry so that they can be read by other HDAs.

The resulting model of the keel, or HDA, can be created inside of a *Houdini* geometry node. This placed a keel with default parameters into the *Houdini* scene file, displaying in the screen viewport. Adjustments to parameters of the keel HDA will have an immediate effect on the model shown in the screen viewport. By adjusting parameters, a user can quickly create a completely different custom keel model, to be used to create an entire ship hull.

The frames were created based on the keel, a feature that will be automatically updated by any changes made to the keel HDA. To describe Oliveira's frames we used three basic shapes; a horizontal line for the flat bottom, a semi-circular futtock arc, and a straight line tangent to the end of the arc to define the top timbers. By describing these shapes, their relationship to each other, and their relationship to the keel, we developed rules which guided the procedural model.

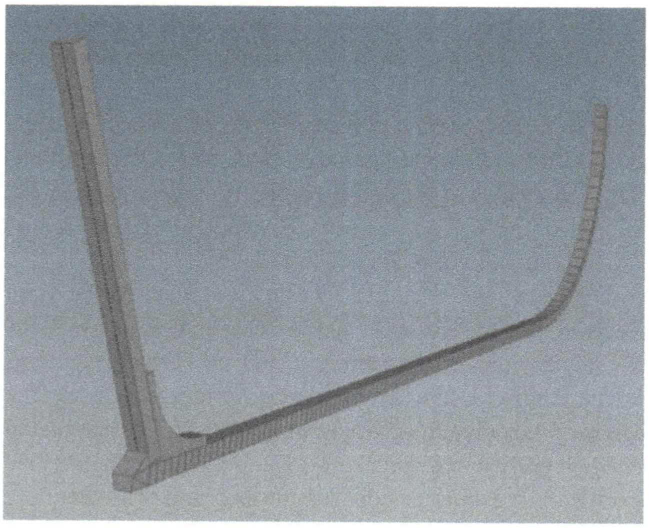

Fig. 8.6 Screen capture showing the model's keel and posts

To create the base curve which will act as a guide for the frame, a horizontal line is first created to represent the flat. The flat of the master frame is the widest of all the frames. For the rest of the frames, the length of the flat can be determined by the amount of its narrowing, and its height from the amount of the rising (Castro 2007; Suarez 2016). The centre of the futtock arcs was calculated geometrically and built into the model. Rising the flat of the frame further above the keel created the Y-frames, which were also defined with parameters and rules. To complete the frame's parameterization, we needed to describe each frame's position along the keel. Oliveira gives us clues to position all frames (Cook 2012, 50). Oliveira's treatise also says that the mast step is placed at half the keel's length. We have calculated the position of the master frame and placed the rest of the frames at similar distances.

Using the parameter values and the rules from the taxonomy developed for the project, we then developed spline curves to control the hull longitudinal shape and act as construction guides for the frame geometry. In the definition of the frames we have considered several ways to create futtocks. This option allowed models to be double-framed, as most ships were after the late seventeenth century, or to have frames composed of floor timbers and futtocks, as they were built before that. Therefore, if the *Futtock* parameter is unchecked, the frame geometry is created by sweeping a rectangular cross section along the guide curve. The dimensions of the rectangular cross section are defined by the parameters. If the futtocks option is turned on, four additional parameters are needed; *First Futtock Height, Second Futtock Height, First Futtock Overlap,* and *Second Futtock Overlap.* The resulting frame model, or HDA, can be created inside of a *Houdini* geometry node. To create a frame, a keel HDA must be used as input. By connecting the keel's output to the frame HDA's input, the keel passes along its attributes and the frame will be created according to its relationship to the keel as described above. A frame can be set to be *Master Frame* or *Custom Frame.* Setting the frame to master frame will automatically create a master frame based on the Oliveira recipe. Frames set to custom will have a procedural relationship with both the keel and the master frame. Adjustments to the parameters of the frame HDA will have an immediate effect on the model shown in the viewport. By adjusting parameters, a user can quickly create any frame in the hull.

The stern panel was defined as a sub-group of timbers within the transversal group, consisting of fashion pieces and transoms. Again, we used Oliveira's recipe and modelled the stern panel as a unit (Suarez 2016). The moulded dimension of the top transom and lower transoms can be defined by two parameters, *Top Transom Sided* and *Lower Transoms Sided.* Again, the resulting model of the stern panel, or HDA, can be created inside of a *Houdini* geometry node.

'Longitudinal reinforcements' was a group of timbers encompassing the whales, stringers and breast hooks. The timbers in this group were determined by the shape of the frames. The whales follow the exterior of the frames and the stringers the interior, and the breast hook follows the interior of the frames at the bow of the ship. To construct the stringers and the whales, there must first be a series of frames which define the hull shape. By using the frames as a guide, two hull shapes can be determined; one by the inside surface and the second by the outside surface of the frames.

'Deck structures' comprised the group of timbers containing both transverse and longitudinal timbers: beams, knees, carlings, clamps and waterways. For this model a deck is imagined as the intersection of a horizontal surface, which curves in three dimensions, and the volume of the hull defined by the frames. The surface has variable longitudinal and transversal curvature to create the cambered shape of a deck surface. The level defined by the surface is the top of the beams, above which the deck planks would be laid. By referencing the intersection surface which represents the deck surface, we could determine the location of all of the timbers in the deck structures group based on their dimensions, relationship to the deck surface, and their relationships to one another (Fig. 8.7).

The placement of the beams in this model is determined by two factors; the height of the top of the beam and its position along the length of the hull. The height of the top of a beam is defined by the deck intersection surface. A beam is connected at either end to a frame. The longitudinal position of a beam is determined by the location of the frames along the keel. The width of each beam can be determined by the distance, at the height of the deck, between each arm of the frame. The bottom surface of a beam can be determined by offsetting from the deck intersection surface by the amount of the beam's moulded dimension. In this model, beneath each beam, connecting the beam to the frame, there is a deck knee. In other models these knees could be placed every other frame or even every three or four frames.

The resulting model of the deck, or HDA, can be created inside of a *Houdini* geometry node. Creating a deck needs three inputs; a keel, stern panel, and a group of frames. The output of the keel, stern panel, and merged frames is input into the deck HDA. This provides the deck HDA with all the necessary information to create a deck. The *Deck Height* parameter defines the deck location in the Y dimension. The *Lateral Bend* and *Longitudinal Bend* parameters control the camber for the deck. Each of the six timbers which make up the deck structure; beams, knees, carlings, clamps, and waterways, which in this model are composed of two timbers, in the Portuguese way, can be turned on or off independently. Each timber's width and sided dimension can be adjusted and all other timbers will update accordingly. Any changes to the keel, frames, or stern structure will update the deck timbers.

The stanchions span from the bottom-most stringer to the bottom deck beams, and then between the beams of each sequential deck. In this model, the stanchions are positioned longitudinally at each frame. The decks HDA also stores the locations at which the carlings meet the deck beams. Given these locations as input into the stanchions HDA, a spline through each set of corresponding points is created. To create the stanchion models, a rectangular cross section is swept along their splines. And once again, the resulting model of the stanchions, or HDA, can be created inside of a *Houdini* geometry node.

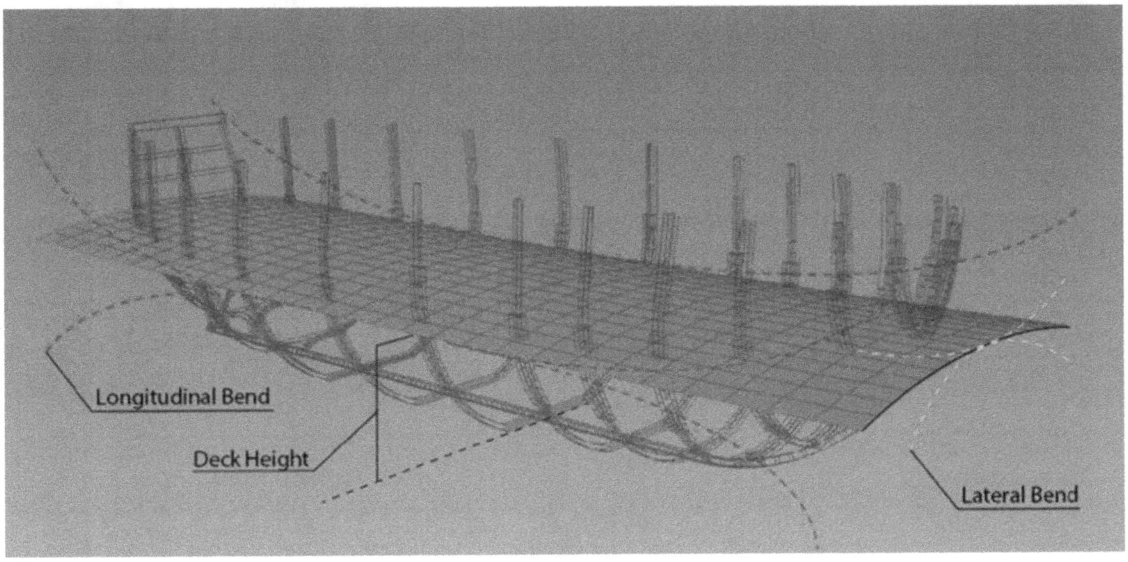

Fig. 8.7 Deck design

A keelson was also considered in this model, with a maststep as an enlarged portion of its section. The keelson sits atop the keel, sandwiching the frames along the flat. It typically starts at the point where the frames change from 'Y-shaped' to 'V-shaped' at the stern, ends forward at the apron, and is notched at each frame so that it locks in over them. The keelson has a rectangular cross section and follows along the lowest point of the frames. The keelson model was also created inside of a *Houdini* geometry node. The keelson is automatically placed along the top of the frames. The start and end locations, relative to the keel length are adjustable via the *Start Distance* and *End Distance* parameters. The *Notch Depth* parameter will adjust the depth of the notches in the keelson, lowering it over the frames by the notched amount. The *Maststep Start, Maststep End, Width,* and *Mortise Dimensions* allow the user full control over the placement and shape of the maststep. Any changes made to the keel or frames will automatically update the keelson HDA.

The planking of a hull is the application of the exterior layer of timbers which form the skin of the ship. The long strips of wood, called strakes, are each custom formed and cut to create the correct flow along the length of the hull. The designing of their shape is largely left up to the eye of the shipwright (Antscherl 2016, 3). The orientation of each strake along its length is determined by the curvature of the frame at that particular point along the hull's length. The plank varies in its moulded dimension along its length to compensate for the hull's 3D curvature. Each strake of the planking can be described by two variables for each frame; the location of the bottom of the strake along the curve of each frame and the moulded dimension of the strake at each frame. The location of the bottom of the strake along the frame, in most cases, can be derived from the strake beneath it since each strake is stacked on the one lying below it (Fig. 8.8).

The planking was the most complex part of this project and missing from this model are the inevitable drop strakes, fillers, and repairs because it proved to be too time consuming and was not essential to the purpose of this experiment. The question of the bevels was similarly not addressed from a theoretical viewpoint. The resulting planking was considered adequate to the objectives of this experiment and is comprised of multiple strake HDAs, which can be created inside of a *Houdini* geometry node. A planking strake needs three inputs; a keel HDA, a stern panel, and a merged group of frames.

The fourth input is optional; it can be used to automatically set the position of the bottom of a strake directly on top of the strake beneath it. This will provide the strake HDA with all the necessary information to create one strake around the volume of the hull. For each frame, the strake HDA will automatically create the *Bottom of Strake #* and *Moulded #* parameters, where the '#' represents the frame number. The *Bottom of Strake #* parameter will slide the current strake along the outside surface of the frame at each respective frame. The *Moulded #* parameter will adjust the strake's moulded dimension at each particular frame. By adjusting these two parameters for each frame, each strake can be designed with a custom shape, to create the desired flow of

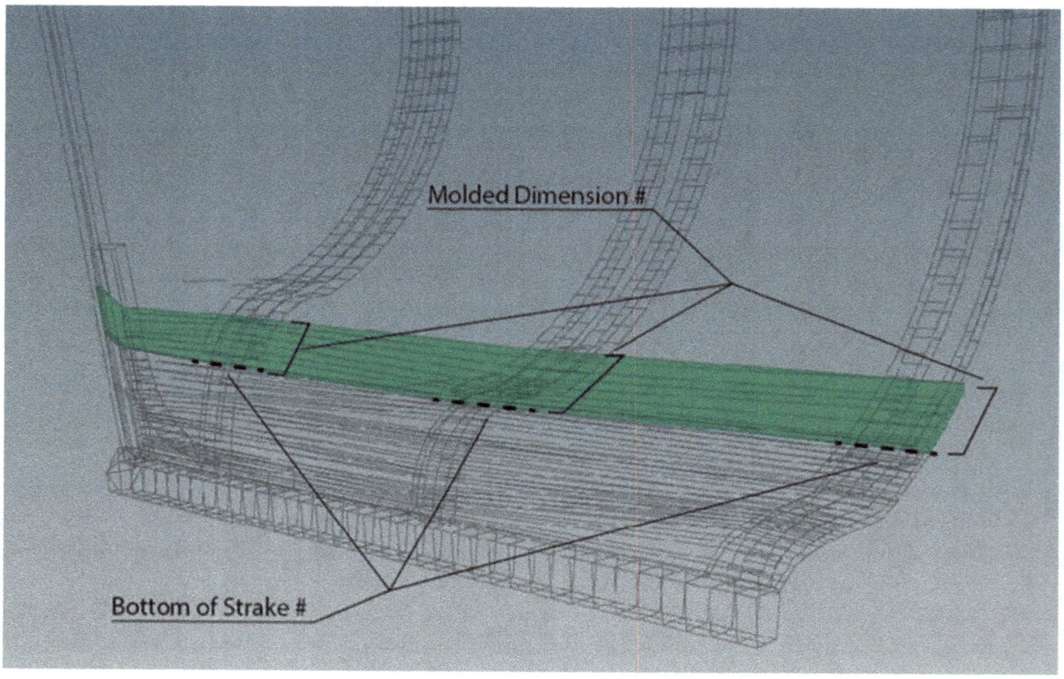

Fig. 8.8 Planking parameters

the hull planking, trying to emulate the methods employed by a shipwright.

The drop strake concept was emulated in the following way: if the *Moulded #* parameter is set to zero for any frame, the strake will not be created at this frame. This feature allows the user to terminate strakes anywhere along the length of the hull, as is often necessary when ships are planked. By using a strake HDA as input the fourth input of another strake HDA, the *Bottom of Strake #* parameter will automatically update to the location of the top of the input strake for each frame.

The ideal workflow for planking the hull is to first create the bottom-most strake (garboard), adjusting the parameters until the desired shape of the first strake is determined. Then, copy and paste the first strake node to create a duplicate of the first strake and use the first strake as input into the new strake. The second strake will maintain the same moulded value but will be automatically placed along top of the first strake. The moulded dimension of the second strake can be adjusted to create a custom shape for the new strake. The new strake can also be extended to frames which the previous strake did not span, by using a non-zero *Moulded #* parameter value at those frames. This process of duplicating and stacking strakes continues up the side of the hull until the entire hull is planked (Fig. 8.9). Any changes made to timbers which affect the hull's shape will automatically update the strake HDAs. If adjustments are made to lower strakes, the strakes above it will automatically adjust to the new shape due to their defined relationship with the strakes below.

8.9 Conclusions and Future Work

This project and its methods proved to be an excellent exploratory reflection on the use of procedural methods to model ship hulls. The results, which were tested by entering the scantlings of a set of timbers from the Belinho 1 shipwreck (Castro et al. 2015), were promising and helped understand the complexity and variability of possible construction solutions in the history of early modern European merchant shipbuilding. *Houdini* proved to be versatile in ways unthinkable with other software packages, such as Autodesk *Maya*, for instance (Yamafune et al. 2016). We have established several directions for future work. Firstly, we have developed a number of additional features and details of the timbers considered for this project that could be included in future procedural models, such as scarves, hatches, stairs, gun ports, or even small details such as fastening patterns. We have also looked at extending the model upwards, and try to model the ship's upper works. But the most interesting direction of research is the definition of more recipes for the construction of ships. We believe that the power of procedural modelling lies precisely on future reflections on the diversity and similarity of shipbuilding recipes. Secondly, this project did not include a user study, which would test the application and user experience of the procedural tools, similar to the research of Chun-Yen Huang and Wen-Kai Tai in *Ting Tools*. Thirdly, we want to look at the possibility of integrating the *Houdini* digital assets created for this project into a game engine, such as *Unity* or *Unreal*, using the *Houdini* engine

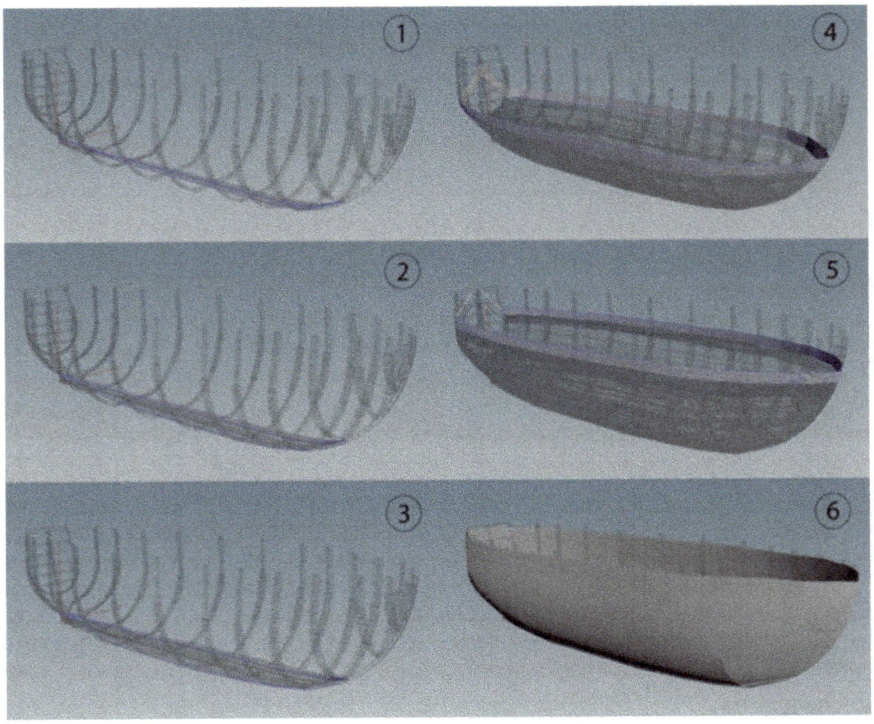

Fig. 8.9 Planking process

plugin. This could create a widely accessible shipbuilding utility tool that could be deployed using *Unity* or *Unreal* engine, which are both free to download and use. We believe that this study firmly established the potential of procedural models to explore research hypotheses and we intend to pursue this avenue of investigation.

Acknowledgements This study would not be possible without the support of the Andre Thomas Family Fellowship, and in fact stemmed from a conversation he initiated between the Texas A&M University Departments of Visualization and Anthropology.

References

Antscherl D (2016) A primer on planking. http://www.admiraltymodels.com/Planking_primer.pdf. Accessed 15 Aug 2016

Castro F (2003) The pepper wreck, an early 17th-century Portuguese Indiaman at the mouth of the Tagus River, Portugal. Int J Naut Archaeol 32(1):6–23. https://doi.org/10.1111/j.1095-9270.2003.tb01428.x

Castro F (2005) Rigging the pepper wreck. Part I: masts and yards. Int J Naut Archaeol 34(1):112–124. https://doi.org/10.1111/j.1095-9270.2005.00048.x

Castro F (2007) Rising and narrowing: 16th-century geometric algorithms used to design the bottom of ships in Portugal. Int J Naut Archaeol 36(1):148–154. https://doi.org/10.1111/j.1095-9270.2006.00116.x

Castro F (2009) Rigging the pepper wreck. Part II: sails. Int J Naut Archaeol 38(1):105–115. https://doi.org/10.1111/j.1095-9270.2005.00048.x

Castro F, Fonseca N, Wells A (2010) Outfitting the pepper wreck. Hist Archaeol 44(2):14–34

Castro F, Almeida A, Bezant J, Biscaia F, Carmo A, Crespo A, Farias I, Gonçalves I, Groenendijk P, Magalhães I, Martins A, Monteiro A, Nayling N, Santos A, Trapaga K (2015) The Belinho 1 timber catalogue. https://www.academia.edu/18771630/Belinho_1_Shipwreck_Timber_Catalogue_2015. Accessed 23 Aug 2018

Cook C (2012) A parametric model of the Portuguese nau. MA Thesis, Texas A&M University

Doran J, Hodson F (1975) Mathematics and computers in archaeology. Harvard UP, Cambridge, MA

Evans TL, Daly PW (2006) Digital archaeology: bridging method and theory. Routledge, London

Fleming A, Clarke DL (1973) Models in archaeology. Man 8(2):315

Haegler S, Mueller P, Gool L (2009) Procedural modeling for digital cultural heritage. EURASIP J Image Video Process 2009:852392. https://doi.org/10.1155/2009/852392

Hazlett AD (2007) The nau of the livro nautico: reconstructing a sixteenth-century Indiaman from texts. PhD dissertation, Texas A&M University

Hill RW (1994) A dynamic context recording and modeling system for archaeology. Int J Naut Archaeol 23(2):141–145. https://doi.org/10.1111/j.1095-9270.1994.tb00453.x

Huang C-Y, Tai W-K (2012) Ting tools: interactive and procedural modeling of Chinese ting. Vis Comput 29(12):1303–1318. https://doi.org/10.1007/s00371-012-0771-3

Lock G (2003) Using computers in archaeology: towards virtual pasts. Routledge, London

Miller P, Richards J (1995) The good, the bad, and the downright misleading: archaeological adoption of computer visualization. In: Huggett J, Ryan N (eds) CAA94. Computer applications and quantitative methods in archaeology 1994, BAR International Series 600. Tempus Reparatum, Oxford, pp 19–22

Mueller P, Vereenooghe T, Wonka P, Paap I, Gool L (2006) Procedural 3D reconstruction of Puuc buildings in Xkipché. In: Ioannides M, Arnold D, Niccolucci F, Mania K (eds) The 7th international symposium on virtual reality, archaeology and cultural heritage (VAST 2006)

Oliveira F (1991) O Liuro da Fabrica das Naus. Academia de Marinha, Lisboa

Richards J (1986) Computer usage in British archaeology, IFA occasional paper no. 1. Institute of Field Archaeologists, Birmingham

Saldaña M (2015) An integrated approach to the procedural modeling of ancient cities and buildings. Digit Scholar Human 30(1):148–163. https://doi.org/10.1093/llc/fqv013

Suarez M (2016) A procedural approach to computer-aided modeling in nautical archaeology. MA thesis, Texas A&M University

Wells A (2008) Virtual reconstruction of a seventeenth-century Portuguese nau. MA thesis, Texas A&M University

Yamafune K, Torres R, Castro F (2016) Multi-Image photogrammetry to record and reconstruct underwater shipwreck sites. J Archaeol Method Theory 24(3):703–725. https://doi.org/10.1007/s10816-016-9283-1

9 Deepwater Archaeological Survey: An Interdisciplinary and Complex Process

Pierre Drap, Odile Papini, Djamal Merad, Jérôme Pasquet, Jean-Philip Royer, Mohamad Motasem Nawaf, Mauro Saccone, Mohamed Ben Ellefi, Bertrand Chemisky, Julien Seinturier, Jean-Christophe Sourisseau, Timmy Gambin, and Filipe Castro

Abstract

This chapter introduces several state of the art techniques that could help to make deep underwater archaeological photogrammetric surveys easier, faster, more accurate, and to provide more visually appealing representations in 2D and 3D for both experts and public. We detail how the 3D captured data is analysed and then represented using ontologies, and how this facilitates interdisciplinary interpretation and cooperation. Towards more automation, we present a new method that adopts a deep learning approach for the detection and the recognition of objects of interest, amphorae for example. In order to provide more readable, direct and clearer illustrations, we describe several techniques that generate different styles of sketches out of orthophotos developed using neural networks. In the same direction, we present the Non-Photorealistic Rendering (NPR) technique, which converts a 3D model into a more readable 2D representation that is more useful to communicate and simplifies the identification of objects of interest. Regarding public dissemination, we demonstrate how recent advances in virtual reality to provide an accurate, high resolution, amusing and appropriate visualization tool that offers the public the possibility to 'visit' an unreachable archaeological site. Finally, we conclude by introducing the plenoptic approach, a new promising technology that can change the future of the photogrammetry by making it easier and less time consuming and that allows a user to create a 3D model using only one camera shot. Here, we introduce the concepts, the developing process, and some results, which we obtained with underwater imaging.

Keywords

Ontology · Machine learning · Non-photorealistic rendering · Virtual reality · Lightfield imaging

9.1 Introduction

Archaeological excavations are irreversibly destructive, and it is thus important to accompany them with detailed, accurate, and relevant documentation, because what is left of a disturbed archaeological complex is the knowledge and record collected. This kind of documentation is mainly iconographic and textual. Reflecting on archaeological sites is almost always done using recording such as drawings, sketches, photographs, maps and sections, topographies, photogrammetry, maps, and a vast array of physical samples that are analyzed for their chemical, physical, and biological characteristics. These records are a core part of the archaeological survey and offer a context and a frame of reference within which artifacts can be analyzed, interpreted, and reconstructed. As noted by Buchsenschutz (2007, 5) in the introduction to the Symposium "*Images et relevés archéologiques, de la preuve à la démonstration*" in Arles in 2007, 'Even when very accurate, drawings only

P. Drap (✉) · O. Papini · D. Merad · J. Pasquet · J.-P. Royer
M. Motasem Nawaf · M. Saccone · M. Ben Ellefi
Aix Marseille Univ, Université de Toulon, CNRS, LIS UMR 7020, Images & Models team, Marseille, France
e-mail: pierre.drap@univ-amu.fr

B. Chemisky · J. Seinturier
COMEX, COmpanie Maritime d'Expertise, Marseille, France

J.-C. Sourisseau
Aix Marseille Univ, CNRS, Ministère de la Culture et de la Communication, CCJ UMR 7299, Aix En Provence, France

T. Gambin
Archaeology Centre, University of Malta, Msida, Malta

F. Castro
Department of Anthropology, Texas A&M University, College Station, TX, USA
e-mail: fvcastro@tamu.edu

J. K. McCarthy et al. (eds.), *3D Recording and Interpretation for Maritime Archaeology*, Coastal Research Library 31,
https://doi.org/10.1007/978-3-030-03635-5_9

retain certain observations to support a demonstration, just as a speech only retains certain arguments, but this selection is not generally explicit.' This is the cornerstone of archaeological work: the survey is both a metric representation of the site and an interpretation of the site by the archaeologist.

In the last century, huge progress was made on collecting 3D data and archaeologists adopted analogue and then digital photography as well as photogrammetry. This was first developed by A. Laussedat in 1849, and the first stereo plotter was built by C. Pulfrich in 1901 (Kraus 1997). Furthermore, we saw the beginning of underwater archaeological photogrammetry (Bass 1966) and finally the dense 3D point cloud point generation based on automatic homologous point description and matching (Lowe 1999). In a way, building a 3D facsimile of an archaeological site is not itself a matter of research even in an underwater context. Creation of a model does not solve the problem of producing a real survey and interpretation of the site according to a certain point of view – a teleological approach able to produce several graphical representations of the same site, according to the final goal of the survey. Indeed, the production of such a survey is a complex process, involving several disciplines and emergent approaches. In this chapter we present the work of an interdisciplinary team, merging photogrammetry and computer vision, knowledge representation, web semantic, deep learning, computational geometry, lightfield cameras dedicated to underwater archaeology (Castro and Drap 2017).

9.1.1 The Archaeological Context

This work is centered on the Xlendi shipwreck, named after its location, found off the Gozo coast in Malta. The shipwreck was located by the Aurora Trust, an expert in deep-sea inspection systems, during a survey campaign in 2008. The shipwreck is located near a coastline known for its limestone cliffs that plunge into the sea and whose foundation rests on a continental shelf at an average depth of 100 m below sea level. The shipwreck itself is therefore exceptional; first due to its configuration and its state of preservation which is particularly well-suited for our experimental 3D modelling project. The examination of the first layer of amphorae also reveals a mixed cargo, consisting of items from Western Phoenicia, and Tyrrhenian-style containers which are both well-dated to the period situated between the end of the eighth and the first half of the seventh centuries BC. The historical interest of this wreck, highlighted by our work, which is the first to be performed on this site, creates added value in terms of innovation and the international profile of the project (Drap et al. 2015).

9.2 Underwater Survey by Photogrammetry

The survey was done using optical sensors: photogrammetry is the best way to collect both accurate 3D data and color information in a full contactless approach and reduced the time on site to the necessary time to take the photographs. The survey had two goals: measuring the entire visible seabed where the wreck is located and extracting known artefacts (amphorae), in order to position them in space and accurately represent them after laboratory study. The photogrammetric system used in 2014 (Drap et al. 2015), was mounted on the submarine *Remora 2000*. In this version, a connection is established between the embedded sub system, fixed on the submarine and the pilot (inside the submarine) to ensure that the survey is fully controlled by the pilot. The photogrammetric system uses a synchronized acquisition of high and low-resolution images by video cameras forming a trifocal system. The three cameras are independently mounted in separate waterproof housings. This requires two separate calibration phases; the first one is carried out on each set of camera/housing in order to compute intrinsic parameters and the second one is done to determine the relative position of the three cameras which are securely mounted on a rigid platform. The second calibration can be done easily before each mission and it affects the final 3D model scale. This allow us to obtain a 3D model at the right scale without any interaction on site. More in detail, the trifocal system is composed of one high-resolution, full-frame camera synchronized at 2 Hz and two low-resolution cameras synchronized at 10 Hz (Drap 2016).

The lighting, a crucial part in photogrammetry, must meet two criteria: the homogeneity of exposure for each image and its consistency between images (Drap et al. 2013). Of course, using only one vehicle, the lights are fixed on the submarine as far as possible from the camera. Hydrargyrum medium-arc iodide (HMI) lamps were used with an appropriate diffuser (note: in the current version overvolted LEDs are used as they are significantly more energy efficient). The trifocal system has two different goals: The first one is the real-time computation of system pose and the 3D reconstruction of the zone of seabed visible from the cameras. The operator can pilot the submarine using a dedicated application that displays the position of the vehicle in real time. A remote video connection also enables the operator to see the live images captured by the cameras. Using the available data, the operator can assist the pilot to ensure the complete coverage of the zone to be surveyed. The pose is estimated based on the movement of the vehicle between two consecutive frames. We developed a system for computing visual odometry in real time and producing a sparse point cloud of 3D points on the fly (Nawaf et al. 2016, 2017).

The second goal is to perform an offline 3D reconstruction of a high-resolution metric model. This process uses the high-resolution images to produce a dense model, scaled based on baseline distances. We developed a set of tools to bridge our visual odometry software to the commercial software *Agisoft Photoscan/Metashape* in order to use the densification capabilities. After this step we obtained a dense pointcloud and a set of oriented high-resolution photographs describing accurately the entire site. This is enough to produce a high resolution orthophoto of the site (1 pixel/0.5 mm), as well as accurate 3D models. The ultimate goal of this process is to study the cargo, hull remains, and remaining artifact collection.

The second problem is to extract known objects for these data. We defined the amphorae typology and the corresponding theoretical 3D models. The recognition process is composed of two different phases: the first one is the artifact detection; and then the position and orientation estimation of each artefact is undertaken in order to calculate the exact size and location. The amphorae detection is done in 2D using the full orthophoto. We used a deep learning approach and obtained 98% of good results (Pasquet et al. 2017). This allows us to extract the relevant part of the 3D data where the artifact is located. We then apply a 3D matching approach to compute the position, orientation and dimension of the known artifact.

It is important to note that during the last decade several excellent works have been done in this context for both underwater archaeology and marine archaeology (Aragón et al. 2018; Balletti et al. 2016; Bodenmann et al. 2017; Bruno et al. 2015; Capra et al. 2015; Martorelli et al. 2014; McCarthy and Benjamin 2014; Pizarro et al. 2017; Secci 2017). Indeed, more technical details of this survey have been published (Drap et al. 2015). More generally, we can observe that entire workshops are now dedicated to photogrammetric survey for underwater archaeology (for example the workshop organized by CIPA/ISPRS, entitled 'Underwater 3D recording and modeling' in Sorrento, Italy in April 2015), hundreds of articles are written on photogrammetric underwater survey for archaeology and substantial research is done on technical aspects, such as calibration (Shortis 2015; Telem and Filin 2010), stereo system (O'Byrne et al. 2018; Shortis et al. 2009) using structured light (Bruno et al. 2011; Roman et al. 2010) or more generally on underwater image processing (Ancuti et al. 2017; Chen et al. 2018; Hu et al. 2018; Yang et al. 2017a, b). In the last few years this discipline has attracted the attention of the industrial world and has been used to record and analyse complex objects of large dimensions (Menna et al. 2015; Moisan et al. 2015). The new challenge for tomorrow is producing accurate and detailed surveys from ROV and AUV in complex environments (Ozog et al. 2015; Zapata-Ramírez et al. 2016).

9.3 The Use of Ontologies

9.3.1 In Underwater Archaeology

The main focus behind this research is the link between measurement and knowledge. All underwater archaeological survey is based on the study of a well-established corpus of knowledge, in a discipline which continues to redefine its standard practices. The knowledge formalization approach is based on ontologies; the survey approach proposed here implies a formalization of the existing practice, which will drive the survey process.

The photogrammetric survey was done with the help of a specific instrumental infrastructure provided by COMEX, a partner in the GROPLAN project (Drap et al. 2015; GROPLAN 2018). Both this photogrammetry process and the body of surveyed objects were ontologically formalized and expressed in OWL2. The use of ontologies to manage cultural heritage advances every year and generates interesting perspectives for its continued study (Bing et al. 2014; Lodi et al. 2017; Niang et al. 2017; Noardo 2017). The ontology developed within the framework of this project takes into account the manufactured items surveyed and the photogrammetry process which is used to measure them. Each modelled item is therefore represented from a measurement point of view and linked to all the photogrammetric data that contributed to the measurement process. To this extent we developed two ontologies: one dedicated to photogrammetric measurement and georeferencing the measured items, and another dedicated to the measured items, principally the archaeological artefacts. The latter describes their dimensional properties, ratios between main dimensions, and default values. Within this project, these two ontologies were aligned in order to provide one common ontology that covers the two topics at the same time. The development architecture of these ontologies was performed with a close link to the Java class data structure, which manages the photogrammetric process as well as the measured items. Each concept or relationship in the ontology has a counterpart in Java (the opposite is not necessarily true). Moreover, the surveyed resources are archaeological items studied and possibly managed by archaeologists or conservators in a museum. It is therefore important to be able to connect the knowledge acquired when measuring the item with the ontology designed to manage the associated archaeological knowledge.

The modelling work of our ontology started from the premise that collections of measured items are marred by a lack of precision concerning their measurement, assumptions about their reconstruction, their age, and origin. It was therefore important to ensure the coherence of the measured items and potentially propose a possible revision. This collection work was presented in a previous study in the context of underwater archaeology with similar problems (Curé et al. 2010; Hué et al. 2011; Seinturier 2007; Serayet 2010; Serayet et al. 2009).

Amongst the advantages of the photogrammetric process is the possibility of providing several 2D representations of the measured artefacts. Our ontology makes use of this advantage to represent the concepts used in photogrammetry, and to be able to use an ontology reasoner on the ABox representing photogrammetric data. In other words, this photogrammetric survey is expressed as an ontology describing the photogrammetric process, as well as the measured objects, and that was populated both by the measurements of each artefact and by a set of corresponding data. In this context, we developed a mapping from an Object Oriented (OO) formalism to a Description Logic (DL). This mapping is relatively easy to accomplish because we have to map a poor semantic formalism toward a richer one (Roy and Yan 2012). We need to manage both the computational aspects (often heavy in photogrammetry) implanted in the artefacts measurable by photogrammetry, and the ontological representation of the same photogrammetric process and surveyed artefacts.

The architecture of the developed framework is based on a close link between, on the one hand, the software engineering aspects and the operative modelling of the photogrammetry process, artefacts measured by photogrammetry in the context of this project and, on the other hand, with the ontological conceptualization of the same photogrammetry process and surveyed artefacts. The present implementation is based on a double formalism, JAVA, used for computation, photogrammetric algorithms, 3D visualization of photogrammetric models, and cultural heritage objects, and then for the definition of ontologies describing the concepts involved in this photogrammetric process, as well as on the surveyed artefacts.

To implement our ontological model, we opted for OWL2 (Web Ontology Language), which has been used for decades as a standard for the implementation of ontologies (McGuinness and Harmelen 2004). This web ontology language allows for modelling concepts (classes), instances (individuals), attributes (data properties) and relations (object properties). In fact, the main concern during the modelling process is the representation of accurate knowledge from a measurement point of view for each concept in the ontology. On the other hand, the same issue presides over the elaboration of the JAVA taxonomy, where we have to manage constraints involving differences in the two hierarchies of concepts within the engineering software side. For this purpose, we developed a procedural attachment method for each concept in the ontology. This homologous aspect of our architecture leads to the fact that each individual of the ontology can produce a JAVA instance since each concept present in the ontology has a homologous class in the JAVA tree. Note here that the adoption of an automatic binding between the ontology construction in OWL and the JAVA taxonomy cannot be produced automatically in our case. Hence, we have abandoned an automatic mapping using JAVA annotation and JAVA beans for a manual extraction, even if this is a common way in literature (Horridge et al. 2004; Ježek and Mouček 2015; Kalyanpur et al. 2004; Roy and Yan 2012; Stevenson and Dobson 2011).

The current implementation is based on a two aspects: JAVA, used for computation, photogrammetric algorithms, 3D visualization of photogrammetric data and patrimonial objects, and OWL for the definition of ontologies describing the concepts involved in the measurement process and the link with the measured objects. In this way, reading an XML file used to serialize a JAVA instance set representing a statement can immediately (upon reading) populate the ontology; similarly reading an OWL file can generate a set of JAVA instance counterparts of the individuals present in the ontology. Furthermore, the link between individuals and instances persists and it can be used dynamically. The huge advantage of this approach is that it is possible to perform logical queries for both aspects of the ontology and the JAVA representation, i.e. to perform semantic queries over ontology instances while benefiting from the computational capabilities in the homologous JAVA side. We can thus read the ontology, visualize in 3D the artefacts present in the ontology, and graphically visualize the result of SQWRL queries in the JAVA viewer.

A further step, after developing and populating the ontology, is to find target ontologies to link to, following semantic web recommendations (Bizer et al. 2009); linking the newly published ontology to other existent ontologies in the web in order to allow ontologies sharing, exchanging and reusing information between them. In cultural heritage contexts, CIDOC CRM is our main target ontology since it is now well adopted by CH actors from theoretical point of view (Gaitanou et al. 2016; Niccolucci 2016; Niccolucci and Hermon 2016) as well as applicative works (Araújo et al. 2018) and an interesting direction toward GIS application based on some connection with photogrammetric survey (Hiebel et al. 2014, 2016).

Several methodologies can be chosen regarding mapping these two ontologies. For example, Amico et al. (2013) choose to model the survey location with an activity (*E7*) in CRM. They also developed a formalism for the digital survey tool mapping the digital camera definition with (*D7 Digital Machine Event*). We see here that the mapping problem is close to an alignment problem, which is an issue in this case. Aligning two ontologies dealing with digital camera definition is not obvious; a simple observation of the lack of interoperability between photogrammetric software shows the scale of the problem. We are currently working on an alignment/extension process with Sensor ML which is an ontology dedicated to sensors. Although some work has already yielded results (Hiebel et al. 2010; Xueming et al. 2010), it is not enough to support the close-range photogrammetry process, from image measurement to artefact representation.

Linking our ontology to CIDOC-CRM can provide more integrity between cultural heritage datasets and will allow more flexibility for performing federated queries cross different datasets in this community. Being a generic ontology, however, the current state of CIDOC-CRM does not support the items that it represents from a photogrammetric point of view, a simple mapping would not be sufficient and an extension with new concepts and new relationships would be necessary. Our extension of the CIDOC-CRM ontology is structured around the triple <*E18_Physical_Thing, P53_has_former_or_current_location, E53_Place*>, which provides a description of an instance of *E53_Place* as the former or current location of an instance of *E18_Physical_Thing*. The current version of this extension relates only to the TBox part of the two ontologies where we used the hierarchical properties *rdfs:subClassOf* and *rdfs:subPropertyOf* to extend the triple <*SpatialObject, hasTransformation3D, transformation3D*>, developed in this project. Note that the mapping operation is done in JAVA by interpreting a set of data held by the JAVA classes as a current identification of the object: 3D bounding box, specific dimension. These attributes are then computed in order to express the right CRM properties.

Our architecture is based on the procedural attachment where the ontology is considered as a homologous side of the JAVA class structure that manages the photogrammetric survey and the measurement of artefacts. This approach ensures that all the measured artefacts are linked with all the observations used to measure and identify them.

A further advantage of adopting ontology is to benefit from the reasoning over the semantic of its intentional and extensional knowledge. For this purpose, the approach that we adopted so far, using the OWLAPI and the Pellet reasoner, allows for performing SQWRL queries using an extension of SWRL Built-In (O'Connor and Das 2006) packages. SWRL provides a powerful extension mechanism that allows for implementing user-defined methods in the rules (Keßler et al. 2009). For this purpose, we have built some spatial operators allowing us to express spatial queries in SWRL (Arpenteur 2018), as for example the operator *isCloseTo* with three arguments which allows for selecting all the amphorae present in a sphere centred on a specific amphora and belonging to a certain typology. A representation of the artefacts measured on the Xlendi wreck, as well as an answer to a SWRL query, is shown in Fig. 9.1.

Our ontology also provides a spatial description as georeferencing of each artefact and all the archaeological knowledge, including relationships provided by archaeologists. Based on our procedural attachment approach, we built a mechanism which allows for the evaluation and visualization of spatial queries from SWRL rules. We are currently extending this approach in a 3D information system dedicated to archaeological survey based on photogrammetric survey and knowledge representation for spatial reasoning.

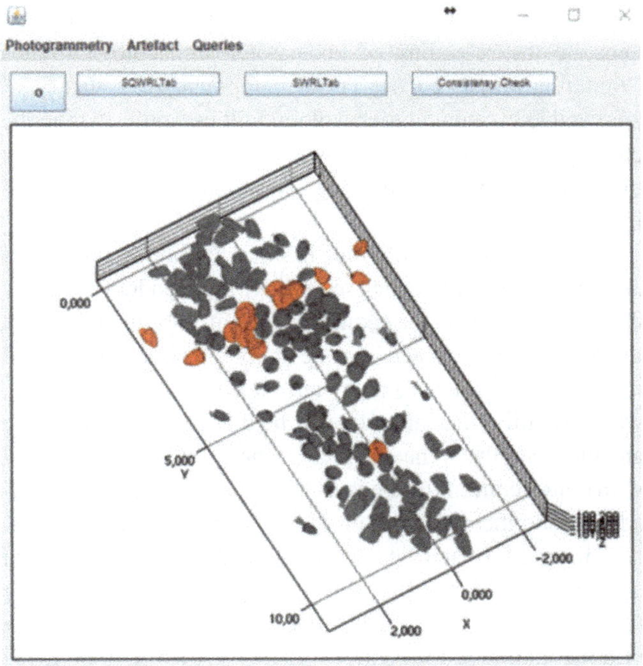

Fig. 9.1 3D visualization of a spatial resquest in SWRL: *Amphorae(?a) ^ swrlArp:isCloseTo(?a, "IdTargetAmphora", 6.2) ^ hasTypologyName(?a, "Pitecusse_365") -> sqwrl:select(?a).* Means select all amphorae with the typology Pitecusse_365 and at a maximum distance of 6.2 m from the amphorae labelled IdTargetAmphora

Finally, we draw the reader's attention to the fact that our ontology has been recently published in the Linked Open Vocabulary (LOV 2018; Vandenbussche et al. 2017) which offers users a keywords search service indexing more than 600 vocabularies in its current version. The LOV indexed all terms in our ontology and provides an online profile metadata ARP (2018) that offers our work a better visibility and allows terms reuse for the ontology meta-designers in different communities.

9.3.2 Application in Nautical Archaeology

The applications to query, visualize, and evaluate survey data acquired through photogrammetric survey methods have the potential to revolutionize nautical archaeology (as evidenced by many chapters in this volume). Even the simplest utilization of off-the-shelf photogrammetry software combined with consumer-grade portable computers and without the need for special graphics hardware, can greatly simplify and expedite the recording of underwater archaeological sites (Yamafune et al. 2016). The development of a theory of knowledge in nautical archaeology will certainly change the paradigm of this discipline, which is hardly half a century old and is still struggling to define its aims and methodologies.

The first and most obvious implication of any process of automation of underwater recording, independent from the operating depths, is the economic benefit. Automated survey methods save time and can be more precise. The second implication is the treatment and storing of primary data, which is still artisanal and performed according to the taste and means of the archaeologists in charge; it is seldom stored with consideration for the inevitable reanalysis that new paradigms and the development of new equipment will eventually dictate. Primary data are traditionally treated in nautical archaeology as propriety of the principal investigator and are often lost, as archaeologists move on, retire, or die. The lack of a methodology to record shipwrecks (Castro et al. 2017) makes it difficult to develop comparative studies, aiming at finding patterns in trade, sailing techniques, and shipbuilding, to cite only three examples where data from shipwreck excavations are almost always truncated or unpublished.

The application of an ontology and a set of logical rules for the identification, definition, and classification of measurable objects is a promising methodology to assess, gather, classify, relate, and analyse large sets of data. Nautical archaeology is a recent sub-discipline of archaeology and attempts to record shipwrecks under the standards commonly used in land archaeology started after 1960. The earliest steps of this discipline were concerned with recording methodology and accuracy in underwater environments. These environments are difficult and impose a number of practical constraints, such as reduced bottom working time, long decompression periods, cold, low visibility, a narrower field of vision, surge, current, or depth. Since the inception of nautical archaeology, theoretical studies aimed at identifying patterns and attempting to address larger anthropological questions related to culture change have emerged. The number of shipwrecks excavated and published, however, makes the sample sizes too small to allow for broad generalizations. Few seafaring cultures have been studied and understood well enough to allow a deep understanding of their history, culture, and development. Classical and Viking seafaring cultures provide two European examples where data from land excavations and historical documents help archaeologists to understand the ships and cargoes excavated, but there is a lack of an organized body of data pertaining to most maritime landscapes and cultures, and a lack of organization of the material culture in relational libraries. Through our work we hope to provide researchers from different marine sciences with both better archives and appropriate tools for the facilitation of their work. This in turn will improve the conditions to the development of broader anthropological studies, for instance, evaluating and relating cultural change in particular areas and time periods.

The study of the history of seafaring is the study of the relations of humans with rivers, lakes, and seas, which started in the Palaeolithic. An understanding of this part of our past entails the recovery, analysis, and publication of large amounts of data, mostly through non-intrusive survey methods. The methodology proposed in GROPLAN aims at simplifying the collection and analysis of archaeological data, and at developing relations between measurable objects and concepts. It builds upon the work of Steffy (1994), who in the mid-1990s developed a database of ship components. This shipbuilding information, segmented in units of knowledge, tried to encompass a wide array of western shipbuilding traditions—which developed through time and space—and establish relations between conception and construction traits in a manner that allowed comparisons between objects and concepts. Around a decade later Carlos Monroy transformed Steffy's database into an ontological representation in RDF-OWL, and expanded its scope to potentially include other archaeological materials (Monroy 2010; Monroy et al. 2011). After establishing a preliminary ontology, completed through a number of interviews with naval and maritime archaeologists, Monroy combined the database with a multi-lingual glossary and built a series of relational links to textual evidence that aimed at contextualizing the archaeological information contained in the database. His work proposed the development of a digital library that combined a body of texts on early modem shipbuilding technology, tools to analyse and tag illustrations, a multi-lingual glossary, and a set of informatics tools to query and retrieve data (Monroy 2010; Monroy et al. 2006, 2007, 2008, 2009, 2011).

Our approach extends these efforts into the collection of data, expands the analysis of measurable objects, and lays the base for the construction of extensive taxonomies of archaeological items. The applications of this theoretical approach are obvious. It simplifies the acquisition, analysis, storage, and sharing of data in a rigorous and logically supported framework. These two advantages are particularly relevant in the present political and economic world context, brought about by the so-called globalization and the general trend it entailed to reduce public spending in cultural heritage projects. The immediate future of naval and maritime archaeology depends on a paradigm change. Archaeology is no longer the activity of a few elected scholars with the means and the power to define their own publication agendas. The survival of the discipline depends more than ever on the public recognition of its social value. Cost, accuracy, reliability (for instance established through the sharing of primary data), and its relationship with society's values, memories and amnesias, are already influencing the amount of resources available for research in this area. Archaeologists construct and deconstruct past narratives and have the power to impact society by making narratives available that illustrate the diversity of the human experience in a world that is less diverse and more dependent on the needs of world commerce, labour, and capital.

The main objective of this work is the development of an information system based on ontologies and capable of establishing a methodology to acquire, integrate, analyse, generate, and share numeric contents and associated knowledge in a standardized and homogenous form. In 2001 the UNESCO Convention on the Protection of the Underwater Cultural Heritage established the necessity of making all archaeological data available to the public (UNESCO 2001). According to UNESCO around 97% of the children of the planet are in school, and 50% have some access to the internet. In one generation archaeology ceases to be a closed discipline and it is likely that a diverse pool of archaeologists from all over the world will multiply our narratives of the past and enrich our experience with new viewpoints and better values. This work is fully built upon this philosophy, and shows a way to share and analyse archaeological data widely and in an organized manner.

9.4 Artefact Recognition: The Use of Deep Learning

9.4.1 The Overall Process Using a Deep Learning Approach

This work aims to detect amphorae on the orthophoto of the Xlendi shipwreck. This image, however, does not contain a lot of examples and so it is complicated to train a machine learning model. Moreover, we cannot easily train the model from another shipwreck to learn an amphorae model because it is difficult to find another orthophoto which contains amphorae with the same topology. So, we propose to use a deep learning approach that is proving its worth in many research fields and shows the best performance on different competitions as ImageNet (Russakovsky et al. 2014) with deep networks (He et al. 2015; Simonyan and Zisserman 2014; Szegedy et al. 2014). We use a Convolution Neural Network (CNN) in order to train the shape of various and different amphorae and the context of the ground. Then we propose to use a transfer learning process to fine-tune our model over the Xlendi shipwreck amphorae. This approach allows us to train the model using a small part of the Xlendi database. Underwater objects are rarely found in a perfect state. Indeed, they can be covered by sediments or biological growth, or by another object, and they are often broken. It is common that amphora necks are separated from an amphora's body. We want to detect all the amphora pieces by performing a pixel segmentation which consists of adopting a pixel-wise classification approach on the orthophoto (Badrinarayanan et al. 2015; Shelhamer et al. 2016). To improve the model, we define three classes: the underground, the body of the amphora and the head of the amphora; which are the rim, the neck and the handles respectively. After the pixel segmentation, we group pixels with similar probabilities together to get an object segmentation.

9.4.2 The Proposed Convolution Neural Network

The CNN is composed of a series of layers in which each layer takes as input the output of the previous layer. The first layer is named the *input layer* and takes as input the testing or the training image. The last layer is the output of the network and gives a prediction map. The output of a layer, noted l in the network, is called a feature map and is noted f_l. In this work, we use four different types of layers: convolution layers, pooling layers, normalization layers and deconvolution layers. We explain the different types of layers in the following:

Convolution layers are composed of convolutional neurons. Each convolutional neuron applies the sum of 2D convolutions between the input feature maps and its kernel. In the simple case where only one feature map is passed to the input convolutional neuron, the 2D convolution between the kernel noted Kof size w × h and the input feature map $I \in R^2$ is I ∗ K and is defined as:

$$\left(I * K\right)_{x,y} = \sum_{x+w/2}^{x'=x-w/2} \sum_{y+h/2}^{y'=y-h/2} I_{x',y'} . K_{x'+w/2-x,y'+h/2-y}$$

where (x,y) are the coordinates of a given pixel into the output feature map.

In the case of neural convolutional neural networks, a neuron takes as input each of p feature maps of the previously layer noted I^l with $I \in \{0..p\}$. The resulting feature map is the sum of p2D convolutions between the kernel $K^l$and the map $I^l$and is defined as:

$$\left(I \otimes K\right) = \sum_{p}^{i=0} I^i * K^i$$

Basically, we apply a nonlinear transformation after the convolution step in order to solve nonlinear classification problems. The most commonly used function is the Rectify Linear Unit (ReLU) which is defined by $f(x) = max(x, 0)$ (Krizhevsky et al. 2012).

Pooling layers quantify the information while reducing the data volume. They apply a sliding window on the image which processes a specific operation. The two most used methods consist in selecting only the maximum or the mean value between different data in the sliding window.

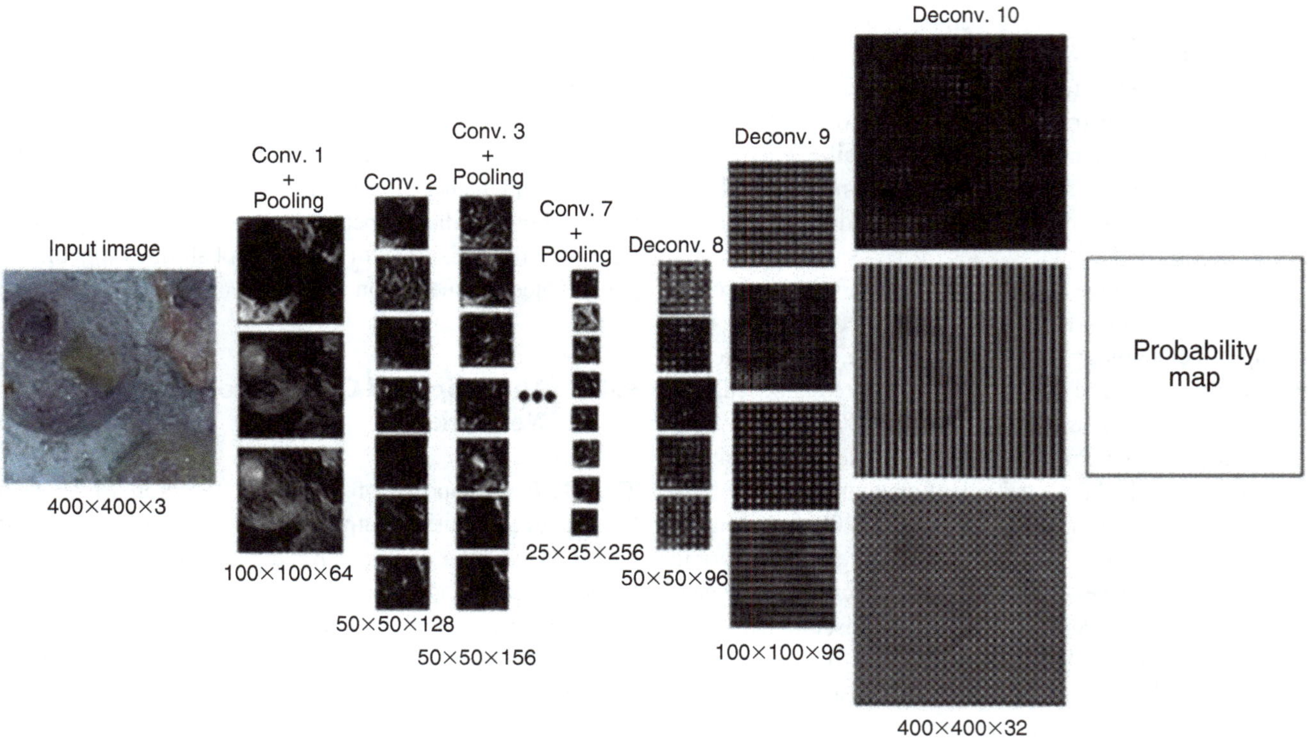

Fig. 9.2 Representation of the architecture that we are proposing, using an example to activate the feature maps

Normalization layers scale the feature maps. The two most used methods dedicated to normalizing are the Local Response Normalization (LRN) and the Batch Normalization (BN) (Ioffe and Szegedy 2015). In this work we use the batch normalization which scales each value of the feature maps depending on the mean and the variance of this value in the batch of images. Moreover, the batch normalization has two hyperparameters which are learned during the training step to scale the importance of the different feature maps in a layer and to add a bias value to the feature map.

Deconvolution layers are the transpose of the convolution layers.

To perform the learning step, parameters of convolution and deconvolution are tuned using a stochastic gradient descent (Bottou 1998). This optimization process is costly but is compatible with parallel processing. In this work we use the Caffe (Jia et al. 2014) framework to train our CNN. The results are obtained using a GTX 1080 card, packed in 2560 cores with a 1.733 GHz base. Our CNN architecture is composed of seven convolution layers, three pooling layers and three deconvolution layers.

9.4.3 Classification Results

We train our CNN on images coming from another site and then we use a small part of the Xlendi image to fine-tune the weights of the CNN. On the Xlendi Image we have only used 20 amphorae as training examples. Results are given on Fig. 9.2 where we can see that all the amphorae in the testing image are detected. The false positives are mainly located on the grind stones. This error is due to the small size of the training database. Indeed, during the pre-training step there are not grinding stone examples in the used images, then during the tuning step only a few grind stone examples are represented. On the segmentation pixel image, the recall is around 57% and the precision around 71%. The recall is low because the edges of the amphorae are rarely detected since the probability is the highest at the middle of each amphora and then it decreases rapidly toward the edges. For the object detection map, the noise is removed and so the recall is close to 100% and the precision is around 80% (Fig. 9.3).

9.5 2D Representation: From Orthophoto to Metric Sketch

Cultural heritage representation has been completely transformed in the last 40 years. Computer graphics introduced in this field, high resolution 3D survey, and 3D modelling and image synthesis, built on surveys results nearly indistinguishable from reality. But even if this huge production of photorealism increased during the last decade, photorealism is far from a drawing made by an expert; the interpretation phase is missing even if it is accurate and detailed. On the other

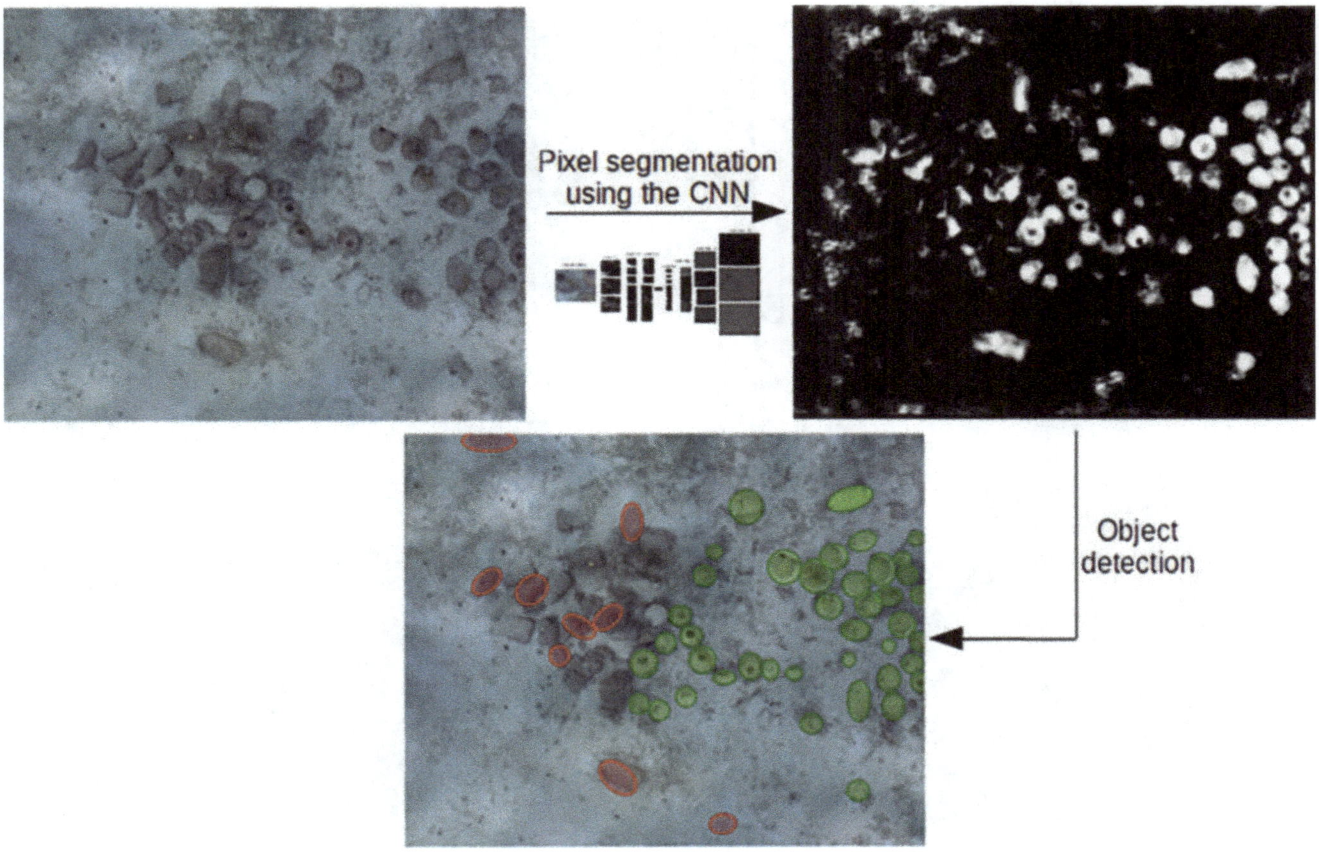

Fig. 9.3 Pixel segmentation on the testing part of the Xlendi orthophoto. On the probability map, the wither the pixel, the higher the probability to be an amphora is. On the object detection map, the green circles represent the correctly detected amphorae and the red circles the false positive detections

hand, 2D representations are still important, easy to manipulate, transfer, publish and annotate. We are working on producing 2D accurate and detailed documents, easily accessible even if they have an important resolution. For example the high resolution orthophoto of the Xlendi site is accessible through the GROPLAN web site (Drap 2016) using IIPImage (Pitzalis and Pillay 2009).

Even if photorealistic 3D or 2D documents are still providing better quality images, we want to replicate the effects of hand-drawn documents, in 2D. These are designed to look like documents traditionally produced by archaeologists and merge two important qualities: they are detailed and accurate, and they respect a common graphical convention used by the archaeological community. These kinds of images, produced with Non-Photorealistic Rendering or Deep Learning as detailed below, have an extra meaning: 'selection.' Photorealism provides the same relevance to all objects in the scene. Hand-drawings and NPR images pick-up only objects with specific properties, hence a selection is made by knowledge. Here we present a work-in-progress and two research directions aimed to produce such products'; one uses 2D documents as orthophoto and the other is based on dense 3D models.

9.5.1 Style Transfer to Sketch the Orthophoto

Image style here is defined by the way of drawing without the content. The style transfer applies the style of a given image to another image. In deep learning this kind of method is well known, to transfer the style of an artist to a real image, see Fig. 9.4. Using the approach of (Gatys et al. 2016) we apply a sketch style to the image. The style A draws different patches of leafs on the output. The resulting image obtained with the style B is blurred and the edges are sheared because the sketch is composed of points, but not of a solid line.

The style D creates some horizontal patterns which are similar to the waves. The style C gives the best visual result, even if the representation of grind stones on the top of the image disappears. This type of image, however, is unusable by archaeologists because it is an artistic vision which does not spotlight the interest regions.

We propose to learn the relevant sketch using a machine learning process on a part of the Xlendi orthophoto. Then we propose to use an architecture similar to the previous one to learn the sketch process. The sketch of Xlendi created by the CNN is given on Fig. 9.7 (bottom right). The weakness of this approach is evident when objects that are not known by the

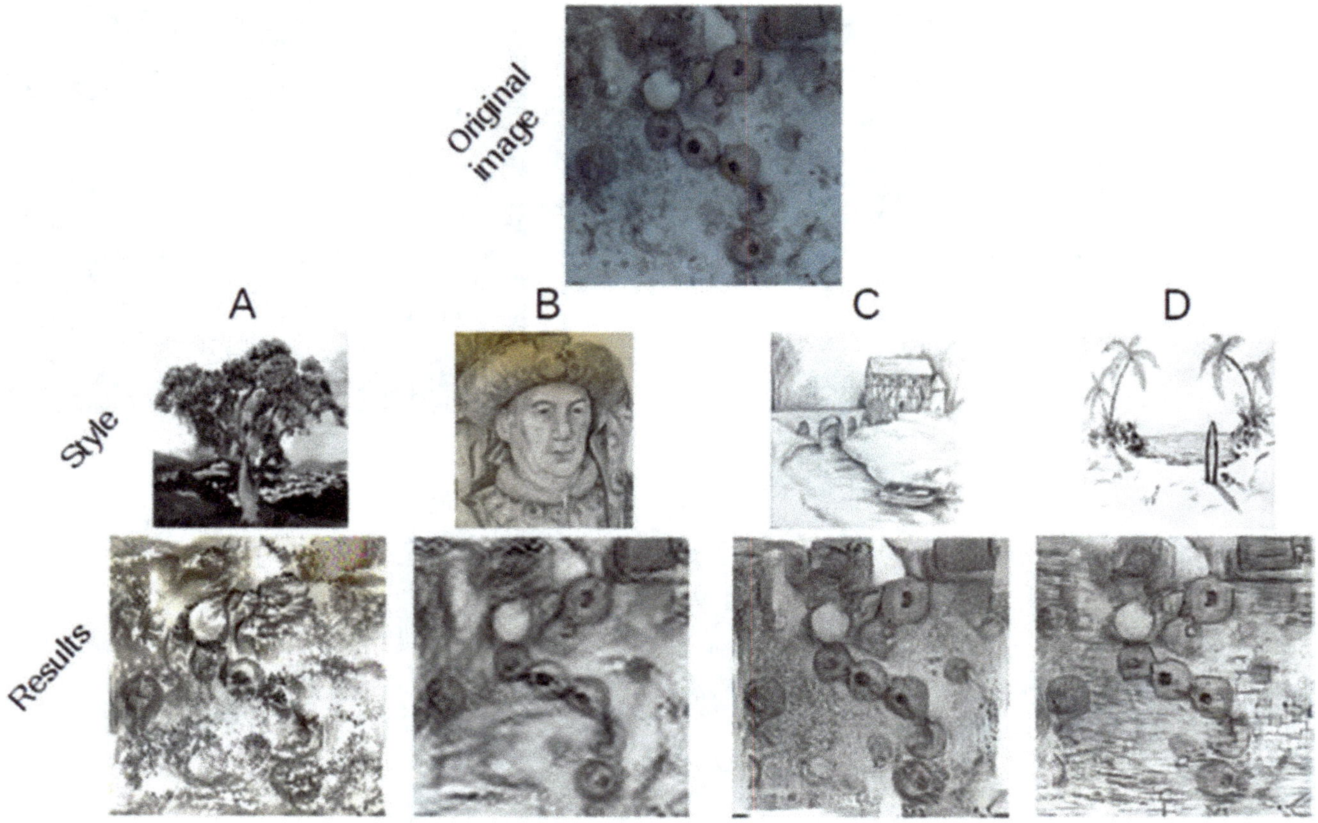

Fig. 9.4 Examples of different styles applied to the same image (C. Nigon)

model are present, because they are not in the training database. Figure 9.5 is focused on a grind stone with a yellow starfish above. Since there is no starfish in our database, the algorithm counts it as a rim of amphora. To avoid this error, we increase the number of examples in the training database.

9.5.2 From 3D Models to NPR: Non-photorealistic Rendering

Several archaeological representation styles found in technical manuals (Bianchini 2008) and published articles (Sousa et al. 2003) are in fact close to the NPR results. These kinds of representation are useful to communicate and to make illustration more readable, direct and clearer.

As of a few years ago, researchers have been trying to extract features from 3D models, such as lines, contours, ridges and valleys, apparent contours and so on, and to compare their results with sketches. This kind of representation, NPR, has been developed for illustration but it is not so much used for archaeology and cultural heritage. Several typologies of representations for 3D archaeological artefacts have been studied during recent years (DeCarlo et al. 2003; Jardim and de Figueiredo 2010; Judd et al. 2007; Raskar 2001; Roussou and Drettakis 2003; Tao et al. 2009; Xie et al. 2014). An interesting open source approach, called *Suggestive Contour Software* (DeCarlo and Rusinkiewicz 2007), allows visualization of 3D models built with mesh by NPR and gives several options to modify final results with various parameters (occlusion contours, suggestive contours, ridges and valley, etc.). The link between accuracy and sketch is a question considered by several scientists in the field of computer vision and virtual reality (Bénard et al. 2014). More recent research in this field now uses deep learning and AI (Bylinskii et al. 2017; Gatys et al. 2016).

Concerning cultural heritage and archaeological artefacts, some morphological properties of amphorae are well represented with NPR rendering (Fig. 9.6). It is simple to identify handles, rims and necks when we extract 'contours' or 'ridges and valleys' because archaeologists and algorithms are looking at the same properties. The curvature of the body is substantial for an amphora and is one of the keys to identify its typology, but it is also a geometrical property enhanced by NPR algorithms. In contrast to image contour extraction, the NPR renderings studied here work on 3D models, which enable the use of the normal surface to accurately extract contours from overlapping shapes.

An orthophoto of an excavation area has too much extra information, and the goal is to extract the important informa-

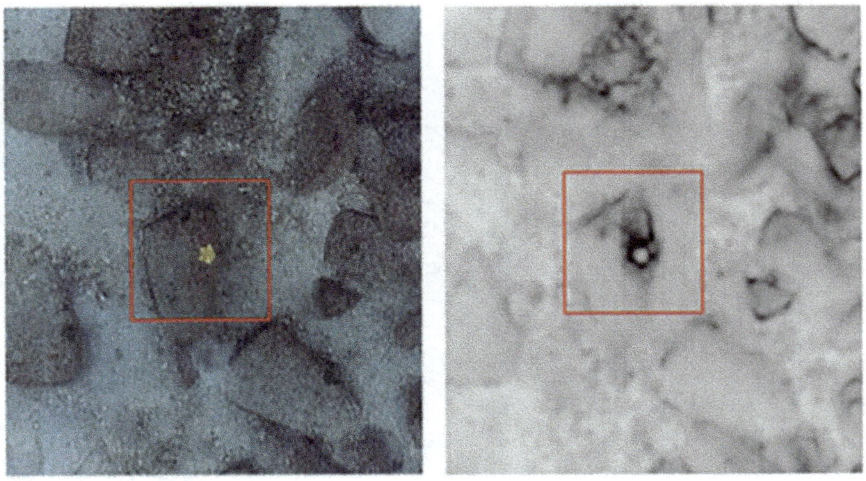

Fig. 9.5 A part of Xlendi centered on a grind stone with a starfish (left), and the sketch of the image based on a CNN (right)

Fig. 9.6 Some amphorae on Xlendi site. On the left, an NPR image made using the called Suggestive Contour Software (DeCarlo et al. 2003). On the right, the same portion of Xlendi site mesh model, acquired with photogrammetric survey

tion contained in the image and to leave out the rest. NPR algorithms can be useful for that because NPR acts as a filter, using information that is not available within the images: the surface normals. Beginning from a 3D model and a point of view, NPR rendering produces an image that is fairly close those obtained through expert interpretation (Fig. 9.7).

Nevertheless, these positive results from NPR are partially caused by a specific configuration of the scene: in Xlendi, amphorae are lying on a flat seabed or are partially covered by sediment. The seabed is mainly uniform and flat so the normal analysis done by the NPR process focus on protruding amphorae and tend to minimize the seabed. These morphological differences can be enhanced by tuning the numerous NPR parameters to highlight the relevant part of the 3D model. It seems that each kind of scene requires a bespoke set of parameters, so the most promising results need to involve knowledge of the process and the experiments in deep learning appear to present an opportunity to progress these ideas.

9.6 Virtual Reality for the General Public

Public dissemination is also important. Once we are able to offer an accurate and appropriate visualization tool to experts, with the correct graphical form, NPR for example, a visualization tool for the general public is possible. The collected data on site and photogrammetry can be used both to extract relevant knowledge, and to produce a facsimile of the site. Even if this approach does not generate specific archaeological knowledge, it offers the possibility to 'visit' an unreachable archaeological site. This can be an interesting feature for the general public, but of course it also allows experts from around the world to have access to an exceptional underwater archaeological site by means of a high resolution and accurate 3D modelling. We also propose to that users visualize and explore an archaeological site using Virtual Reality (VR) technology. This visualization is made of both photorealistic and semantic representations of the observed objects including amphorae and other archaeological material. A user can freely navigate the site, switch from one representation to another, and interact with the objects by means of dedicated controllers. The different representations that are visualized in the tool, can be seen as queries mixing geometry, photorealistic rendering, and knowledge (for example: display the amphorae colorized by type on the site).

Although software rendering packages such as *Unity*, *Sketchfab* and *Unreal Engine*, or APIs such as *OpenVR, HTC Vive, SDK* already exist, we have chosen to develop our own solution.

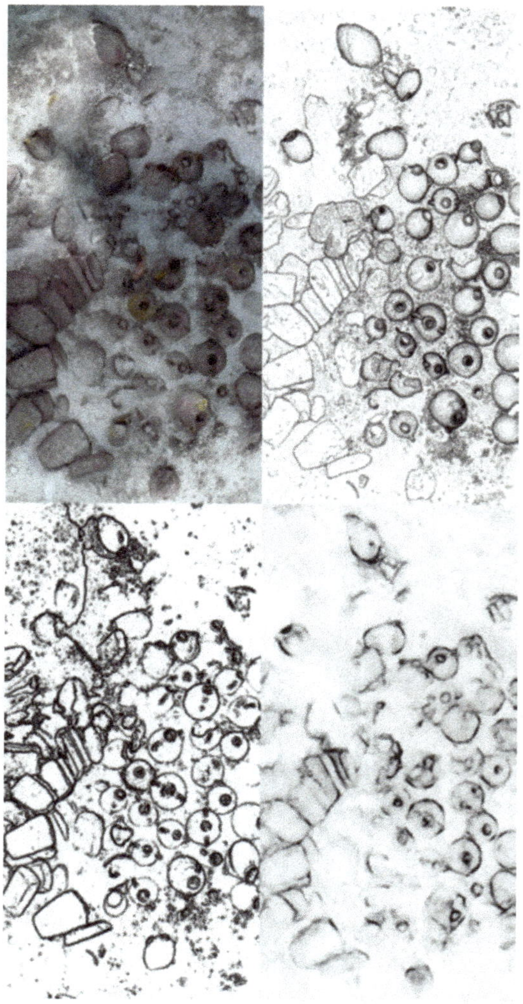

Fig. 9.7 The Xlendi wreck in Malta: (top left) orthophoto, (top right) hand-made design by Gina De Angelis (University of Rome III, Rome, Italy) on the orthophoto. (bottom left) NPR generated from the 3D mesh. (bottom right) A sketch of Xlendi created by the CNN

The work presented here is based on the Arpenteur project (2018; Drap 2017) that provides geometrical computation capabilities, photogrammetric features and representation and processing of knowledge within ontologies. We chose to use *jMonkeyEngine* (jME 2018) as a basis for our development, which is a game engine made especially for modern 3D development, as it uses shader technology. 3D games can be written for a large set of devices using this engine. *jMonkeyEngine* is written in Java and uses Lightweight Java Game Library (LWJGL 2018) as its renderer. Indeed, it provides high level functionalities for scene description while retaining a power of display comparable to native libraries.

A first experiment of such a tool was presented during a CNRS symposium in Marseilles in May 2017. The innovative exhibition (IE 2018) can be seen on Fig. 9.8. This demonstrator enables the display of an underwater archaeological site in Malta. Inside the demonstrator, the user can view the site in a photorealistic mode. It can even simulate underwater conditions using a filter and for example visualize the site as if it was underwater or as if it was outdoors. The user can also display the cargo of the wreck as described by the archaeologists and mix the photorealism and the NPR. This demonstrator will be enhanced for visualizing the result of ontology-based queries and allowing an archaeologist to formulate and verify hypotheses with the impression of being on the site. For the general public, the immersion capacity can be used to visit inaccessible sites.

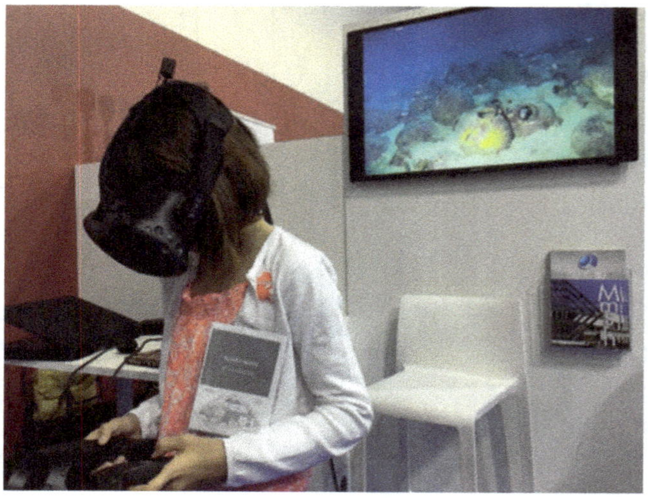

Fig. 9.8 Immersive experience on the Xlendi wreck for general public

9.7 New 3D Technologies: The Plenoptic Approach

In traditional photogrammetry, it is necessary to completely scan the target area with many redundant photos taken from different points of view. This procedure is time consuming and frustrating for large sites. For example, thousands of photos have been captured to cover Xlendi shipwreck. Whereas many 3D reconstruction technologies are developed for terrestrial sites relying mostly on active sensors such as laser scanners, solutions for underwater sites are limited. The idea of plenoptic camera (also called lightfield camera) is to simulate a 2D array of aligned tiny cameras. In practice, this can be achieved by placing an array of a micro-lenses between the image sensor and main lens as shown in Fig. 9.9. In this way, the raw image obtained contains information about the position and direction of all light beams present in the image field, so that the scene can be refocused at any depth plane, but it is also possible to obtain different views from a single image after capturing. The first demonstration of this technique was published by Ng et al. (2005).

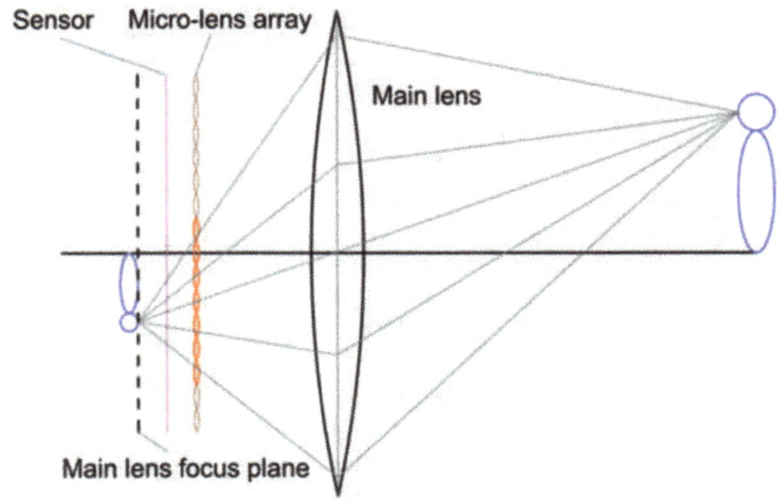

Fig. 9.9 Diagram showing plenoptic camera internal design

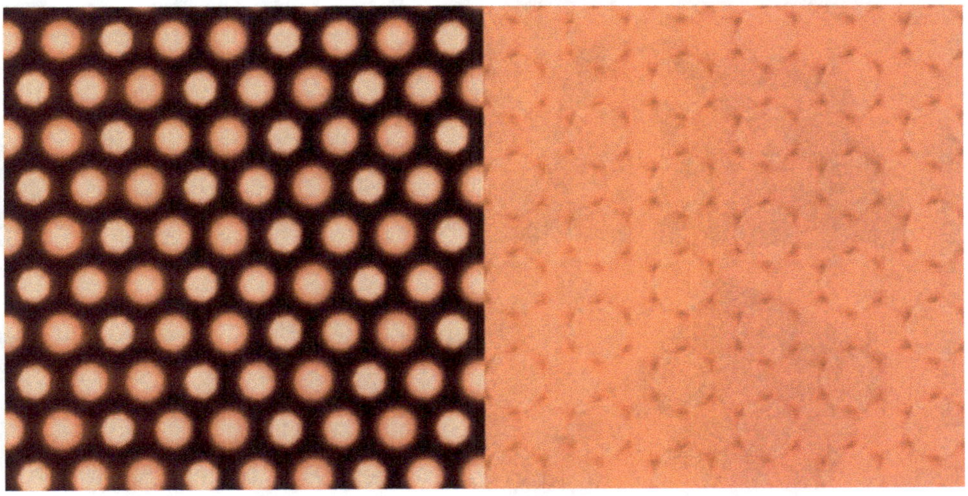

Fig. 9.10 Two examples of plenoptic zoomed images showing micro images. Taken at two different aperture settings (f/13 left, f/9 right) in the presence of a white filter

In a plenoptic camera, pairs of *close* micro lenses can be considered as a stereo pair under the condition of having parts of the scene appearing in both acquired micro images.[1] By applying a stereo feature points matching to those micro images it is possible to estimate a corresponding depth map. It is essential during image acquisition to respect a certain distance to the scene to remain within the working depth range. This working depth range can be enlarged but at a price of depth levels accuracy.[2] In practice, the main lens focus should be set beyond the image sensor, in this case, each micro lens will produce a slightly different micro image for the object with respect to its neighbours. This allows the operator to have a depth estimation of the object.

Recent advances in digital imaging resulted in the development and manufacture of high quality commercial plenoptic cameras. The main manufacturers are Raytrix (2017) and Lytro (Lytro 2017). While the latter is concentrating on image refocusing and off-line enhancement, the first is focusing on 3D reconstruction and modelling. The plenoptic camera used in this project is a modified version of Nikon D800 by Raytrix. A layer that is composed of around 18,000 micro-lenses is placed in front of the 36.3 mega pixels CMOS original image sensor. The micro-lenses are of three types, which differ in their focal distance. This helps to enlarge the working depth range, by combining three zones that correspond to each lens type; namely; near, middle and far range lens. The projection size of micro lens (micro images) can be controlled by changing the aperture of the camera. In our case, the maximum diameter of a micro image is around 38–45 pixels depending on lens type. Figure 9.10 illustrates two examples of captured plenoptic images showing micro images taken at two different aperture settings.

Our goal is to make this approach work in the marine environment so that only one shot is enough to obtain a properly scaled 3D model, which is a great advantage underwater,

[1]There is no common term in literature used for the projection of a single micro lens. In this book, we refer to it as micro image.

[2]Despite the usage of subpixel accuracy, recovered depth is roughly discrete.

Fig. 9.11 Underwater images taken using a plenoptic camera (first row), it shows also the repetition of an edge in the scene which is essential to perform the depth estimation. The processed total focus image (second row) and reconstructed 3D models (third row)

where diving time and battery capacity are limited. The procedure to work with a plenoptic camera is described in the following. Figure 9.11 shows examples of plenoptic underwater images and reconstructed 3D model.

Micro-lens Array (MLA)—Calibration step: computes the camera's intrinsic parameters, which include the position and the alignment of the MLA. The lightfield camera returns an uncalibrated raw image containing the pixels' intensity values read by the sensor during shooting. The position of each individual micro lens is to be identified in the raw image. To localize the exact position of each the micro lens in the raw image, a calibration image is taken using a white diffusive filter generating continuous illumination to highlight the edges and vignetting of each micro lens as shown in Fig. 9.10 (left). Using simple image processing techniques, it is possible to localize the exact position of each micro-lens which corresponds to its optical centre up to subpixel accuracy. In the same way, by taking another calibration image after changing the aperture of the camera so that the exterior edges of micro-lens are touching (Fig. 9.10), we could detect the micro-lens outer edges by circle fitting with the help of the computed centres positions. Hence, for any new image without the filter it is possible to extract micro images easily. Finally, a metric calibration must also be performed to convert from pixels to metric units. Here, the use of a calibration grid enables to determine accurately the internal geometry of the micro-lens array. The calibration results provide many parameters: intrinsic parameters, orientation of MLA, tilt of the MLA with respect to image sensor as well as distortion parameters.

MLA Imaging—each micro lens will produce an image for a small part of the scene, from the opposite side, every part of the captured scene is projected to several micro lenses at different angles. Hence, it is possible to produce a synthesized image with a focus plane considered at some distance. This is done by defining for each point in the object space the micro lenses where a certain point is projected. By selecting the projected points in each micro image, the average pixel value is considered the final pixel colour in the focused image.

Total focus processing—in the used plenoptic camera from Raytrix (this holds also for other brands such as Lytro) there are three types of micro lenses with different focal throughout the MLA. Each type of micro lens is considered as a sub-array and it has certain depth of focus. So that the three depths of focus associated to the three types are added together for more depth range. Here, it is possible to reconstruct a full focus image with a large field of depth using the all types of lenses to compute a single sharp image. For more details on the total focus process, we refer to Perwass and Wietzke (2012).

Depth map computation—a brute force stereo points matching is performed among neighbouring micro-lenses. The neighbourhood is an input parameter that defines the maximum distance on the MLA between two neighbouring micro-lenses. This process can be easily run in parallel. This is achieved using *Raytrix RxLive* software which relies totally on the GPU. The total processing time does not exceed 400 milliseconds per image. Next, using the computed calibration parameters, each matched point is triangulated in 3D in order to obtain sparse point cloud.

Dense point cloud and surface reconstruction—in this step, all pixel data in the image are used to reconstruct the surface in 3D using iterative filling and bilateral filtering for smoother depth intervals.

9.8 Conclusions

This chapter addresses a number of relevant problems related to the acquisition, analysis, and dissemination of archaeological data from underwater contexts. Underwater archaeology is expensive and computers are streamlining its processes and, perhaps more importantly, promise to increase the accuracy of the recording process and make it available to a wider number of scholars. This trend is changing the enduring individualistic paradigm and pushing archaeology to a team-based discipline, where knowledge is acquired and narratives are constructed in a continuous, iterative process, more similar to that of the hard sciences. The circulation of primary data is a fundamental step in this trend, expected to transform archaeological interpretations into something closer to community projects, where narratives are constructed and deconstructed in a much more exciting and dynamic process than the traditional ones, where publications often took decades to appear. We have presented several techniques that have the potential to make successful underwater archaeological surveys quicker, cheaper, and more accurate.

Data acquisition and processing using photogrammetry allow the capture of an impressive amount of underwater site features and details. The representation of photogrammetry data using ontologies has two main benefits. The first is to facilitate data sharing between researchers with different backgrounds, such as archaeologists and computer scientists. The second is to improve and expand data analysis and to identify patterns or to generate different statistics using a simple query language that is close to natural language. The proposed set of tools also allows researchers to create sketched images that are close to what is commonly used and produced by archaeologists. The proposed automatic detection and recognition method, using deep learning, promises to be tremendously useful, particularly at larger sites, given the amount of effort it saves. Our experiment with plenoptic cameras are is one of few attempts found in literature so far to apply this technique to underwater archaeology and appears fruitful. This is an avenue of research that we intend to pursue in the near future.

Finally, using virtual reality to visualize the 3D data has produced a countless number of applications, both for pedagogical purposes and as a means to share archaeological discoveries with the public, inviting a wider audience to participate in the production process, and promoting and raising the awareness of the underwater heritage. Survey and representation are always guided by intentions, like archaeological excavation: we try to find, or find out, to measure, and record. This human action is based on choices and selections, even if they appear to be unconscious. The goal is not objectivity, but how we can guide and make those choices and selections explicit. Our answer is to enlarge the knowledge base using several resources: ontologies to create relations between measurable objects and concepts, improve analysis and sharing knowledge, and NPR and Deep Learning to improve object recognition and representation of artefacts.

Acknowledgements This work is partially done in the framework of the project GROPLAN (2018, ANR-13-CORD-0014) funded by the French Agency for Scientific Research (ANR). We would like to thank the artist Cosette Nigon for providing sketches in the deep learning, style transfer approach (Fig. 9.4). We would like to thank also Gina De Angelis and Alessandra Sprega, two architects, graduates from the University of Rome III, Rome, Italy for the archaeological design made from an orthophoto of underwater wreck for training the CNN (Fig. 9.7, top right).

References

Amico N, Ronzino P, Felicetti A, Niccolucci F (2013) Quality management of 3D cultural heritage replicas with CIDOC-CRM. In: Paper presented at the CEUR workshop, La Valetta, Malta, 26 September 2013, vol 1117, pp 61–69. CEUR-WS.org

Ancuti CO, Ancuti C, De Vleeschouwer C, Garcia R (2017) A semi-global color correction for underwater image restoration. ACM SIGGRAPH 2017 Posters, Los Angeles

Aragón E, Munar S, Rodríguez J, Yamafune K (2018) Underwater photogrammetric monitoring techniques for mid-depth shipwrecks. J Cult Herit. https://doi.org/10.1016/j.culher.2017.12.007. (in press)

Araújo C, Martini RG, Rangel Henriques P, Almeida JJ (2018) Annotated documents and expanded CIDOC-CRM ontology in the automatic construction of a virtual museum. In: Rocha Á, Reis LP (eds) Developments and advances in intelligent systems and applications. Studies in computational intelligence, vol 718. Springer, pp 91–110. https://doi.org/10.1007/978-3-319-58965-7

ARP (2018) Arpenteur ontology, linked open vocabularies. http://lov.okfn.org/dataset/lov/vocabs/arp. Accessed 6 Sept 2018

Arpenteur (2018) Built-in operators. http://www.arpenteur.org/ontology/ArpenteurBuiltInLibrary.owl. Accessed 6 Sept 2018

Badrinarayanan V, Kendall A, Cipolla R (2015) SegNet: a deep convolutional encoder-decoder architecture for image segmentation. arXiv:1511.00561 [cs]

Balletti C, Beltrame C, Costa E, Guerra F, Vernier P (2016) 3D reconstruction of marble shipwreck cargoes based on underwater multi-image photogrammetry. Digit Appl Archaeol Cult Herit 3(1):1–8. https://doi.org/10.1016/j.daach.2015.11.003

Bass GF (1966) Archaeology under water. Thames and Hudson, Bristol

Bénard P, Hertzmann A, Kass M (2014) Computing smooth surface contours with accurate topology. ACM Trans Graph 33(2):19. https://doi.org/10.1145/2558307

Bianchini M (2008) Manuale di rilievo e di documentazione digitale in archeologia. Aracne

Bing L, Chan KCC, Carr L (2014) Using aligned ontology model to convert cultural heritage resources into semantic web. In: Paper presented at the 2014 IEEE international conference on Semantic Computing, Newport Beach, CA, 16–18 June 2014, pp 120–123. https://doi.org/10.1109/ICSC.2014.39

Bizer C, Heath T, Berners-Lee T (2009) Linked data-the story so far. In: Sheth A (ed) Semantic services, interoperability and web applications: emerging concepts. Kno.e.sis Center, Wright State University, Dayton, pp 205–227

Bodenmann A, Thornton B, Nakajima R, Ura T (2017) Methods for quantitative studies of seafloor hydrothermal systems using 3D visual reconstructions. ROBOMECH J 4:22. https://doi.org/10.1186/s40648-017-0091-5

Bottou L (1998) On-line learning and stochastic approximations. In: Saad D (ed) On-line learning in neural networks. Cambridge University Press, Cambridge, pp 9–42

Bruno F, Bianco G, Muzzupappa M, Barone S, Razionale AV (2011) Experimentation of structured light and stereo vision for underwater 3D reconstruction. ISPRS J Photogramm Remote Sens 66(4):508–518

Bruno F, Lagudi A, Gallo A, Muzzupappa M, Davidde Petriaggi B, Passaro S (2015) 3D documentation of archaeological remains in the underwater park of Baiae. Int Arch Photogramm, Remote Sens Spat Inf Sci XL-5/W5:41–46. https://doi.org/10.5194/isprsarchives-XL-5-W5-7-2015

Buchsenschutz O (2007) Images et relevés archéologiques, de la preuve à la démonstration. Papers presented at the 132e congrès national des sociétés historiques et scientifiques, Arles. Les éditions du comité des travaux historiques et scientifiques

Bylinskii Z, Wook Kim N, O'Donovan P, Alsheikh S, Madan S, Pfister H, Durand F, Russell B, Hertzmann A (2017) Learning visual importance for graphic designs and data visualizations. In: Paper presented at the 30th Annual ACM symposium on user interface software & technology. arXiv:1708.02660 [cs.HC] https://doi.org/10.1145/3126594.3126653

Capra A, Dubbini M, Bertacchini E, Castagnetti C, Mancini F (2015) 3d reconstruction of an underwater archaeological site: comparison between low cost cameras. Int Arch Photogramm Remote Sens Spat Inf Sci XL-5/W5:67–72. https://doi.org/10.5194/isprsarchives-XL-5-W5-67-2015

Castro F, Drap P (2017) A arqueologia marítima e o future/Maritime archaeology and the future. VESTÍGIOS—Rev Lat-Am Arqueologia Hist 11(1):40–55

Castro F, Bendig C, Bérubé M, Borrero R, Budsberg N, Dostal C, Monteiro A, Smith C, Torres R, Yamafune K (2017) Recording, publishing, and reconstructing wooden shipwrecks. J Marit Archaeol 13(1):55–66. https://doi.org/10.1007/s11457-017-9185-8

Chen Z, Zhang Z, Bu Y, Dai F, Fan T, Wang H (2018) Underwater object segmentation based on optical features. Sensors 18(1):196. https://doi.org/10.3390/s18010196

Curé O, Sérayet M, Papini O, Drap P (2010) Toward a novel application of CIDOC CRM to underwater archaeological surveys. In: Paper presented at the 4th IEEE international conference on Semantic Computing, ICSC 2010, Pittsburgh, 22–24 September 2010, pp 519–524. https://doi.org/10.1109/ICSC.2010.104

Decarlo D, Rusinkiewicz S (2007) Highlight lines for conveying shape. In: Paper presented at the international symposium on Non-Photorealistic Animation and Rendering (NPAR), August 2007

Decarlo D, Finkelstein A, Rusinkiewicz S, Santella A (2003) Suggestive contours for conveying shape. ACM Trans Graph (Proc SIGGRAPH) 22(3):848–855

Drap P, Merad DD, Mahiddine A, Seinturier J, Peloso D, Boï J-M, Chemisky B, Long L (2013) Underwater photogrammetry for archaeology: what will be the next step? Int J Herit Digit Era 2(3):375–394. https://doi.org/10.1260/2047-4970.2.3.375

Drap P, Merad D, Hijazi B, Gaoua L, Saccone MMNM, Chemisky B, Seinturier J, Sourisseau J-C, Gambin T, Castro F (2015) Underwater photogrammetry and object modeling: a case study of Xlendi wreck in Malta. Sensors 15:30351–30384. https://doi.org/10.3390/s151229802

Drap P (2016) GROPLAN web site: GROPLAN ontology and photogrammetry; Generalizing surveys in underwater and nautical archaeology. The GROPLAN project web site: http://www.groplan.eu. Web page available: http://www.groplan.eu. Last Access date: 2018-12-06

Drap P (2017) ARPENTEUR, an Architectural PhotogrammEtry Network Tool for EdUcation and Research, The ARPENTEUR project web site: http://www.arpenteur.org. Web page available: http://www.arpenteur.org. Last Access date: 2018-12-06

Gaitanou P, Gergatsoulis M, Spanoudakis D, Bountouri L, Papatheodorou C (2016) Mapping the hierarchy of EAD to VRA Core 4.0 through CIDOC CRM. In: Garoufallou E, Subirats Coll I, Stellato A, Greenberg J (eds) Metadata and semantics research: proceedings of the 10th international conference, MTSR 2016, Göttingen, Germany, 22–25 November 2016. Springer, pp 193–204. https://doi.org/10.1007/978-3-319-49157-8

Gatys LA, Ecker AS, Bethge M, Hertzmann A, Shechtman E (2016) Controlling perceptual factors in neural style transfer. arXiv:1611.07865 [cs.CV]

GROPLAN (2018) GROPLAN: Généralisation du relevé, avec ontologies et photogrammétrie, pour l'archéologie navale et sous-marine. http://www.groplan.eu. Accessed 6 Sept 2018

He K, Zhang X, Ren S, Sun J (2015) Deep residual learning for image recognition. arXiv:1512.03385 [cs]

Hiebel G, Hanke K, Hayek I (2010) Methodology for CIDOC CRM based data integration with spatial data. Contreras F, Melero FJ (eds) CAA'2010 fusion of cultures: proceedings of the 38th conference on computer applications and quantitative methods in archaeology, Granada, April 2010, pp 1–8

Hiebel G, Doerr M, Hanke K, Masur A (2014) How to put archaeological geometric data into context? Representing mining history research with CIDOC CRM and extensions. Int J Herit Digit Era 3(3):557–578. https://doi.org/10.1260/2047-4970.3.3.557

Hiebel G, Doerr M, Eide Ø (2016) CRMgeo: a spatiotemporal extension of CIDOC-CRM. Int J Digit Libr 18(4):271–279. https://doi.org/10.1007/s00799-016-0192-4

Horridge M, Knublauch H, Rector A, Stevens R, Wroe C (2004) A practical guide to building OWL ontologies using the protege-OWL plugin and CO-ODE tools edition 1.0

Hu H, Zhao L, Li X, Wang H, Liu T (2018) Underwater image recovery under the nonuniform optical field based on polarimetric imaging. IEEE Photon J 10(1):1–9. https://doi.org/10.1109/JPHOT.2018.2791517

Hué J, Sérayet M, Drap P, Papini O, Würbel E (2011) Underwater archaeological 3D surveys validation within the removed sets framework. In: Liu W (ed) Symbolic and quantitative approaches to reasoning with uncertainty, ECSQARU 2011. Lecture notes in computer science, vol 6717. Springer, Berlin, pp 663–674. https://doi.org/10.1007/978-3-642-22152-1_56

IE (2018) Innovative exhibition. http://innovatives.cnrs.fr/. Accessed 6 Sept 2018

Ioffe S, Szegedy C (2015) Batch normalization: accelerating deep network training by reducing internal covariate shift. arXiv:1502.03167 [cs]

jME (2018) jMonkeyEngine. http://jmonkeyengine.org/. Accessed 6 Sept 2018

Jardim E, De Figueiredo LH (2010) A hybrid method for computing apparent ridges. In: 23rd conference on graphics, patterns and images (SIBGRAPI), Gramado, 30 August–3 September 2010. https://doi.org/10.1109/SIBGRAPI.2010.24

Ježek P, Mouček R (2015) Semantic framework for mapping object-oriented model to semantic web languages. Front Neuroinform 9(3):1–15. https://doi.org/10.3389/fninf.2015.00003

Jia Y, Shelhamer E, Donahue J, Karayev S, Long J, Girshick R, Guadarrama S, Darrell T (2014) Caffe: convolutional architecture for fast feature embedding. In: MM'14 proceedings of the 22nd ACM international conference on Multimedia. Orlando, 3–7 November 2014. ACM, New York, pp 675–678. https://doi.org/10.1145/2647868.2654889

Judd T, Durand F, Adelson E (2007) Apparent ridges for line drawing. ACM Trans Graph 26(3):19. https://doi.org/10.1145/1276377.1276401

Kalyanpur A, Pastor DJ, Battle S, Padget JA (2004) Automatic mapping of OWL ontologies into java. In: Paper presented at the SEKE, pp 98–103

Keßler C, Raubal M, Wosniok C (2009) Semantic rules for context-aware geographical information retrieval. In: Barnaghi P, Moessner K, Presser M, Meissner S (eds) Proceedings of smart sensing and context: 4th European conference, EuroSSC 2009, Guildford, UK, 16–18 September 2009. Springer, Berlin, pp 77–92

Kraus K (1997) Photogrammetry: advanced methods and applications (vols 1 and 2). Dummlerbush, Bonn

Krizhevsky A, Sutskever I, Hinton GE (2012) ImageNet classification with deep convolutional neural networks. In: Pereira F, Burges CJC, Bottou L, Weinberger KQ (eds) NIPS'12 Proceedings of the 25th international conference on neural information processing systems, vol 1. Curran Associates, pp 1097–1105

Lodi G, Asprino L, Nuzzolese AG, Presutti V, Gangemi A, Recupero DR, Veninata C, Orsini A (2017) Semantic web for cultural heritage valorisation. In: Hai-Jew S (ed) Data analytics in digital humanities. Multimedia systems and applications. Springer, pp. 3–37. https://doi.org/10.1007/978-3-319-54499-1

LOV (2018) Linked open vocabularies. http://lov.okfn.org/dataset/lov/. Accessed 6 Sept 2018

Lowe D (1999) Object recognition from local scale-invariant features. In: Paper presented at the international conference on computer vision, Corfu, Greece, September 1999, pp 1150–1157

LWJGL (2018) Lightweight Java Game Library. https://www.lwjgl.org/. Accessed 6 Sept 2018

Lytro (2017) Lytro ILLUM https://www.lytro.com/. Accessed 17 Aug 2017

Martorelli M, Pensa C, Speranza D (2014) Digital photogrammetry for documentation of maritime heritage. J Marit Archaeol 9(1):81–93. https://doi.org/10.1007/s11457-014-9124-x

McCarthy JK, Benjamin J (2014) Multi-image photogrammetry for underwater archaeological site recording: an accessible, diver-based approach. J Marit Archaeol 9(1):95–114. https://doi.org/10.1007/s11457-014-9127-7

McGuinness DL, van Harmelen F (2004) OWL web ontology language overview. http://www.w3.org/TR/owl-features/, February 2004. World Wide Web Consortium (W3C) recommendation

Menna F, Nocerino E, Troisi S, Remondino F (2015) Joint alignment of underwater and above—the photogrammetric 3d models by independent models adjustment. International archives of the photogrammetry, remote sensing and spatial information sciences, XL-5/W5:143–151. https://doi.org/10.5194/isprsarchives-XL-5-W5-143-2015

Moisan E, Charbonnier P, Foucher P, Grussenmeyer P, Guillemin S, Koehl M (2015) Adjustment of sonar and laser acquisition data for building the 3D reference model of a canal tunnel. In: Paper presented at the Sensors

Monroy C (2010) A digital library approach to the reconstruction of ancient sunken ships. PhD dissertation, Texas A&M University

Monroy C, Parks N, Furuta R, Castro F (2006) The nautical archaeology digital library. In: Proceedings of the 10th European conference on research and advanced technology for digital libraries, Alicante, Spain

Monroy C, Furuta R, Castro F (2007) A multilingual approach to technical manuscripts: 16th- and 17th-century Portuguese shipbuilding treatises. In: Proceedings of the 7th ACM/IEEE-CS joint conference on Digital libraries, Vancouver, BC, Canada, 8–23 June 2007, pp 413–414. https://doi.org/10.1145/1255175.1255258

Monroy C, Furuta R, Castro F (2008) Design of a computer-based frame to store, manage, and divulge information from underwater archaeological excavations: the Pepper wreck case. In: Paper presented at the Society for Historical Archaeology annual meeting, Sacramento

Monroy C, Furuta R, Castro F (2009) Ask not what your text can do for you. Ask what you can do for your text (a dictionary's perspective). Digital Humanities, pp 344–347

Monroy C, Furuta R, Castro F (2011) Synthesizing and storing maritime archaeological data for assisting in ship reconstruction. In: Ford CB, Alexis (eds) Oxford handbook of maritime archaeology. Oxford University Press (Pub.), pp 327–346

Nawaf MM, Hijazi B, Merad D, Drap P (2016) Guided underwater survey using semi-global visual odometry. In: Paper presented at the COMPIT 15th international conference on computer applications and information technology in the maritime industries, Lecce, pp 287–301

Nawaf MM, Drap P, Royer J-P, Saccone M, Merad D (2017) Towards guided underwater survey using light visual odometry. Int Arch Photogramm, Remote Sens Spat Inf Sci Arch XLII-2/W3:527–533. https://doi.org/10.5194/isprs-archives-XLII-2-W3-527-2017

Ng R, Levoy M, Brédif M, Duval G, Horowitz M, Hanrahan P (2005) Light field photography with a hand-held plenoptic camera. Stanford University Computer Science Tech Report CSTR 2005-02:1–10

Niang C, Marinica C, Markhoff B, Leboucher E, Malavergne O, Bouiller L, Darrieumerlou C, Francois Laissus F (2017) Supporting semantic interoperability in conservation-restoration domain: the PARCOURS project. J Comput Cult Herit 10(3):1–20. Special Issue on Digital Infrastructure for Cultural Heritage, Part 2. https://doi.org/10.1145/3097571

Niccolucci F (2016) Documenting archaeological science with CIDOC CRM. Int J Digit Libr 18(3):223–231. https://doi.org/10.1007/s00799-016-0199-x

Niccolucci F, Hermon S (2016) Expressing reliability with CIDOC CRM. Int J Digit Libr 18(4):281–287. https://doi.org/10.1007/s00799-016-0195-1

Noardo F (2017) A spatial ontology for architectural heritage information. In: Grueau C, Laurini R, Rocha JG (eds) Geographical information systems theory, applications and management: second international conference, GISTAM 2016, Rome, Italy, 26–27 April 2016, Revised Selected Papers. Communications in computer and information science book series, vol 741. Springer, pp 143–163. https://doi.org/10.1007/978-3-319-62618-5

O'Byrne M, Pakrashi V, Schoefs F, Ghosh B (2018) A stereo-matching technique for recovering 3D information from underwater inspection imagery. Comput Aided Civ Inf Eng 33(3):193–208. https://doi.org/10.1111/mice.12307

O'Connor MJ, Das A (2006) A mechanism to define and execute SWRL Built-ins in Protégé-OWL

Ozog P, Troni G, Kaess M, Eustice RM, Johnson-Roberson M (2015) Building 3D mosaics from an autonomous underwater vehicle, doppler velocity log, and 2D imaging sonar. In: 2015 IEEE international conference on robotics and automation (ICRA), Seattle, WA, 26–30 May 2015, pp 1137–1143. https://doi.org/10.1109/ICRA.2015.7139334

Pasquet J, Demesticha S, Skarlatos D, Merad D, Drap P (2017) Amphora detection based on a gradient weighted error in a convolution neuronal network. In: Paper presented at the IMEKO international conference on Metrology for Archaeology and Cultural Heritage, Lecce, Italy, 23–25 October 2017, pp 691–695

Perwass C, Wietzke L (2012) Single lens 3D-camera with extended depth-of-field. In: Paper presented at the Human Vision and Electronic Imaging

Pitzalis D, Pillay R (2009) Il sistema IIPImage: un nuovo concetto di esplorazione di immagini ad alta risoluzione. Archeol Calcolatori Suppl 2:239–244

Pizarro O, Friedman A, Bryson M, Williams SB, Madin J (2017) A simple, fast, and repeatable survey method for underwater visual 3D benthic mapping and monitoring. Ecol Evol 7(6):1770–1782. https://doi.org/10.1002/ece3.2701

Raskar R (2001) Hardware support for non-photorealistic rendering. ACM Trans Graph/Proc ACM SIGGRAPH 2001:41–47. https://doi.org/10.1145/383507.383525

Raytrix (2017) Raytrix, 3D light field camera technology. https://www.raytrix.de/. Accessed 17 Aug 2017

Roman C, Inglis G, Rutter J (2010) Application of structured light imaging for high resolution mapping of underwater archaeological sites. In: Paper presented at the OCEANS 2010 IEEE, Sydney, 24–27 May 2010, pp 1–9

Roussou M, Drettakis G (2003) Photorealism and non-photorealism in virtual heritage representation. In: Proceedings of the 4th international conference on virtual reality, archaeology and intelligent cultural heritage, Brighton

Roy S, Yan MF (2012) Method and system for creating owl ontology from java. In: Limited infosys technologies editor, Google Patents

Russakovsky O, Deng J, Su H, Krause J, Satheesh S, Ma S, Huang Z, Karpathy A, Khosla A, Bernstein M, Berg AC, Fei-Fei L (2014) ImageNet large scale visual recognition challenge. arXiv:1409.0575 [cs]

Secci M (2017) Survey and recording technologies in Italian underwater cultural heritage: research and public access within the framework of the 2001 UNESCO Convention. J Marit Archaeol 12:109–123. https://doi.org/10.1007/s11457-017-9174-y

Seinturier J (2007) Fusion de connaissances: applications aux relevés photogrammétriques de fouilles archéologiques sous-marines. PhD dissertation, Université du Sud Toulon Var

Sérayet M (2010) Raisonnement à partir d'information structurées et hiérarchisées: application à l'information archéologique. PhD dissertation, Université de la Méditerranée

Sérayet M, Drap P, Papini O (2009) Encoding the revision of partially preordered information in answer set programming. In: Sossai C, Chemello G (eds) Symbolic and quantitative approaches to reasoning with uncertainty: 10th European Conference, ECSQARU 2009, Verona, Italy, 1–3 July 2009, ECSQARU 2009. Lecture notes in computer science, vol 5590. Springer, Berlin, pp 421–433. https://doi.org/10.1007/978-3-642-02906-6_37

Shelhamer E, Long J, Darrell T (2016) Fully convolutional networks for semantic segmentation. arXiv:1605.06211 [cs]

Shortis MR (2015) Calibration techniques for accurate measurements by underwater camera systems. Sensors 15(12):30810–30826. https://doi.org/10.3390/s151229831

Shortis MR, Harvey ES, Abdo DA (2009) A review of underwater stereo-image measurement for marine biology and ecology applications. In: Gibson RN, Atkinson RJA, Gordon JDM (eds) Oceanography and marine biology: an annual review, vol 47. CRC Press, Boca Raton, pp 257–292

Simonyan K, Zisserman A (2014) Very deep convolutional networks for large-scale image recognition. arXiv:1409.1556 [cs]

Sousa MC, Foster K, Wyvill B, Samavati F (2003) Precise ink drawing of 3D models. Comput Graph Forum 22:369–379

Steffy JR (1994) Wooden ship building and the interpretation of shipwrecks. Texas A&M University Press, College Station

Stevenson G, Dobson S (2011) Sapphire: generating Java Runtime artefacts from OWL ontologies. In: Salinesi C, Pastor O (eds) Proceedings of the advanced information systems engineering workshops: CAiSE 2011 International Workshops, London, UK, 20–24 June 2011. Springer, Berlin, pp 425–436

Szegedy C, Liu W, Jia Y, Sermanet P, Reed S, Anguelov D, Erhan D, Vanhoucke V, Rabinovich A (2014) Going deeper with convolutions. arXiv:1409.4842 [cs]

Tao L, Renju L, Hongbin Z (2009) 3D line drawing for archaeological illustration. In: Paper presented at the computer vision workshops (ICCV Workshops), 2009 IEEE 12th international conference, 27 September–4 October 2009, pp 907–914

Telem G, Filin S (2010) Photogrammetric modeling of underwater environments. ISPRS J Photogramm Remote Sens 65(5):433–444

UNESCO (2001) Convention on the protection of the underwater cultural heritage http://www.unesco.org/new/en/culture/themes/underwater-cultural-heritage/. Accessed 6 Sept 2018

Vandenbussche P-Y, Atemezing GA, Poveda-Villalón M, Vatant B (2017) Linked Open Vocabularies (LOV): a gateway to reusable semantic vocabularies on the web. Semant Web 8:437–452

Xie J, Hertzmann A, Li W, Winnemeller H (2014) PortraitSketch: face sketching assistance for novices. In: Proceedings of the 27th annual ACM symposium on user interface software and technology, Honolulu, Hawaii, 5–8 October 2014. ACM, New York, pp 407–417. https://doi.org/10.1145/2642918.2647399

Xueming P, Beckman P, Havemann S, Tzompanaki K, Doerr M, Fellner DW (2010) A distributed object repository for cultural heritage. In: Paper presented at the VAST 2010, Paris, September 2010, pp 105–114

Yamafune K, Torres R, Castro F (2016) Multi-Image photogrammetry to record and reconstruct underwater shipwreck sites. J Archaeol Method Theory 24(3):703–725. https://doi.org/10.1007/s10816-016-9283-1

Yang X, Liu Z-Y, Li C, Wang J-J, Qiao H (2017a) Robust underwater image stitching based on graph matching. In: Sun Y, Lu H, Zhang L, Yang J, Huang H (eds) Intelligence science and big data engineering. IScIDE 2017, Lecture notes in computer science, vol 10559. Springer, Cham, pp 521–529. https://doi.org/10.1007/978-3-319-67777-4_46

Yang X, Liu Z-Y, Qiao H, Song Y-B, Ren S-N, Ji D-X, Zheng S-W (2017b) Underwater image matching by incorporating structural constraints. Int J Adv Robot Syst 14(6):1–10. https://doi.org/10.1177/1729881417738100

Zapata-Ramírez PA, Huete-Stauffer C, Scaradozzi D, Marconi M, Cerrano C (2016) Testing methods to support management decisions in coralligenous and cave environments: a case study at Portofino MPA. Mar Environ Res 118:45–56

Quantifying Depth of Burial and Composition of Shallow Buried Archaeological Material: Integrated Sub-bottom Profiling and 3D Survey Approaches

10 Quantifying Depth of Burial and Composition of Shallow Buried Archaeological Material: Integrated Sub-bottom Profiling and 3D Survey Approaches

Trevor Winton

Abstract

This chapter presents proof of concept results from a program of in situ experimental and shipwreck survey measurements using non-linear (parametric) sub-bottom profiler (SBP) acoustic technology. Currently adopted acoustic methods have practical limitations for in situ management purposes for underwater sites with buried archaeological material. Sidescan and multibeam sensors do not quantify material buried below the seabed; linear SBP surveys are challenging to operate in very shallow water and have difficulties with respect to interpretation in the top 30 cm of the seabed; and confidence estimates for parametric SBP depth of burial measurements have yet to be published. The prime purposes of this research, consequently, are: to quantify shallow buried archaeological sites in 3D with confidence estimates, by measuring the depth of sediment cover, thickness and lateral extent of buried archaeological material; and to investigate relationships between acoustic waveform parameters and the type and degradation condition of that buried material. This improved measurement and interpretation capability, when combined with the other geophysical search tools such as multibeam echo sounders and magnetometers, will also aid in the assessment of the archaeological research potential of underwater sites.

Keywords

Marine geophysics · In situ management · Parametric sonar · Underwater cultural heritage

10.1 Introduction

The trend in maritime archaeology to favour in situ preservation over more destructive methods, has led to an increased need to measure burial depths and composition of shallow-buried material using non-invasive methods. This has been discussed widely within the archaeological community (Bergstrand and Godfrey 2007; Gregory 2007; Richards 2011a; Richards et al. 2014; Shefi and Veth 2015) and continues to evolve pragmatically as technological improvements result in the increased discovery of UCH. Simultaneously, excavation, conservation, storage, and display costs continue to increase. Furthermore, there is usually a significant time gap between discovery and potential site excavation, thus many sites awaiting investigation may require protection in the interim period in order to maintain the quality of the archaeological material (Manders et al. 2008). The protection of UCH through in situ preservation as a first option has also been consistently emphasized for preserving submerged and waterlogged cultural heritage for future generations, and politically galvanized by UNESCO in Article 2 of the 2001 *Convention on the Protection of the Underwater Cultural Heritage* (UNESCO 2001).

The ultimate in situ preservation goal for many heritage managers is the ability to maintain or create a stable, protective environment (Manders et al. 2008; Ortmann et al. 2010) to conserve as much as possible of the currently remaining archaeological material. Decisions on how to achieve this goal need to be based on the understanding of the true site extent, the types of material present and their state of degradation, and the potential exposure of archaeological material on a site (both which is visible on or above the seabed, and which may lie beneath the seabed—and in many situations, is neither visible nor known) (Gregory and Matthiesen 2012; Richards et al. 2014; Wheeler 2002; Winton 2015).

Burial depth, and the continuity of this sediment coverage through time, is the single most important site-specific influence on the rate of degradation of shipwreck material

T. Winton (✉)
Maritime Archaeology Program, Flinders University, Adelaide, SA, Australia
e-mail: wint0062@flinders.edu.au

J. K. McCarthy et al. (eds.), *3D Recording and Interpretation for Maritime Archaeology*, Coastal Research Library 31,
https://doi.org/10.1007/978-3-030-03635-5_10

(Stewart 1999; Winton 2015). Studies have shown that the extent of biological degradation of organic materials decreases considerably with burial depths greater than 50 cm, where anaerobic conditions limits the effects of marine borers, fungi and most bacteria, and would be suitable for the preservation of timbers (Björdal et al. 2000; Gregory 1998; Richards 2011b; Shefi and Veth 2015). If electrically isolated metal components are buried in anaerobic sediments, then aerobic corrosion mechanisms are also avoided. By contrast, materials with shallower depths of sediment coverage are subject to much higher aerobic microbiological and chemical degradation rates and can also be exposed to combined physical and biological degradation processes if the shallow protective sediment layer is eroded, or purposefully excavated.

Effective in situ management of maritime archaeological sites consequently requires a priori 3D information to identify: if archaeological material is buried below the seabed (or riverbed or lakebed); its lateral extent and depth of burial, especially if this depth is less than 50 cm; the major material types both exposed and potentially buried; and their state of deterioration. Currently this information is gained from the outcomes of long-term or episodic seabed erosion, or purposeful excavation. As both expose previously buried anaerobic material to aerobic conditions, however, a proven non-invasive method to measure and identify shallow-buried material is desired.

10.2 Non-invasive Geophysical Measurements

A variety of marine seismic reflection techniques, including single beam and Multibeam echo sounders (SBES, MBES), sidescan sonar (SSS), synthetic aperture sonar (SAS) and sub-bottom profilers (SBP) have been progressively used since the early 1950s to investigate a range of submerged geomorphological and archaeological sites (Bjørnø 2017b, c; Dix et al. 2008; Quinn 2012). SBES, MBES, SSS and SAS devices have traditionally been used for bathymetric mapping and visualizing the seabed and objects on, or above, the seabed. Single beam acoustic ground discrimination systems (AGDS) are based on SBES and used to classify seabed type and map submerged archaeological materials lying on the seabed (Lawrence and Bates 2001). Sub-bottom imaging is carried out with high-frequency seismic profiling systems. Significant advances in the development of SBP technology, including analysis and imaging software (Bull et al. 2005; Missiaen et al. 2005; Müller et al. 2005; Plets et al. 2009; Wunderlich and Müller 2003) led in situ managers (Manders et al. 2008) to the view that 'SBP instruments provide a non-intrusive view of material below the seabed and that sites could then be managed with material still in their protective burial environment.'

Nearly all the published SBP surveys of underwater shipwreck sites have used Chirp SBPs which were first developed in 1981. From 2001 this system was optimized and enhanced as 2D systems, and subsequently 3D, at the University of Southampton. Descriptions of these applications is given by Arnott et al. (2005), Cvikel et al. (2017), Dix et al. (2008), Forrest et al. (2005); Grøn and Boldreel (2013), Grøn et al. (2015), Lafferty et al. (2006), Plets et al. (2005, 2008, 2009), Quinn et al. (1997a, b, c, 1998a, b), and Vardy et al. (2008). The earlier SBP surveys were effectively used to help locate or map site extent, and qualitatively improve understanding of site formation processes on the *Invincible*, *Mary Rose*, *La Surveillante* and *Pandora* shipwreck sites (Forrest et al. 2005; Quinn et al. 1997a, b, c, 1998a, b, 2002). Following improvements in data processing and data interpretation processes, quantitative analysis of shipwreck sites became possible whereby derived reflection coefficients were used to predict the degradation state of the buried ship's timbers (Arnott et al. 2005; Bull et al. 1998; Quinn et al. 1997b) as well as the 3D shape of the buried ship remains (Plets et al. 2008, 2009).

Despite these successful applications, there are difficulties for heritage managers to use (linear) Chirp SBP systems for in situ management purposes, especially in shallow (<5 m water depths) owing to vessel-induced bubble turbulence, restricted acoustic geometry of the system, wide acoustic beam patterns and inability to discriminate in the top 30 cm. While Chirp systems can be pulled by divers to avoid boat noise interference (Plets et al. 2007, 2009) and data processing techniques can be used to correct for geometry and optimize the processing of the collected data, their field operability remains difficult (Bjørnø 2017c). Chirp SBPs use wide acoustic beam patterns (20–30°) which limits horizontal resolution. Although instrument technical improvements have progressively improved resolution in shallow water depths from approximately 2–3 m (Plets et al. 2008) to 0.4–0.7 m resolution (Plets et al. 2009) and 'decimeter resolution' (Gutowski et al. 2015), finer resolution to 0.25 cm (horizontal) and around 4.5 cm (vertical) can only be achieved in 3D by expert use of post-processing software (Dix, personal communication, 22 December 2017). Bull et al. (1998) report that small lateral variations in the very-near surface sediments have a profound effect on Chirp acoustic returns in the top 30 cm of the seabed, resulting in high uncertainty in very shallow sub-bottom measurement, unfortunately at precisely in the depth range of maximum importance.

SBPs based on nonlinear acoustic phenomena have advanced from early experimental acoustic arrays developed in the mid-1980s, and due to their inherently different acoustic wave characteristics, have stronger in situ management potential for sub-bottom profiling of shallow buried archaeological material. Nonlinear (parametric) SBPs produce low-frequency pulses (secondary difference-frequencies) as an outcome of the interaction between two simultaneously generated high sound pressure, higher-frequency (primary) sound waves transmitted at slightly different frequencies. Advantageous (seabed penetrating) pulse qualities includes narrow (±2°) beam width with consequential high horizontal resolution; very low side-lobe levels, which reduce clutter and signal-to-noise ratios and enhance the separation of backscattering from seafloor and sub-surface reflectors, leading to improved ability to detect very shallow and acoustically weak reflectors. They also include high pulse repetition rates allowing more 'hits' per target and higher boat survey speeds and a smaller combined transmitter/receiver array which significantly improves field operability as it can be vertically mounted from a vessel, rather than towed in an array (Bjørnø 2017a, c; Caiti et al. 1999; Wunderlich et al. 2005a, b).

Commercially available parametric SBPS include Atlas Hydrographic GmbH (*Parasound*), Germany; Kongsberg Defence Systems (*TOPAS* PS systems), Norway; and Innomar Technologie GmbH, Germany (SES-2000) (Bjørnø 2017a). Both the TOPAS PS and SES-2000 systems are available in different models optimized to operate in different water depths from very shallow water to full ocean depth, and while the TOPAS is best known in the parametric SBP range, it is less popular due to its high cost (Kozaczka et al. 2013).

Missiaen et al. (2008) used a parametric SBP in a comparative assessment of different shallow geophysical methods and, as this technique demonstrated the highest measurement resolution, it was then subsequently used to record complex geomorphological structures in a shallow tidal estuary. Parametric SBPs have also been used to detect surface and shallow buried steel pipelines, cables, spheres and steel canisters (Kozaczka et al. 2013; Vasudevan et al. 2006; von Deimling et al. 2016). Their application to maritime archaeological sites has also been demonstrated through an experimental deployment of an SBP on a ROV in deep water in the eastern Mediterranean to map two Phoenician ships (Mindell and Bingham 2001) as well as a trial to identify a narrow 0.2 m diameter wooden post and other embedded wooden archaeological objects in the Baltic (Müller and Wunderlich 2003; Wunderlich et al. 2005b). Other examples include case studies of the identification of possible remains of Roman dykes and human activities including salt/peat exploitation in prehistoric tidal gullies (Ostend, Belgium) and of an exposed shipwreck on the Buiten Ratel sandbank (Belgium) (Missiaen 2010); and seismic imaging of the scattered remains of the Dutch East Indiaman *'t Vliegent Hart* (Missiaen et al. 2012). More recently Innomar Technologie GmbH introduced a multi-transducer sub-bottom profiler to capture very high data density in shallow waters (SES-2000 *quattro*) which can be subsequently viewed in 3D using gridding and visualization software. Missiaen et al. (2017) conducted 3D seismic surveys using the SES-2000 *quattro* across shallow intertidal areas at the coastal site of Ostend-Raversijde, Belgium. This complemented the previous 2D parametric surveys on this site and provided an image of the complex peat exploitation patterns, the features of which matched with old aerial photographs.

With the exception of the Kozaczka et al. (2013) buried canister trial and the post-measurement dredging and recovery of measured cylinders and poles (Gutowski et al. 2015; Vardy et al. 2008), there has been no reported quantitative verification of linear or non-linear SBP performance (e.g., accuracy of depth of cover estimates, Type I and Type II errors associated with the correct identification/interpretation of a buried reflector, assessment of different reflector material types) through testing against previously surveyed and reburied shipwreck materials. Hence, there is clearly a need for such a trial to assess SBP performance, and particularly for the reported more favorable performance characteristics of the parametric SBP, in mapping shallow buried archaeological material. The following sections describe a research plan and field survey (proof of concept) results to assess the performance and data interpretation capabilities of a parametric SBP for in situ management purposes, and using complimentary tools, to identify and characterize shallow buried archaeological material.

10.3 Parametric SBP Surveys

This research quantifies the accuracy and variability associated with the non-invasive parametric SBP measurements of shallow depths of sediment burial, as well as investigating the potential relationships between acoustic wave parameters and types and condition of a variety of buried material. This was achieved through both in situ experimental burial and comparative in situ wreck-site surveys.

10.3.1 In Situ Experimental Burial Survey

The in situ experimental component involved shallow burial of timber beams ('sleepers') at different burial depths (10, 30 and 50 cm) with different grain orientations and in different sediment types. Following a period of reconsolidation of the sediments (after underwater excavation and back-filling the holes around and over the buried timbers), these sleepers were measured with an Innomar parametric SES-2000 compact SBP. The experimental parameters, listed in Table 10.1, were chosen to be representative of equipment measurement capabilities and in situ conditions on a range of wreck sites. Mid–coarse-grained sands and fine silts-muds represent the typical endpoints in both the range of sediment environments

Table 10.1 In situ experimental parameters

Parameter	Included within in situ experimental burial survey
Sediment environment	Mid-coarse grained sands and
	Fine grained silty sands[a]
Timber types	European Oak (*Quercus robur*)
	Pine (*Pinus radiata*)
	Australian hardwood—Jarrah (*Eucalyptus marginate*)[a]
Timber sample size (nominal)	50 × 12.5 × 12.5 cm (sleepers)
	12.5 × 12.5 × 12.5 cm (blocks)[b]
Burial depths/depths of sediment cover	10 cm, 30 cm, 50 cm
Replication	Triplicates
Grain orientation	Longitudinal grain horizontal and vertical
Timber stacking	Single, 10 + 30 cm, 10 + 30 + 50 cm

[a]Not yet installed

[b]For ease of removal, blocks with depths of burial of 30 cm and 50 cm were cut with a 45° taper on top. Blocks with 10 cm depth of burial remained with flat top, otherwise taper would protrude above seabed surface

in which maritime archaeological material lies buried, and the sediment penetration range for the SBP. To date, sleepers have been buried adjacent to the *James Matthews* (1841) wreck site on the northern side of Woodman Point, approximately 7 km south of Fremantle, Western Australia (Fig. 10.1). Here water depths range 1.5–2.8 m and the calcareous sediments have been characterized as medium sands with some coarse-grained skeletal material (Richards et al. 2009). European oak and pine represent timbers commonly used in European shipbuilding (Zisi 2016), and Australian hardwood was commonly used in Australian colonial-period shipbuilding (O'Reilly 2007; Pemberton 1979; Staniforth and Shefi 2014). The cross-sectional dimensions of the timber samples were based on the theoretical measurement resolution of the SBP (<5 cm vertical and 5–10 cm horizontal, respectively, in water depths 1.5–2.8 m and burial depths 0.1–0.5 m (Bergersen 2016 pers. comm. 29 July). Timber grain orientation reflects the different acoustic properties of timber (Arnott et al. 2005) and timber orientations likely to be found on shipwreck sites (Zisi 2016).

The replicated experimental design permits statistical analysis of the accuracy and variability associated with SBP measurements of timber sleepers buried at three shallow depths (10 cm, 30 cm and 50 cm), the ability to characterize any acoustic 'signatures' from different timber species (currently pine and oak) and to assess whether-or-not timber grain orientation is a significant variable in in situ acoustic measurements. Stacking of the timbers replicates what might be found on some wreck sites and may reveal the ability and acoustic strength of the SBP system to measure multiple layers of timbers. It is intended to leave the buried sleepers in place for at least 5 years, enabling repeated SBP measurements on an annual basis to determine if the acoustic 'signatures' change through time due to microbial degradation processes. Representative numbers of the small blocks will be removed at the time of each of these subsequent annual SBP surveys, and following laboratory analysis to measure percent moisture content, this data will be used to assess comparative changes in timber density, at different burial depths, through time (Table 10.2).

Air-dried European Oak (*Quercus robur*), originally sourced from Poland, was provided to size by the Western Australian Museum (WAM) Conservation Department from left over timber used to build the 1999 replica Dutch ship *Duyfken*. Freshly sawn green pine (*Pinus radiata*) was purchased from a sawmill in the south-western of Western Australia, then cut to length. All sleepers were fabricated as a single beam, except where two and three beams were respectively vertically stacked with 7.5 cm gaps in between the timber beams (Fig. 10.2). Three pine sleepers were each cut into 16.5 cm lengths, each section then rotated through 90° such that the end grain was vertical, then these were drilled and pinned using pine dowels and PVA timber glue to reform the 50 cm long vertical grain sleepers. Endplates for the sleepers and blocks were cut from inert 12 mm PVC sheeting, with dowel holes and slots drilled and cut to measure. These endplates were securely attached by driving two 25 mm diameter PVC dowels into slightly undersized drilled holes in the ends of each timber beam and block. The varying length of each PVC endplate, from the upper surface of the timber sleeper to the underside of the pre-cut slot, enabled accurate placement below the seabed and subsequent measurement of actual depth of sediment cover over the top face of the timber sleepers and blocks. Unique labels for each sleeper and block were engraved and blackened into the PVC endplates using a soldering iron and engraved colour-coded PVC cattle tags were also attached via nylon cable ties.

Burial of the sleepers adjacent to the *James Matthews* wreck site was achieved through diver-operated water dredging. Together with staff, the WAM Departments of Archaeology and Conservation provided their new dive and research vessel *Dirk Hartog* as surface support, including surface supplied air, SCUBA, water dredge and dive platform. Burial of the 38 pine and oak sleepers and blocks was accomplished in approximately 30 h of dive time, during 16 dives, over a 5-day period in mid-February 2017. An additional 18 dives were needed for site preparation, layout, recording and trialing prior to and following the dredging activities. Two 30-m long pre-installed parallel measuring tapes, set 50 cm apart and tied off to permanently installed star pickets, guided the positioning of the sleepers in a straight line (the sleepers were buried at right angles to the tapes with each endplate touching one of the tapes). Blocks were installed in a grid with locations identified using two perpendicular tapes. Each sleeper/block hole was dredged by the diver operated suction

Fig. 10.1 Location map of the *James Matthews* wreck-site (Google Maps)

head until the required burial depth was achieved, with sand stockpiled on the side. Each sleeper/block was randomly selected from the vessel, additional temporary weights placed on the timber to overcome its natural buoyancy, swum to the seabed and a long flat plank inserted through the endplate slot/s. The sleeper/block was then placed in the dredged hole and if the horizontal plank rested on the natural seabed surface at both ends of the excavated hole, then the correct burial depth had been achieved. In this situation the dredge head was reversed, and the stockpiled sand dredged back into the

Table 10.2 In situ experimental design

Timber type	Burial depth	Grain orientation	Stacking	Sleeper/block
Pine	3 × 10 cm, 3 × 30 cm, 3 × 50 cm	Horizontal	Single	Sleeper
	3 × 30 cm	Vertical	Single	Sleeper
	1 × (10 + 30 cm)	Horizontal	Stacked	Sleeper
	1 × (10 + 30 + 50 cm)			
	6 × 10 cm, 6 × 30 cm, 6 × 50 cm	Vertical	Single	Block
Oak	3 × 30 cm	Horizontal	Single	Sleeper
	3 × 30 cm	Vertical	Single	Block
Jarrah[a]	3 × 30 cm	Horizontal	Single	Sleeper
	3 × 30 cm	Vertical	Single	Block

[a]Not yet installed

hole, burying the sleeper/block (Fig. 10.3). Temporary weights were progressively removed during the backfilling. Following completion of all dredge backfilling a weighted horticultural rake was used by an over-weighted diver to smooth seabed irregularities around each sleeper/block and to restore seabed levels to the underside of each endplate slot (Fig. 10.4). Table 10.3 lists the order and separation distances of all buried sleepers.

The Innomar SES-2000 *compact* SBP was selected for the sub-bottom measurements as it operates in very shallow coastal waters, from around 50 cm to 400 m water depths (Innomar 2018). It is designed to be pole mounted on a survey or autonomous vessel, forward of propeller wash to avoid acoustic noise; has a sampling ping rate of up to 40 pings/second and data acquisition rate of 70 kHz, allowing for high survey vessel speeds of 2 m/s; and a very narrow transmit beam width (−3 dB) of ±2° which in shallow water depths(<2.8 m) and for shallow (<50 cm) buried timbers results in an acoustic foot of <10 cm.

Field data collection occurred on 7 and 8 June 2017 using the Innomar SES-2000 *compact* SBP and associated SESWIN software, together with a Trimble POS MV Surfmaster GNSS G2 real time satellite positioning antenna and heave correction sensor (IMU) (Applanix 2018) mounted on the WAM's research vessel *Dirk Hartog* (Fig. 10.5). The SBP transducer head was positioned 50 cm below sea surface level. Offsets from each sensor mounting position relative to the center of the SBP transducer were recorded and included into the positioning calculations. Fugro Satellite Positioning Pty. Ltd. supplied Marinestar positioning solution which enabled real time position tracking to approximately 15–20 cm in both the horizontal (x, y) and vertical (z) directions, and with post-processing, 2 cm accuracy in the horizontal and vertical position. Surface marker buoys were tethered at each end and midway along the 30 m long line of sleepers, and multiple measurement runs were made with the coxswain guided by the surface buoys.

10.3.2 *James Matthews* Comparative In Situ Surveys

The *James Matthews* was a copper-sheathed wooden snow brig of 107 tons, constructed with iron deck knees and assembled with copper and iron fasteners and wooden treenails. The significance and history of this ship, which was wrecked in 1841 on the northern side of Woodman Point, Western Australia (Fig. 10.1) is described by Henderson (2009). The wreck covers an area 26 × 7 m and is mostly buried to a depth of 1.5–2.0 m in medium grained (phi = 1.5) calcareous sand with the starboard side preserved to the bulwarks by the sand cover (Richards 2001; Richards et al. 2009).

Following its discovery in July 1973, WAM undertook multiple maritime archaeological excavations on the *James Matthews* site between 1974 and 1977, and during the extensive 1975–1976 excavation, surveyed and recorded the entire remaining ship's structure using a 3D recording grid frame and plumblines (Baker and Henderson 1979; Henderson 1977). The relative x, y, and z positions of almost 5000 points of interest were recorded by hand on underwater plastic film, from which a 2D scale plan was drawn. In 2000 a conservation survey was also undertaken by WAM (Richards 2001) and identified timber type and its degradation condition at six test trench locations. The original 1975–1976 survey data sheets, together with the conservation survey data, have recently been extracted from WAM's archives, digitized, converted into a point cloud of data and used to digitally reconstruct a 3D *AutoCAD* model of the buried shipwreck remains (Fig. 10.6).

The conversion into 3D digital format of the in situ remains of degraded and non-degraded timbers of the keel/keelson, ribs and planking (beech, white oak and elm, respectively), the concreted iron deck knees and iron bars, slate, timber (pine) cargo and stone ballast, provides an opportune baseline for comparative analyses with SBP data. In June 2017, immediately following the SBP survey of the buried sleepers adjacent

Fig. 10.2 Single and multiple stacked timber sleepers, each 50 × 12.5 × 12.5 cm in dimension, showing assembly for 50 cm burial depth (right) measured from top face of timber surface to underside of slot

Fig. 10.3 Author burying timber sleepers (J. Carpenter, WAM)

to this wreck site, a total of 89 SBP transects (77 east-west and 12 north-south) were run with an average 1-m spacing across the *James Matthews* wreck-site (Fig. 10.7). These transects were collected using WAM's *Dirk Hartog*, traveling at an average speed of 2 m/s, with the same onboard SBP and real time positioning and motion correction equipment that was used to record the nearby buried sleepers.

Fig. 10.4 Buried sleepers, 2 weeks after burial, showing endplates and slot (covered in growth) standing 50 cm apart above seabed

Table 10.3 Sleeper burial details

Sleeper number	Sleeper ID	Nominal burial depth (cm)	Distance from northern end (m)	Actual burial depth prior to SBP measurements (cm)
	Star picket (north)		Tape tied off at 0.5	
1	Pup30	30	4.2	28
2	P30	30	5.2	29
3	P30	30	6.55	27
4	P10	10	7.8	7
5	P50	50	8.85	41
6	P30	30	9.65	29
7	P10	10	10.76	10
8	O30	30	11.66	27
9	Pup30	30	12.76	28
10	O30	30	14.35	30
11	Pup30	30	15.10	30
12	P50	50	16.83	49
13	P50	50	17.75	45
14	O30	30	19.10	29
15	P10/30	10/30	20.22	10
16	P10/30/50	10/30/50	22.05	9
17	P10	10	23.73	10
	Star picket (south)		Tape tied off at 29.9	

Legend: P pine, O oak, Pup pine with vertical grain; 10/30/50-nominal burial depth (cm)

10.4 Results and Discussion

10.4.1 In Situ Experimental Burial Survey

Figure 10.8 depicts a typical annotated SBP echo plot (Innomar ISE2 software) approximately 20 m in length showing the collective acoustic traces recorded along the line of buried sleepers (A). These data were collected via vessel mounted transducer/receiver, and despite best efforts to maintain a central track within an accuracy of ±25 cm along the full 30 m length of the buried sleepers, this southwest–northeast run mapped only the sleepers from approximately midway to the northern end, as identified by 'Besser Blocks' placed on the seabed at these two locations. Image (A) shows the raw echo plot without any post-processing of the acoustic data. The locations and depths of the buried sleepers are shown by the black hyperbolae (horseshoe shaped) lines. The lower image (B) shows the raw acoustic wave amplitude and phase of the individual acoustic Trace 271 through the water column and seabed (vertical scale exaggerated and plotted in Excel for greater visual clarity).

The SBP measures the two-way travel time taken for an emitted acoustic wave generated from the transmitter to travel through the water column and reflect from the seabed surface, and through the water column and sediment column to reflect

Fig. 10.5 WAM research vessel *Dirk Hartog* showing mounting locations of SES-2000 *compact* SBP transducer, Trimble GNSS antenna and applanix IMU sensors

from any buried reflector, and then return to the receiver head. Using a default setting for the speed of sound in water (1500 m/s) the respective depths below the transmitter/receiver (transducer) head are simply calculated by multiplying the sound velocity by 0.5 × two-way travel time. For Trace 271 in Fig. 10.8, these correspond to 2.61 and 3.01 m, respectively, resulting in a calculated 40.0 cm depth of burial (DoB). At the time of measurement, seawater temperature was 19 °C and salinity 35 ppt which would result in an actual speed of sound in seawater of 1517 m/s (Lovett 1978). A sensitivity assessment of the variability in assumed/actual speed of sound in the water column indicated that the estimated depths of seabed and reflector surface buried up to 50 cm below the seabed would increase by 2–3 cm. Despite these increases, the variability of the difference between the two simultaneous depth estimates (i.e. DoB) ranges only from 0 to 1 cm for reflector surfaces buried 10–50 cm. However, the assumed speed of sound through the sediment column has a greater effect on DoB estimates. Based on Richards et al. (2009) characterization of the sediments over the adjacent *James Matthews* wreck site, a sediment velocity correction factor of 1.195 can be applied to the speed of sound in the water column. This correction factor was derived by Robb et al. (2005) from in situ measurements of the speed of sound in intertidal medium grained sands. An increase in the speed of sound in the sediment from 1500 to 1813 m/s results in an increase in DoB of approximately 2–8.3 cm for sleepers buried 10–50 cm.

Depths of burial for all identified buried reflectors were initially identified by the locations of hyperbolae from unprocessed SBP data in echo plot Runs 025025.RAW and 024600.RAW (Fig. 10.9) and their positions matched to the known locations of sleepers. At each of these locations the depth and amplitude values of the central and two adjacent traces on either side (at 2 m/s vessel speed and 40 pings/s sampling frequency, a 12.5 cm wide sleeper would be theoretically insonified (hit) by three acoustic waves) were exported from the ISE software, tabulated into an Excel spreadsheet, then plotted (see Fig. 10.8b for one such trace). The seabed surface was determined as the depth corresponding to the zero intercept between the two maxima seabed amplitudes, and the depth of the upper surface of

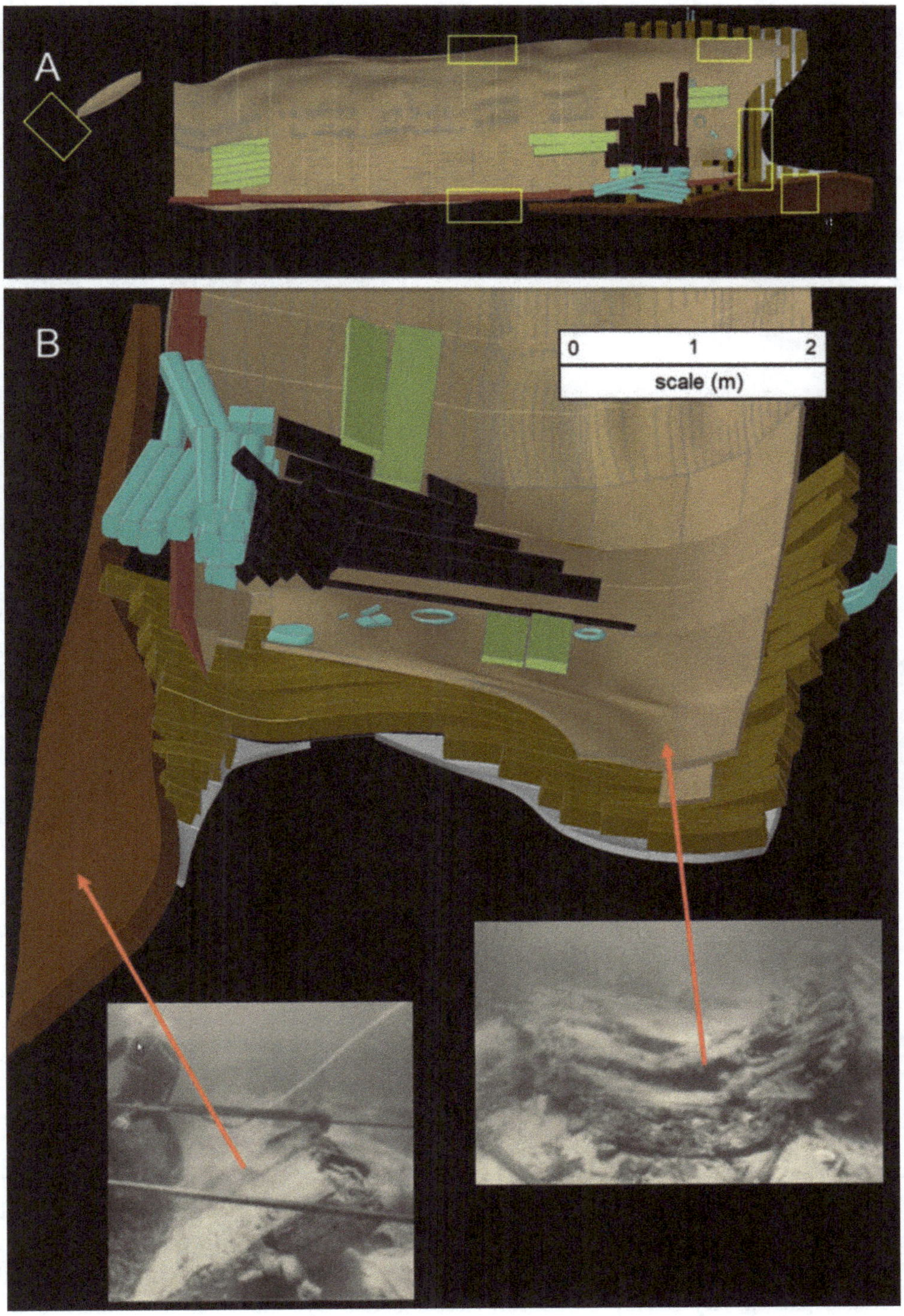

Fig. 10.6 Plan view (**a**) and expanded oblique stern view of remaining buried starboard hull (**b**) showing partially complete 3D digital model of the *James Matthews* shipwreck remains with corresponding survey photos. Total length of buried remains is 26 m (**a**). Keelson and keel (dark browns); ceiling planking (tan); ribs (mid-brown); outer planking (light grey); iron ballast and curved deck knees (blue); remaining slate mound (black); pine timber cargo (dull yellow) and 2000 test trench locations (bright yellow boxes). Excavation survey photos courtesy WAM (P. Baker, WAM)

the buried sleepers were likewise determined at the zero intercept between the two maxima sediment/timber interface amplitudes. In Fig. 10.8b, these respectively correspond to −2.61 and −3.01 m giving a recorded DoB of 40 cm. The trace selected to best represent the central location of each buried sleeper was the one with the shallowest interface depth (corresponding to the most vertical radiated and reflected acoustic waves).

Actual sleeper burial depths were calculated by divers recording the average height of the endplate slots of each sleeper above/below seabed level. This distance was then subtracted from the known precise distance from the slot to each respective upper sleeper face. For example, trace 271 in Fig. 10.8 was recorded at the location of one of the pine (P50) sleepers, which was assembled with a 50 cm gap between the bottom face of the slot and the upper face of the timber. The average depth of the slot above the seabed for this sleeper was measured to be nine cm and hence the actual a depth of sediment cover was 41 cm. Note that when initially buried, this sleeper

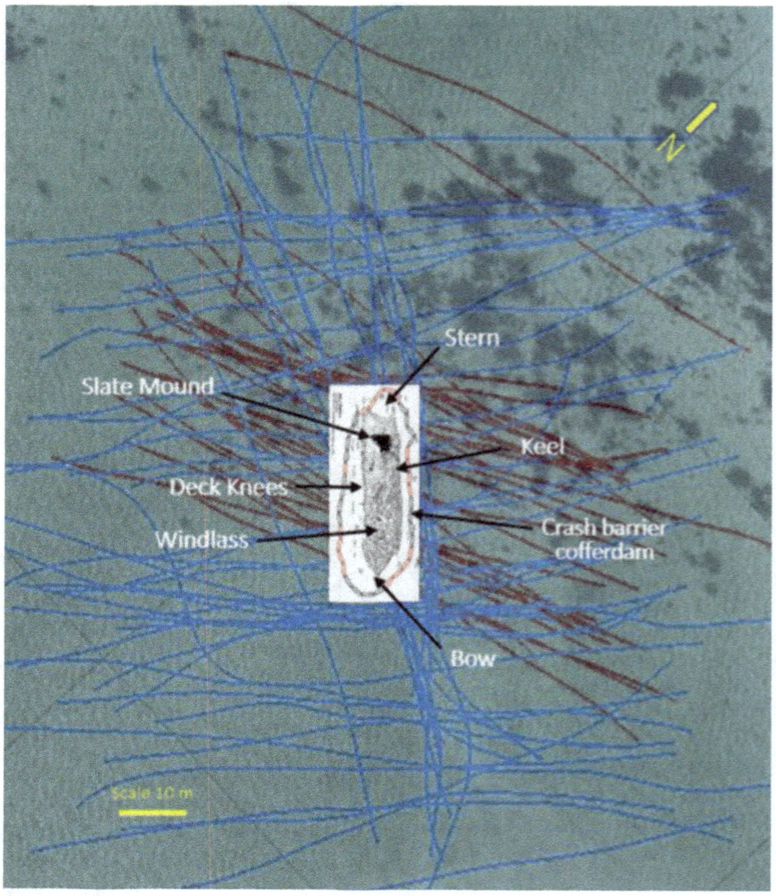

Fig. 10.7 SBP track lines collected over *James Matthews* shipwreck site collected on 7 June (red) and 8 June 2017 (blue)

had a nominal 50 cm depth of burial (Table 10.3), however localized seabed erosion had subsequently scoured the surface and reduced the sleeper's actual depth of sediment cover.

The DoB results extracted from selected traces for each of 12 sleepers (three of which were measured on both runs) in echo plot Runs 025025.RAW and 024600.RAW were compared to diver measured actual burial depths using XLSTAT 2017.4 software to produce a scatter plot as shown in Fig. 10.9. This figure shows the variability in the 15 individual SBP estimates of DoB, and a linear trend line through this data. Uncorrected (for sediment velocity) SBP measurements are shown to: overestimate burial depths in very shallow (7–10 cm) sleeper burials by 3–7 cm; underestimate burial depths in the mid (26–40 cm) burial depth range by 3–4 cm; and underestimate (by 2–5 cm) the measurement of the deepest (50 cm) burial depths. When the speed of sound is corrected to better represent the actual acoustic wave speed in the water column and up to 50 cm through the sediment, all estimates are increased by between 2 and 9 cm corresponding to burial depths 10–50 cm, respectively. Following adjustments to the speed of sound, Fig. 10.9 shows that the SBP linear trend line is closer to the 1:1 relationship between estimated and actual burial depth, especially for sleepers buried deeper than 25–30 cm.

The potential relationship between acoustic wave parameters and types and condition of a variety of buried material was assessed using the original reflection coefficient method by Warner (1990) and reworked in Plets et al. (2007). A reflection coefficient (K_R) is the numerical expression for the strength of the reflection of the acoustic wave from a boundary (seabed surface, the interface between two sedimentary layers or a buried object) and relates to the ratio of the amount of energy reflected to the amount of energy transmitted across the boundary (Telford et al. 1990). A portion of the energy of acoustic sound waves reflect from a boundary if a contrast exists between the elastic properties (acoustic impedances) of the two media that form the boundary, and the remaining portion is transmitted across the boundary. The acoustic impedance of each media is simply the product of its density, ρ, and its compressional P-wave velocity, V_p. In terms of the material properties of each media

$$K_R = \left(\rho_2 V_{p2} - \rho_1 V_{p1}\right) / \left(\rho_2 V_{p2} + \rho_1 V_{p1}\right) \tag{10.1}$$

In Appendix A of Plets et al. (2008) the reflection coefficient for a deeper reflector can be calculated based on acoustic trace properties and known/assumed compressional P-wave velocity values for sediment and the deeper (timber) reflector:

$$K_{DR} = A_{DR}\left[v_w\left(TWT_p/2\right) + v_{DR}\left(TWT_{DR} - TWT_p\right)/2\right]/x \tag{10.2}$$

where x is a calibration coefficient

$$x = \left[A_p v_w\left(TWT_p/2\right)\right]/K_p \tag{10.3}$$

and

$$K_p = \left[A_m TWT_m\right]/\left[A_p TWT_p\right] \tag{10.4}$$

K_{DR} – reflection coefficient of deeper reflector
K_p – reflection coefficient of primary (seabed) reflector
v_{DR} – sound velocity in sediment
v_w – sound velocity through water
$A_{DR/p/m}$ – amplitude of deeper reflector/seabed/seabed 1st multiple
$TWT_{DR/p/m}$ – two-way travel time to deeper reflector/seabed/seabed 1st multiple
now, knowing
$TWT_{DR/p/m} = 2d_{DR/p/m}/v_{DR/p/w}$
d_{DR} – depth from seabed surface to deeper reflector
d_p – depth from water surface to seabed
$v_p = v_w = 1517$ m/s and $v_{DR} = 1813$ m/s
and $TWT_{DR} - TWT_p$ is TWT in seabed $= 2d_{DR}/v_{DR}$

then by substituting (10.4) into (10.3)

$$x = \left[A_p^2 d_p^2\right]/\left[A_m d_m\right] \tag{10.5}$$

whereby Eq. (10.2) simplifies to:

$$K_{DR} = A_{DR}\left[d_{DR} + d_p\right]/x \tag{10.6}$$

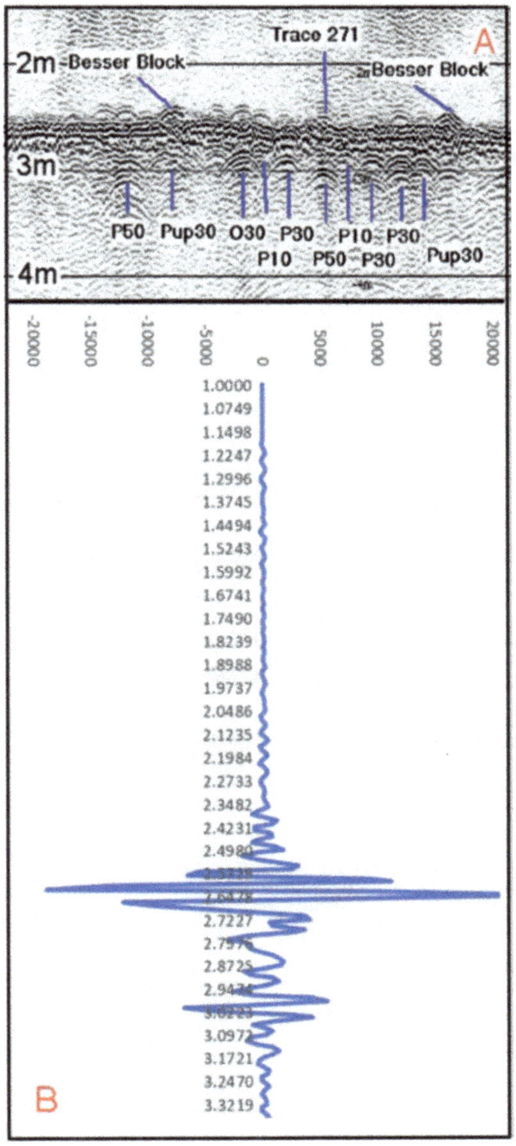

Fig. 10.8 SBP Echo plot 025024.RAW showing unprocessed curtain data recorded at 12 kHz in a SW–NE direction along the line of buried sleepers each buried approximately 1 m apart (**a**) and individual echo Trace 271 plotted in Excel (**b**). Horizontal distance shown in (**a**) is 19.4 m and horizontal black lines show depths in meter increments. Horizontal scale in (**b**) is wave amplitude, vertical scale is meters below transducer head: seabed surface is at −2.61 m and top sleeper interface is at −3.01 m

Amplitudes for the seabed, deeper reflector and seabed first multiple, together with their respective depths were tabulated from the Excel plots created for each buried sleeper identified in Sect. 10.4.1 above, as well as at locations away from the buried sleepers. Tabulated amplitude and depth data for the seabed and first seabed multiple at 39 trace locations away from the sleepers were used in Eq. (10.5) to calculate individual x values. The calibration factor x (1,500,000) was calculated as the 50th percentile from the combined frequency distribution derived from the 39 individual x values (Fig. 10.10) as per Plets et al. (2007).

Reflection coefficients for all sleepers identified in Runs 025025.RAW and 024600.RAW were then calculated using Eq. (10.6) and are shown as a scatter plot in Fig. 10.11. This plot shows identification and separation of buried sleepers by timber type, with little influence by their burial depth. It also shows that grain orientation (pine horizontal vs pine vertical) is not a strong influence on in situ SBP acoustic wave form measurements as they cannot be separated in the scatter plot. There was uncertainty associated with identification of a sleeper along Run 0246000.RAW. It was tentatively identified as an oak sleeper (O?) with a nominal burial depth of 30 cm. This same sleeper was identified as oak on Run 025025.RAW. The reflection coefficient for the sleeper on Run 025025.RAW is consistent with the reflection coefficient value for the other confirmed oak sleeper, while the reflection coefficient for uncertain sleeper on Run 0246000.RAW matched those from other pine sleepers. This suggests

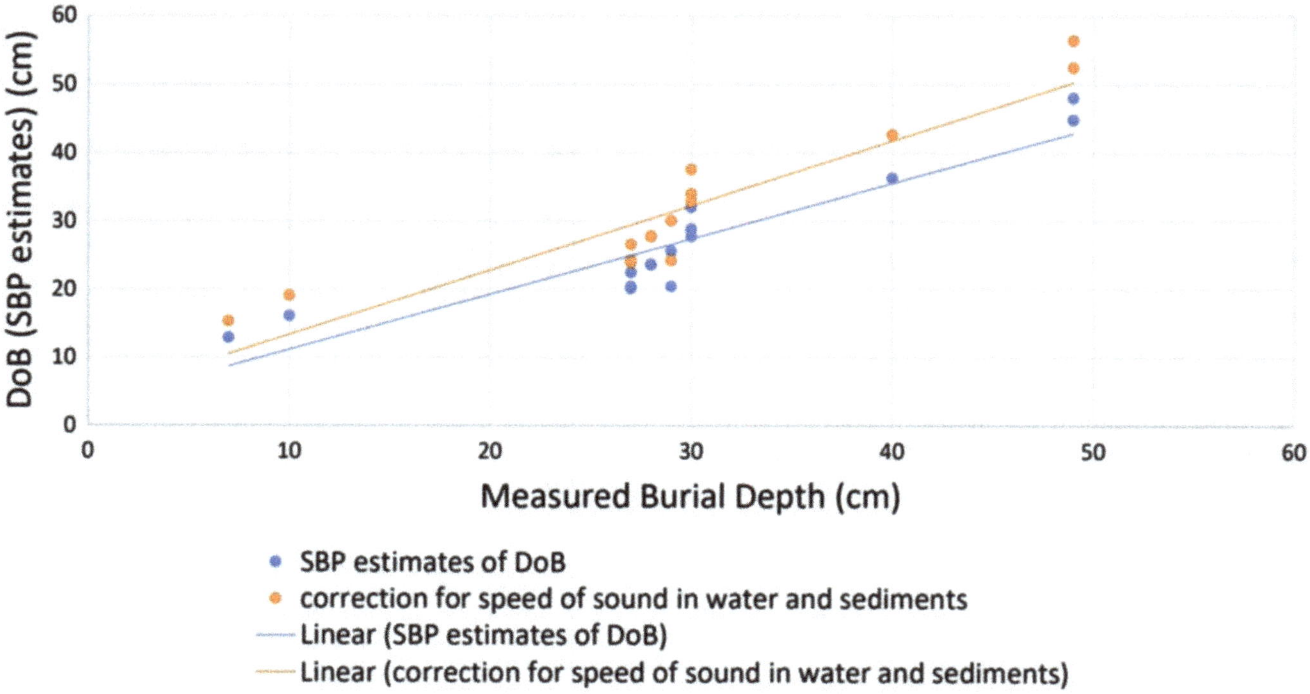

Fig. 10.9 Scatter plot of sleeper burial depths measured by SBP vs actual burial depths

Fig. 10.10 Relative frequency distribution for calibration factor x

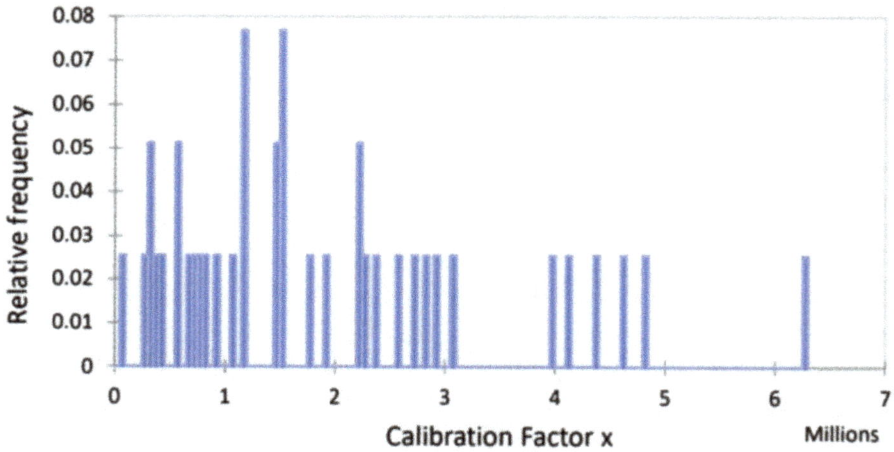

that the tentative identification of the sleeper may have been wrong. The reflection coefficient for one nominal 30 cm deep pine sleeper was higher than those from the oak sleepers, and this may also have resulted from an incorrect sleeper identification, or possibly from partial insonification or non-homogeneous sediment cover.

While the calculated reflection coefficient for each buried sleeper shows promise as a means of identifying different buried material types, the magnitude of these values (c. 0.038 oak and 0.013 for pine) appear to be low compared with previous field and theoretical results (Arnott et al. 2005; Plets et al. 2009). As can be seen by Eq. (10.1), K_{DR} values approach zero when the acoustic impedance for each layer are similar (Eq. 10.7). Changes in value of either the density (ρ) of sediment or timber and/or the assumed compressional acoustic wave speed (v_p) in the sediment or timber can significantly change the absolute magnitude of the derived reflection coefficient.

$$\rho_{\text{timber}} \, v_{p\,\text{timber}} = \rho_{\text{sediment}} \, v_{p\,\text{sediment}} \tag{10.7}$$

In the early work by Quinn et al. (1997c) reflection coefficients were calculated for a range of sediment types (sand, sand-silt-clay, clay) and for 11 different timbers using known densities and theoretical velocity values. Sediment densities ranged from 2100 to 1450 kg/m^3 and corresponding velocity values ranged from 1734 to 1496 m/s. Timber

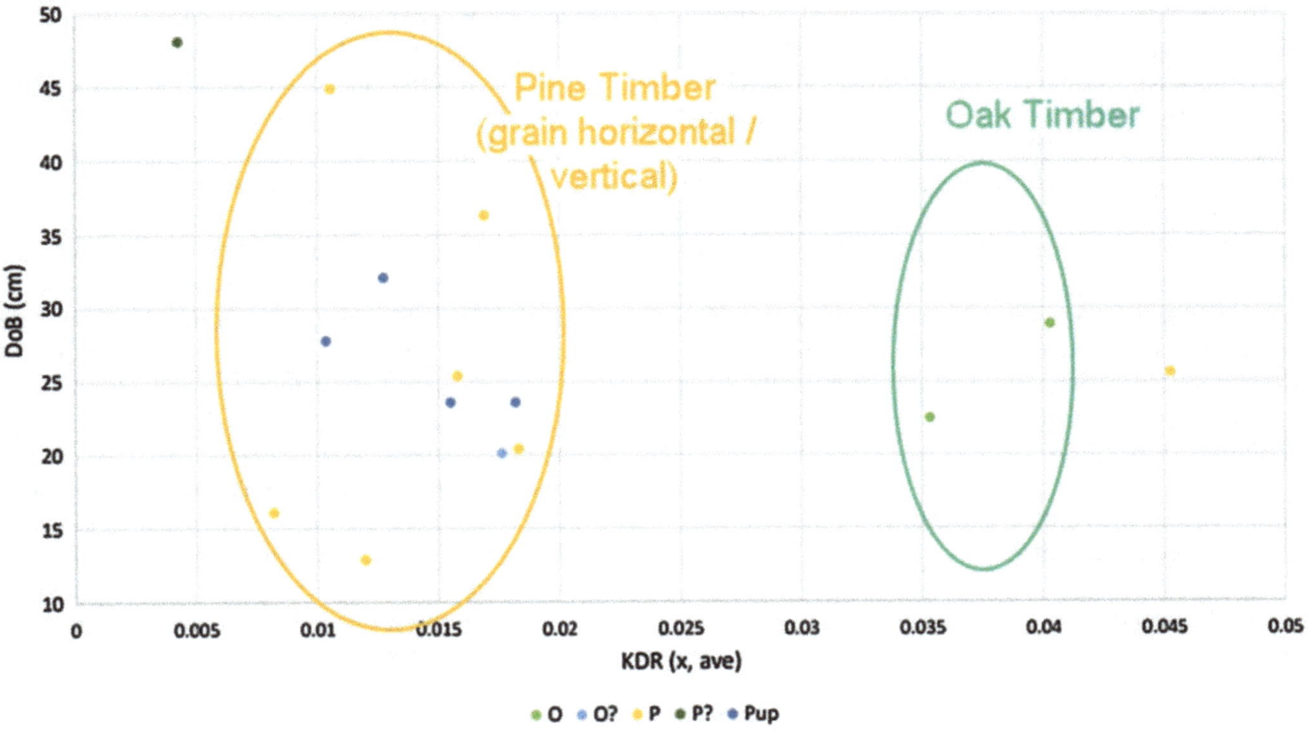

Fig. 10.11 Scatter plot of calculated reflection coefficients against depth of burial

densities for oak and Scotts pine were 660 and 580 kg/m^3, respectively, and theoretical $v_{p\ timber}$ values ranged from 3120 to 1230 m/s (oak) and 6010 to 1000 m/s (Scotts pine) for longitudinal, radial and tangential grain orientations. Theoretical reflection coefficients calculated for oak in sand ranged from −0.28 to −0.47 (longitudinal–radial) and in clay from −0.03 to 0.25. Likewise, for Scotts pine and other softwoods, reflection coefficients ranged from <0.02 to <0.25. Dix et al. (2001) demonstrated that the timber density values used in this earlier work were based on air dried values and did not represent timber in its fully saturated state. This difference results in a 25–43% reduction in the timber's compressional wave speeds, and consequently, for reflection coefficients. For example, the K_{DR} values for saturated oak in sand reduce from −0.28 to −0.04 (longitudinal) and from −0.47 to −0.31 (radial).

Arnott et al. (2005) explored the theoretical relationships between timber degradation and reflection coefficients by exposing oak and pine in seawater for durations up to 9 months and by laboratory measurements of density and compressional wave velocities in those timbers with different states of degradation. The authors plotted (in Fig. 10.2a) theoretical K_R values against timber density ranging from 280 to 580 kg/m^3 for highly degraded–undegraded oak in sand, sand-silt-clay and clay sediments based on tangential and radial velocity measurements in the timber. Extrapolating this linear relationship for higher oak densities from 630 and 700 kg/m^3 would result in theoretical K_R (radial and tangential) values ranging from −0.045 to 0.04 in sand. Similarly, theoretical K_R relationships were derived for highly degraded–undegraded pine (densities 165–525 kg/m^3) however a non-linear relationship was established. With lower water temperatures, Arnott et al. (2005) used 1522 and 1734 m/s for compressional wave speeds in water and sand, respectively, and the same sediment densities as used by Quinn et al. (1997b).

In situ derivation of K_R values were undertaken by Plets et al. (2007, 2009) through interpretation of Chirp data collected on *Grace Dieu* (1439), River Hamble, UK. The remaining timbers at this site are heavily degraded oak, and the sediments are fine silts (phi = 4.7). Compressional wave velocities used were 1484 and 1517 m/s for water and sediments, respectively, and sediment density calculated as 1766 kg/m^3. Derived K_R (radial and tangential) values for timbers in the wreck ranged from 0.07 to 0.19.

Presently the only estimated K_R values for timbers in sand are derived from theoretical analyses using laboratory derived v_p timber values for fresh and degraded oak and pine samples. In the current study green Radiata pine and fully fire-dried oak was used for the sleepers with SBP measurements made 3 months after burial. The density of green Radiata pine is around 1000 kg/m^3 (Forest Products Commission 2018) and the average bulk density of the oak timber was measured to be 642 (n = 6, range 579–751) kg/m^3 using the method by Grattan (1987). The extent to which the moisture content of these buried timbers increased from the time of burial to the time of SBP measurement is currently unknown, but it is likely that they would still be higher than the densities used in the analyses by Arnott et al. (2005). Consequently, the extrapolated K_R (radial and tangential) values (ranging from −0.045 to 0.40 for oak densities 630–700 kg/m^3) may be reasonably indicative for the timber at the time of measurement. In addition, due to the manner of water dredge backfilling, the sand density covering the sleepers may have been substantially lower than the surrounding undisturbed sediments (identified recently when the author noticed significant reduction in penetration resistance to driving in sediment cores) which would also affect the acoustic impedance of the sand covering the sleepers.

10.4.2 *James Matthews* Comparative In Situ Wreck-Site Surveys

SBP track lines over the *James Matthews* shipwreck site have previously been shown in Fig. 10.7, using *QPS Fledermaus* software ver7.7.6.628, together with the 2D plan of the 1976 *James Matthews* excavation survey. Full analyses (i.e. quantitative comparison of SBP trace information and geo-located survey data, together with quasi-3D visualization of all track data) have yet to be completed, but indicative interpretations on several lines have been made. Figure 10.12 shows the longitudinal profile of the buried wreck remains from bow to stern and the unprocessed raw echo plot. The vertical distance between the horizontal black lines on this plot is 1 m. The top of the continuous (horizontal) red/black/red line represents the seabed surface, and the bright outer red hyperbolae (horseshoe) reflectors, sitting approximately 1 m above the seabed and 30 m apart on the left and right-hand sides of the echo plot, depict plastic road crash barriers installed as a cofferdam in December 2013 by WAM as part of the in situ management plan (Richards et al. 2014; Winton and Richards 2005). This figure also depicts one single vertical wave Trace (1154), recorded forward of the slate mound, showing two major reflectors: the top reflector being the seabed surface (−2.57 m) and 0.53 m below this is the upper face of the ship timbers. Below the upper face of the timbers are several more weaker reflectors, most likely additional timber interfaces. Overall this echo plot depicts the upper longitudinal surface (ranging from −2.57 to −2.94 m) and depth of burial (10–53 cm), and possibly the lower limit of buried reflectors (−2.85 to −3.4 m), associated with the remains of the *James Matthews* shipwreck.

Figure 10.13 depicts transverse echo plots from the deck knees to the keel (left to right) across the slate mound and amidships. Trace 1506 through the slate mound (with its upper surface at −2.0 m sitting 70–75 cm above the seabed) shows that nearly all the acoustic wave energy is reflected from the top 45 cm of the slate, with very little energy propagating through to lower levels (below −2.45 m). By contrast with Trace 1412, which is closer to the deck knees and clear of slate, there are strong multiple reflectors located at depth (from around −3.0 to −3.6 m, possibly representing hull timbers). This indicates that the slate is reflecting most acoustic energy and there is an acoustic shadow below the mound where hull timbers were not recorded. The WAM excavation survey also recorded iron bars laying between the slate mound and the keel and like the slate mound, the acoustic wave energy is strongly reflected from this area with little propagation below. Trace 688 at the amidships location, forward of the slate mound, similarly shows multiple reflectors from just below the seabed surface to −3.3 m, probably representing ship's timbers and timber cargo.

The indicative interpretations from the *James Matthews* SBP data runs provide insight into the depth of sediment cover, the total burial depth and potentially different material types associated with the archaeological remains at this site. Further detailed analyses will provide quantitative analyses of sediment thicknesses and ship hull form. Also, the different reflection characteristics qualitatively assessed from slate, iron and timber suggest that the reflection coefficient method applied to the buried sleepers may equally be applicable to characterize different material/timber types and states of degradation on the shipwreck site.

10.5 Future Surveys and Analyses

Further sleeper burials at the current and at a finer-grained site are planned to complete the experimental design. Sleepers must be buried in fine sediments to enable comparison of results with those buried in the coarser sands at the *James Matthews* site. They must be constructed from an Australian hardwood (Jarrah, which was used in Australian colonial period shipbuilding (O'Reilly 2007; Pemberton 1979; Staniforth and Shefi 2014), and need to be buried at both sites to compliment the European timbers (oak and pine). Those already buried at the *James Matthews* site allow comparative analyses of derived reflection coefficients. Oak sleepers need to be buried at 10 and 50 cm at the *James Matthews* site to better assess the depth related relationships with derived reflection coefficients. Iron plates need to be buried at both sites to asses accuracy and variability of measurement of buried iron associated with iron used in ship construction and carried as cargo. Following completion of the installation of these new sleepers, both sites need to be fully surveyed using the same SBP equipment. In addition, further SBP surveys are planned 1, 2 and possibly 5 years after burial to detect any possible influence of biological degradation on acoustic reflection measurements.

All future SBP surveys at both sites will be accomplished using a purpose-developed, wheel-based underwater trans-

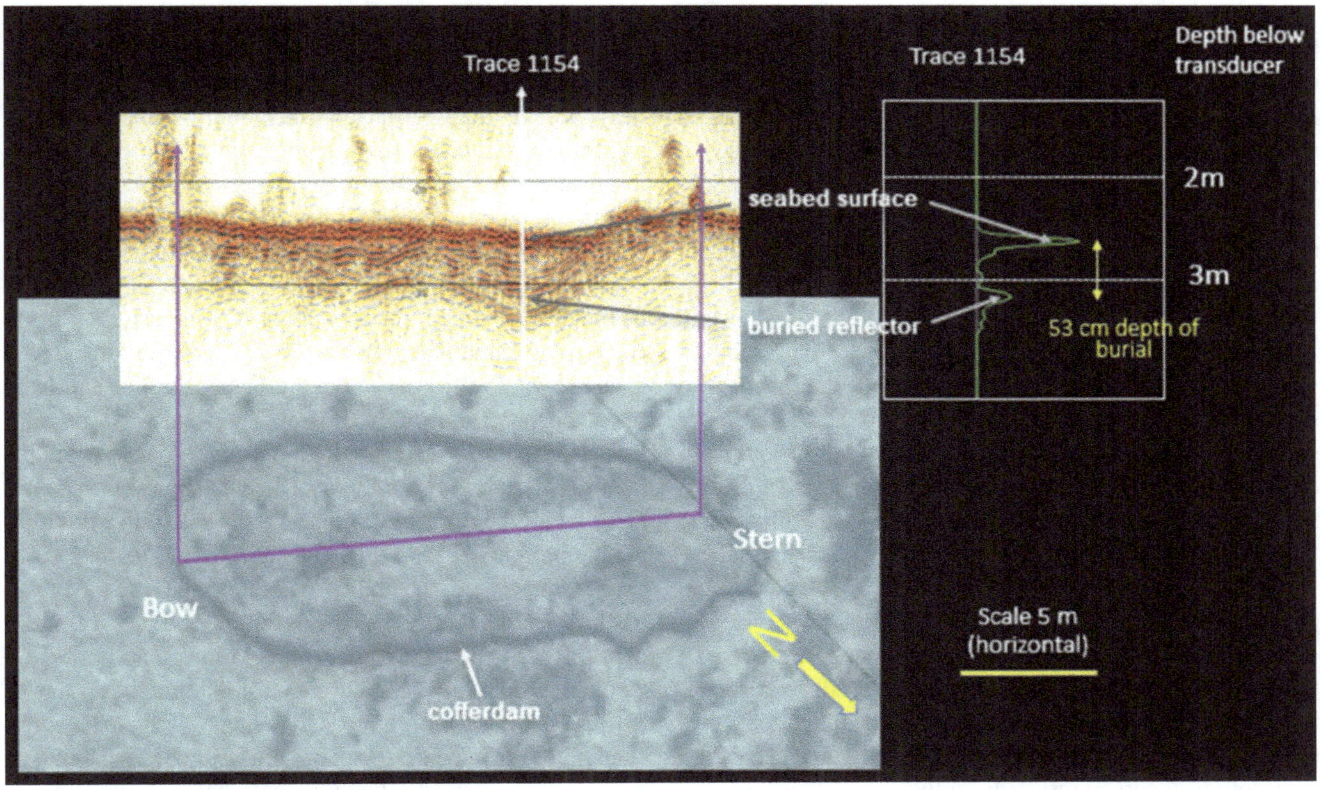

Fig. 10.12 SBP echo curtain data from bow to stern of buried remains of *James Matthews*. Amplitude peaks in Trace 1154 show a buried reflector 53 cm below seabed surface

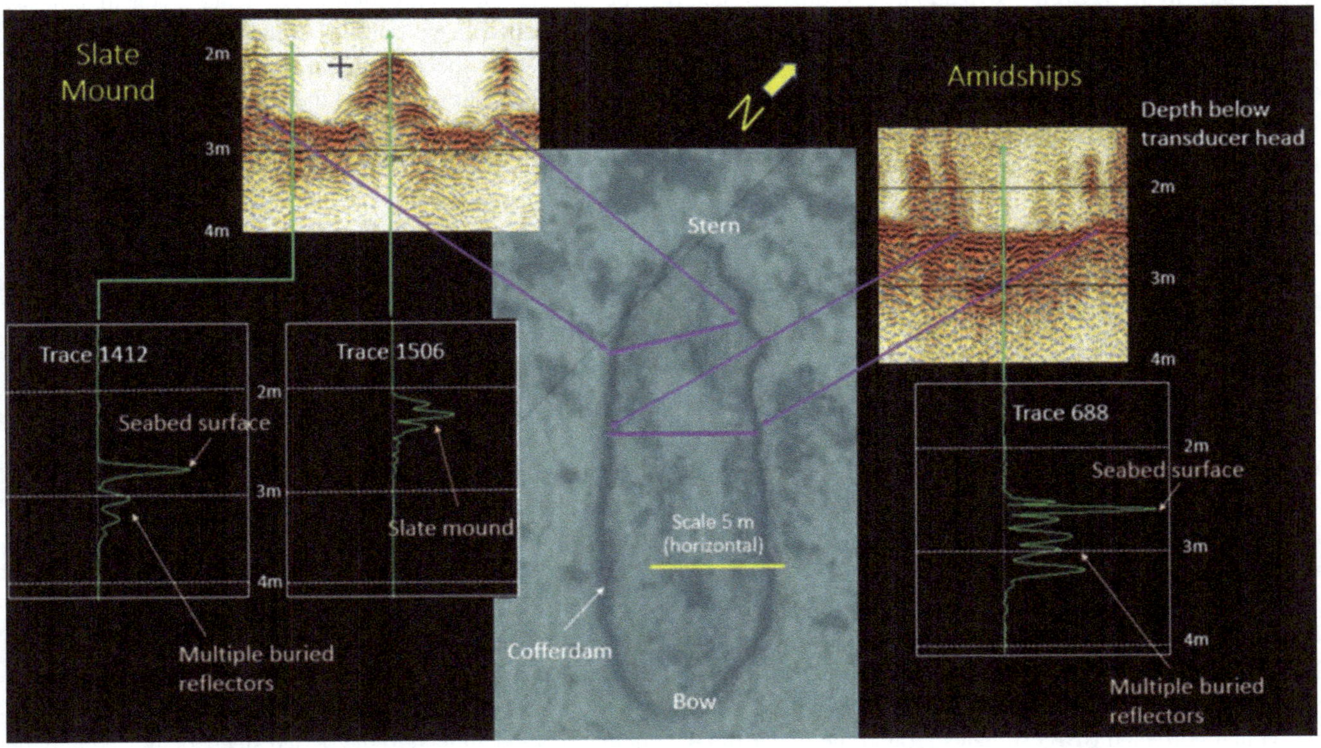

Fig. 10.13 SBP echo curtain data across the slate mound and amidships locations over *James Matthews*. Amplitude response in Trace 1506 (top of slate mound) shows little evidence of deeper reflectors (most acoustic energy absorbed/reflected by slate) compared to Trace 1412, where multiple peaks at depth show buried reflectors. Likewise Trace 688 shows evidence of multiple stacked buried reflectors (most likely timber cargo and hull structure) amidships, forward of the slate mound

ducer sled with a mast mounted RTK GPS sensor, which will be pulled along the seabed to ensure highly controlled positioning of the transducer head relative to each sleeper. Multiple runs back and forth will be undertaken to obtain sufficient measurement data to permit statistical analyses of the accuracy, precision and Type I and II error estimates associated with the measurement of depth of sediment cover, and the variability associated with reflection coefficient estimates.

To assess the confidence in the magnitude of reflection coefficient estimates from the acoustic data, timber samples which were simultaneously buried alongside their respective sleepers will be recovered at the time of each future SBP survey and analysed for bulk and conventional density, percent moisture content and depth of degradation. Replicated sediment cores collected from reference and backfilled locations on both sites will be analysed for in situ and bulk density, grain size distribution and pore water dissolved oxygen, pH and redox potential.

To translate the quantitative measurement performance of the parametric SBP from the individual buried sleeper environment to a complex buried shipwreck site, the direct comparison of the SBP measured data and the corresponding archaeological survey results from the *James Matthews* site is planned. Once the local survey coordinates from the original 1976 archaeological excavation survey and constructed 3D digital model of the *James Matthews* wreck-site are transformed to the WGS 84 navigational coordinate system, then SBP predicted and actual measured depth locations, reflector surfaces and material identified will be assessed. This will provide quantitative evaluation of sediment thickness (DoB) and hull form in 2D profiles, and interpretation in quasi-3D format using gridding and visualization software.

10.6 Conclusions

The proof of concept survey provided initial quantification of accuracy and variability associated with non-invasive parametric SBP measurements of shallow buried maritime archaeological material. The in situ experimental component measured the depth of burial of shallow-buried oak and pine timber beams ('sleepers') at different burial depths (10, 30 and 50 cm) with different grain orientations in coarse sediments using the Innomar parametric SES-2000 *compact* SBP. A linear trendline between sediment and water velocity corrected SBP predictions versus actual measured depth of burial slightly over-estimates (by 1–3 cm) the 1:1 relationship between estimated and actual burial depth. Variability of individual corrected SBP estimates around the linear trendline was 1–6 cm. For three sleepers detected in multiple runs, a difference of up to four cm was identified between runs. Statistical analysis of the variability in individual trace measurements for all sleeper burials will be undertaken with greater confidence once a second set of SBP measurements are undertaken at two sites (different sediment characteristics), with a full set of replicated multi-timber and iron sleeper types at three burial depths, and with a greater control over position of SBP measurements.

Interpretation of SBP survey data recorded from these buried sleepers also identified reflection coefficient relationships between acoustic wave parameters and buried timber, but not with timber grain orientation. The acoustic reflection coefficient signatures, derived from the individual SBP acoustic wave reflections from the seabed surface and from the sleepers with different timber types and grain orientations, were used to identify material type. A scatter plot of depth of burial vs reflection coefficients shows discrimination of material type (pine timber versus oak timber). Within the limitations of currently available data, the scatterplot results suggest that the ability to identify material type from SBP measurements may possibly be independent of burial depth. It also shows that grain orientation (pine horizontal vs pine vertical) does not have a strong influence on in situ SBP acoustic wave form measurements as grain orientation cannot be separated in the scatter plot. However, the magnitude of these derived reflection coefficients is very low (c. 0.038 for oak and 0.013 for pine) and lower than theoretical values calculated by others based on laboratory measurements of the speed of sound through oak and pine timbers. The calculation of reflection coefficients is sensitive to the acoustic impedance properties of the reflector interface (i.e. the density and speed of sound through the overlying sediment and buried timber) and their values used in this study appear to be quite different to and beyond the range previously tested. No conclusive statement can be made regarding the representativeness of the K_R values derived to date in this study, but in situ sediment and timber impedance data will be collected as part of ongoing studies to improve confidence in the results.

Interpretation of SBP measurements over the buried shipwreck material at the *James Matthews* wreck-site demonstrated a proven ability to undertake a non-invasive approach to record depth of sediment cover, the total burial depth and potentially different material types associated with the archaeological remains. The *James Matthews* shipwreck site has been fully excavated, archaeologically recorded and then reburied by maritime archaeologists from the Western Australian Museum in the mid-1970s. A 2D plan from that survey was used to identify key features on selected SBP runs. Different reflection characteristics from known locations of slate, iron and timber suggest that the reflection coefficient method applied to the buried sleepers may equally be applicable to characterize different material types and states of degradation on the shipwreck site.

While still ongoing, this research demonstrates that parametric SBPs can be used for in situ management purposes and the critical quantification of shallow buried archaeological material between 10 and 50+ cm. Further confidence in the measurement accuracy and variance estimates and ability to differentiate material types will be achieved following planned additional sleeper burials and repeated measurements in both fine and coarse-grained sediments, as well as direct 3D spatial comparison with recorded shipwreck material. In combination with other geophysical search tools, these parametric SBP results also reveal the importance of, and opportunity associated with, 3D recording and interpretation of buried maritime archaeological material.

Acknowledgments Special thanks and my fullest appreciation are extended to all staff from the Departments of Maritime Archaeology and Conservation, WAM, and especially to Adjunct Associate Professor Jeremy Green, Adjunct Professor Mike (Mack) McCarthy and Vicky Richards, for their encouragement, personal support in the field and conservation laboratory, and provision of vessels together with diving and dredging equipment. Thanks also to those volunteer members from MAAWA who helped with the diving and other field activities. Special thanks also to Dr. Doug Bergersen, MD Acoustic Imaging Pty Ltd. who provided the Innomar SES-2000 *compact* SBP and SESWIN acquisition software, the GNSS G2 navigational sensors and software, and provided his time to support field data measurements. I would like to acknowledge Fugro Satellite Positioning Pty Ltd. who supplied their Marinestar positioning solution to improve GNSS G2 real time positioning data, and thank Innomar Technologie GmbH and QPS b.v. for the numerous technical discussions and the provision of their respective ISE2 V29533 and Fledermaus ver7.7.6.628 software packages. Lastly, I would also like to thank the anonymous peer reviewers whose comments helped to improve the quality of this chapter.

References

Applanix (2018) POS MV Datasheet. https://www.applanix.com/downloads/products/specs/posmv/ POS-MV-.pdf. Accessed 27 Aug 2018

Arnott S, Dix J, Best A, Gregory D (2005) Imaging of buried archaeological materials: the reflection properties of archaeological wood. Mar Geophys Res 26(2):135–144

Baker P, Henderson G (1979) James Matthews excavation. A second interim report. Int J Naut Archaeol 8(3):225–244

Bergstrand T, Godfrey N (2007) Reburial and analyses of archaeological remains: studies on the effect of reburial on archaeological materials performed in Marstrand, Sweden 2002–2005. The RAAR project

Björdal CG, Daniel G, Nilsson T (2000) Depth of burial, an important factor in controlling bacterial decay of waterlogged archaeological poles. Int Biodeterior Biodegrad 45(1–2):15–26. https://doi.org/10.1016/S0964-8305(00)00035-4

Bjørnø L (2017a) Finite-amplitude waves. In: Neighbors T, Bradley D (eds) Applied underwater acoustics. Elsevier, Amsterdam, pp 857–888

Bjørnø L (2017b) Sonar Systems. In: Neighbors T, Bradley D (eds) Applied underwater acoustics. Elsevier, Amsterdam, pp 587–742

Bjørnø L (2017c) Underwater acoustic measurements and their applications. In: Neighbors T, Bradley D (eds) Applied underwater acoustics. Elsevier, Amsterdam, pp 889–947

Bull JM, Quinn R, Dix JK (1998) Reflection coefficient calculation from marine high resolution seismic reflection (Chirp) data and application to an archaeological case study. Mar Geophys Res 20(1):1–11

Bull JM, Gutowski M, Dix JK, Henstock TJ, Hogarth P, Leighton TG, White PR (2005) Design of a 3D Chirp sub-bottom imaging system. Mar Geophys Res 26(2–4):157–169

Caiti A, Bergem O, Debedal J (1999) Parametric sonars for seafloor characterization. Meas Sci Technol 10:1105–1115

Cvikel D, Grøn O, Boldreel LO (2017) Detecting the Ma'agan Mikhael B shipwreck. Underw Technol 34(2):93–98. https://doi.org/10.3723/ut.34.093

Dix JK, Arnott S, Best AI, Gregory D (2001) The acoustic characteristics of marine archaeological wood. In: Leighton TG, Heald GJ, Griffiths H, Griffiths G (eds) Acoustical oceanography 23(2):299–306. Institute of Acoustics, St Albans, UK

Dix JK, Bastos A, Plets R, Bull J, Henstock T (2008) High resolution sonar for the archaeological investigation of marine aggregate deposits. Project Report, vol 3365

Forest Products Commission WA (2018) Radiata pine. https://www.fpc.wa.gov.au/node/906. Accessed 27 Aug 2018

Forrest J, Homer J, Hooper G (2005) The application of acoustic remote sensing to maritime archaeological site surveys. AIMA Bull 29:1–8

Grattan DW (1987) Waterlogged wood. In: Pearson C (ed) Conservation of marine archaeological objects. Butterworths, London

Gregory D (1998) Re-burial of timbers in the marine environment as a means of their long-term storage: experimental studies in Lynaes Sands, Denmark. Int J Naut Archaeol 27(4):343–358. https://doi.org/10.1111/j.1095-9270.1998.tb00814.x

Gregory D (2007) Environmental monitoring: reburial and analysis of archaeological remains. Studies on the effects of reburial on archaeological materials performed in Marstrand, Sweden 2002–2005. In: Bergstrand T, Godfrey IN (eds) The RAAR Project. Bohuslans Museums, Uddevalla, pp 59–90

Gregory D, Matthiesen H (2012) Nydam Mose: in situ preservation at work. Conserv Manag Archaeol Sites 14(1–4):479–486. https://doi.org/10.1179/1350503312Z.00000000041

Grøn O, Boldreel LO (2013) Sub-bottom profiling for large-scale maritime archaeological survey: an experience-based approach. OCEANS, Bergen, Norway, 10–14 June 2013, MTS/IEEE. In: Paper presented at the OCEANS 2013 MTS/IEEE Bergen: The Challenges of the Northern Dimension. https://doi.org/10.1109/OCEANS-Bergen.2013.6608027

Grøn O, Boldreel LO, Cvikel D, Hermand JP (2015) Subbottom profiling in shallow water: the Akko 4 test case. Rio Acoustics 2015 IEEE/OES Acoustics in Underwater Geosciences Symposium, Rio de Janeiro, Brazil, 29–31 July 2015. https://doi.org/10.1109/RIOAcoustics.2015.7473611

Gutowski M, Malgorn J, Vardy M (2015) 3D sub-bottom profiling—high resolution 3D imaging of shallow subsurface structures and buried objects. MTS/IEEE OCEANS 2015 Genova: Discovering Sustainable Ocean Energy for a New World, Genoa, Italy, 18–21 May 2015. https://doi.org/10.1109/OCEANS-Genova.2015.7271468

Henderson G (1977) Four seasons of excavation on the James Matthews wreck. First southern hemisphere conference on maritime archaeology, Fremantle

Henderson G (2009) Redemption of a slave ship: the James Matthews. Western Australian Museum, Welshpool

Innomar (2018) Innomar SES-2000 compact Sub-Bottom Profiler fact sheet. https:// www.innomar. com/ses2000compact.php. Accessed 10 Sept 2018

Kozaczka E, Grelowska G, Kozaczka S, Szymczak W (2013) Detection of objects buried in the sea bottom with the use of parametric echosounder. Arch Acoust 38(1):99–104

Lafferty B, Quinn R, Breen C (2006) A side-scan sonar and high-resolution Chirp sub-bottom profile study of the natural and anthropogenic sedimentary record of Lower Lough Erne, northwestern Ireland. J Archaeol Sci 33(6):756–766. https://doi.org/10.1016/j.jas.2005.10.007

Lawrence MJ, Bates CR (2001) Acoustic ground discrimination techniques for submerged archaeological site investigation. Mar Technol Soc J 35(4):65–73. https://doi.org/10.4031/002533201788058053

Lovett JR (1978) Merged seawater sound-speed equations. Acoust Soc Am 63(6):1713–1718. https://doi.org/10.1121/1.381909

Manders M, Gregory D, Richards V (2008) The in situ preservation of archaeological sites underwater. an evaluation of some techniques. In: May M, Jones M, Mitchell J (eds) Proceedings of the 2nd heritage, microbiology and science, microbes, monuments and maritime materials conference, 28 June–1 July 2005. RSC Publishing, London, pp 179–203

Mindell DA, Bingham B (2001) A high-frequency, narrow-beam sub bottom profiler for archaeological application. In: Paper presented at the IEEE Oceans 2001, Honolulu

Missiaen T (2010) The potential of seismic imaging in marine archaeological site investigations. Relicta 6:219–236

Missiaen T, Wardell N, Dix J (2005) Subsurface imaging and sediment characterisation in shallow water environments—introduction to the special volume. Mar Geophys Res 26(2):83–85

Missiaen T, Slob E, Donselaar ME (2008) Comparing different shallow geophysical methods in a tidal estuary, Verdronken Land van Saeftinge, Western Scheldt, the Netherlands. Neth J Geosci Geol Mijnb 87(2):151–164

Missiaen T, Demerre I, Verrijken V (2012) Integrated assessment of the buried wreck site of the Dutch East Indiaman 't Vliegent Hart. Relicta 9:191–208

Missiaen T, Evangelinos D, Claerhout C, De Clercq M, Pieters M, Demerre I (2017) Archaeological prospection of the nearshore and intertidal area using ultra-high resolution marine acoustic techniques: results from a test study on the Belgian coast at Ostend-Raversijde. Geoarchaeology 1–15

Müller S, Wunderlich J (2003) Detection of embedded objects using parametric sub-bottom profilers. Int Hydrogr Rev 4(3):76–82

Müller S, Wunderlich J, Hümbs P, Erdmann S (2005) High-resolution sub-bottom profiling for the "shallow survey" common data set using the parametric echosounder SES-2000. In: Paper presented at the Shallow Survey 2005 4th International conference, Plymouth, 12–15 September 2005

O'Reilly R (2007) Australian built wooden sailing vessels of the South Australian intrastate trade. Flinders University, Maritime Archaeology Monograph Series 5, Adelaide

Ortmann N, McKinnon JF, Richards V (2010) In situ preservation and storage: practitioner attitudes and behaviours. AIMA Bull 34:27–44

Pemberton B (1979) Australian coastal shipping. Melbourne University Press, Carlton

Plets RMK, Dix JK, Adams JR, Best AI, Mindell DA (2005) High resolution acoustic imagery from a shallow archaeological site: the Grace Dieu-a case study. In: Proceedings of the international conference Underwater Acoustic Measurements: Technologies & Results, Heraklion

Plets RMK, Dix JK, Best AI (2007) Mapping of the buried Yarmouth Roads wreck, Isle of Wight, UK, using a Chirp Sub-Bottom Profiler. Int J Naut Archaeol 37(2):360–373. https://doi.org/10.1111/j.1095-9270.2007.00176.x

Plets RMK, Dix JK, Adams JR, Best AI (2008) 3D reconstruction of a shallow archaeological site from high-resolution acoustic imagery: the Grace Dieu. Appl Acoust 69(5):399–411

Plets RMK, Dix JK, Adams JR, Bull JM, Henstock TJ, Gutowski M, Best AI (2009) The use of a high-resolution 3D Chirp sub-bottom profiler for the reconstruction of the shallow water archaeological site of the *Grace Dieu* (1439), River Hamble, UK. J Archaeol Sci 36(2):408–418. https://doi.org/10.1016/j.jas.2008.09.026

Quinn R (2012) Acoustic remote sensing in maritime archaeology. In: Catsambis A, Ford B, Hamilton DL (eds) The Oxford handbook of maritime archaeology. Oxford University Press, Oxford, pp 68–89. https://doi.org/10.1093/oxfordhb/9780199336005.013.0003

Quinn R, Bull JM, Dix JK (1997a) Buried scour marks as indicators of palaeo-current direction at the *Mary Rose* wreck site. Mar Geol 140(3–4):405–413

Quinn R, Bull JM, Dix JK (1997b) Imaging wooden artefacts using Chirp sources. Archaeol Prospect 4(1):25–35

Quinn R, Bull JM, Dix JK, Adams JR (1997c) The *Mary Rose* site—geophysical evidence for palaeo-scour marks. Int J Naut Archaeol 26(1):3–16. https://doi.org/10.1111/j.1095-9270.1997.tb01309.x

Quinn R, Adams JR, Dix JK, Bull JM (1998a) The Invincible (1758) site—an integrated geophysical assessment. Int J Naut Archaeol 27(2):126–138. https://doi.org/10.1111/j.1095-9270.1998.tb00796.x

Quinn R, Bull JM, Dix JK (1998b) Optimal processing of marine high-resolution seismic reflection (Chirp) data. Mar Geophys Res 20(1):13–20

Quinn R, Breen C, Forsythe W, Barton K, Rooney S, O'Hara D (2002) Integrated geophysical surveys of the French frigate *La Surveillante* (1797), Bantry Bay, Co. Cork, Ireland. J Archaeol Sci 29(4):413–422. https://doi.org/10.1006/jasc.2002.0732

Richards V (2001) *James Matthews* (1841) conservation pre-disturbance survey report. Department of Materials Conservation, Western Australian Museum, Fremantle

Richards V (2011a) In situ preservation—application of a process-based approach to the management of underwater cultural heritage. Asia-Pacific Regional Conference on Underwater Cultural Heritage, Manila

Richards V (2011b) In situ preservation and reburial of the ex-slave ship *James Matthews*. AICCM Bull 32(1):33–43

Richards V, Godfrey IM, Blanchette RA, Held B, Gregory DJ, Reed E (2009) In situ monitoring and stabilisation of the *James Matthews* shipwreck site. In: Proceedings of the 10th ICOM Group on Wet Organic Archaeological Materials conference, Amsterdam

Richards V, MacLeod I, Veth P (2014) The Australian Historic Shipwreck Preservation Project—in situ preservation and long-term monitoring of the *Clarence* (1850) and *James Matthews* (1841) shipwreck sites. In: Proceedings of the 2nd Asia-Pacific Regional conference on Underwater Cultural Heritage, Honolulu, Hawaii, 13–16 May 2014

Robb GBN, Dix JK, Best AI, Bull JM, Leighton TG, White PR, Seal A (2005) The compressional wave and physical properties of inter-tidal marine sediments. Underwater Acoustic Measurements: Technologies & Results, Heraklion

Shefi D, Veth P (2015) A critical analysis and philosophical review of 'rapid reburial': the *Clarence* project. Int J Naut Archaeol 44(2):371–381. https://doi.org/10.1111/1095-9270.12105

Staniforth M, Shefi D (2014) Shipbuilding in the Australian Colonies before 1850. 2014 ACUA Underwater Archaeology proceedings

Stewart DJ (1999) Formation processes affecting submerged archaeological sites: an overview. Geoarchaeology Int J 14(6):565–587

Telford WM, Geldart LP, Sheriff RE (1990) Applied geophysics, 2nd edn. Cambridge University Press, Cambridge

UNESCO (2001) UNESCO Convention on the Protection of the Underwater Cultural Heritage. http://www.unesco.org/new/en/culture/themes/underwater-cultural-heritage/2001-convention/

Vardy ME, Dix JK, Henstock TJ, Bull JM, Gutowski M (2008) Decimeter-resolution 3D seismic volume in shallow water: a case study in small-object detection. Geophysics 73(2):B33–B40

Vasudevan M, Sivakholundu KM, Venkata Rao D, Kathiroli S (2006) Application of parametric acoustics for shallow-water near-surface geophysical investigations. Oceans 2006—Asia Pacific

Von Deimling JS, Held P, Feldens P, Wilken D (2016) Effects of using inclined parametric echosounding on sub-bottom acoustic imaging and advances in buried object detection. Geo-Marine Letters: 1–7

Warner M (1990) Absolute reflection coefficients from deep seismic reflections. Tectonophysics 173:15–23

Wheeler AJ (2002) Environmental controls on shipwreck preservation: the Irish context. J Archaeol Sci 29(10):1149–1159. https://doi.org/10.1006/jasc.2001.0762

Winton T (2015) Understanding the interactive nature of in situ processes for management of submerged cultural heritage material. AIMA Bull 39:71–83

Winton T, Richards V (2005) In situ containment of sediment for shipwreck reburial projects. In: Proceedings of the 9th ICOM Group on Wet Organic Archaeological Materials conference, Copenhagen, 7–11 June 2004

Wunderlich J, Müller S (2003) Non-linear echo sounders for high-res sub-bottom profiling. Seal Technol 44(9):23–26

Wunderlich J, Müller S, Erdmann S, Buch T, Hümbs P, Endler R (2005a) High-Resolution acoustical site exploration in very shallow water—a case study, Near Surface 2005. EAGE, Palermo

Wunderlich J, Wendt G, Müller S (2005b) High-resolution echo-sounding and detection of embedded archaeological objects with nonlinear sub-bottom profilers. Mar Geophys Res 26(2):123–133

Zisi A (2016) Relationship between wood density and ultrasound propogation velocity: a non-destructive evaluation of waterlogged archaeological wood state of preservation based on its underwater acoustic properties. PhD dissertation, University of Southampton

Check for updates

Resolving Dimensions: A Comparison Between ERT Imaging and 3D Modelling of the Barge *Crowie*, South Australia

11

Kleanthis Simyrdanis, Marian Bailey, Ian Moffat, Amy Roberts, Wendy van Duivenvoorde, Antonis Savvidis, Gianluca Cantoro, Kurt Bennett, and Jarrad Kowlessar

Abstract

Three-dimensional (3D) modelling is becoming a ubiquitous technology for the interpretation of cultural heritage objects. However most 3D models are based on geomatic data such as surveying, laser scanning or photogrammetry and therefore rely on the subject of the study being visible. This chapter presents the case study of *Crowie*, a submerged and partially buried barge wrecked near the town of Morgan in South Australia. *Crowie* was reconstructed using two alternative approaches; one based on a combination of historic photographs and computer graphics and the second based on geophysical data from electrical resistivity tomography (ERT). ERT has been rarely used for maritime archaeology despite providing 3D representation under challenging survey conditions, such as in shallow and turbid water. ERT was particularly successful on *Crowie* for mapping the external metal cladding, which was recognisable based on very low resistivity values. An alternative 3D model was created using historic photographs and dimensions for *Crowie* in combination with information from acoustic geophysical surveys. The excellent correspondence between these models demonstrates the efficacy of ERT in shallow maritime archaeology contexts.

Keywords

Electrical resistivity tomography · Geophysics · Historic shipwreck · Riverine archaeology

11.1 Introduction

This chapter presents recent efforts to map and create a three-dimensional (3D) model of *Crowie*; a wrecked and submerged historic barge located at Morgan on the River Murray in South Australia (Fig. 11.1). *Crowie* was launched in 1911 and sank while at anchor (circa 1950) (Roberts et al. 2017; Simyrdanis et al. 2018). *Crowie* is an important vessel with multiple layers of significance including its substantial economic contribution to the colony of South Australia (e.g., Kenderdine 1993), its large size and, more uniquely, the Aboriginal significance attached to this vessel.

Roberts et al. (2017) undertook the first study of this vessel which was primarily concerned with locating and describing the submerged, but unburied, remains of *Crowie* via multibeam and sidescan imaging and exploring its Aboriginal significance. Subsequent research has sought to improve our knowledge regarding the dimensions and condition of the buried portion of the vessel via electrical resistivity tomography (ERT) and to validate the accuracy of these data by comparing the results to 3D model created by acoustic geophysical methods and historic photographs. ERT can image submerged and buried shipwreck remains in situ without disturbing the site or undertaking expensive recovery projects. This provides exciting new opportunities to create digital content as part of the increasing trend towards virtual muse-

K. Simyrdanis (✉) · G. Cantoro
Laboratory of Geophysical-Satellite Remote Sensing, Institute for Mediterranean Studies, Rethymno, Greece
e-mail: ksimirda@ims.forth.gr; gianluca.cantoro@gmail.comv

M. Bailey · A. Roberts · W. van Duivenvoorde · K. Bennett
J. Kowlessar
Archaeology, College of Humanities, Arts and Social Sciences, Flinders University, Adelaide, SA, Australia
e-mail: bail0164@flinders.edu.au; amy.roberts@flinders.edu.au; wendy.vanduivenvoorde@flinders.edu.au; kurt.bennett@flinders.edu.au; kowl0004@flinders.edu.au

I. Moffat
Archaeology, College of Humanities, Arts and Social Sciences, Flinders University, Adelaide, SA, Australia

McDonald Institute for Archaeological Research, University of Cambridge, Cambridge, UK
e-mail: ian.moffat@flinders.edu.au

A. Savvidis
VR Developer & CGI Illusionist, Thessaloniki, Greece

J. K. McCarthy et al. (eds.), *3D Recording and Interpretation for Maritime Archaeology*, Coastal Research Library 31,
https://doi.org/10.1007/978-3-030-03635-5_11

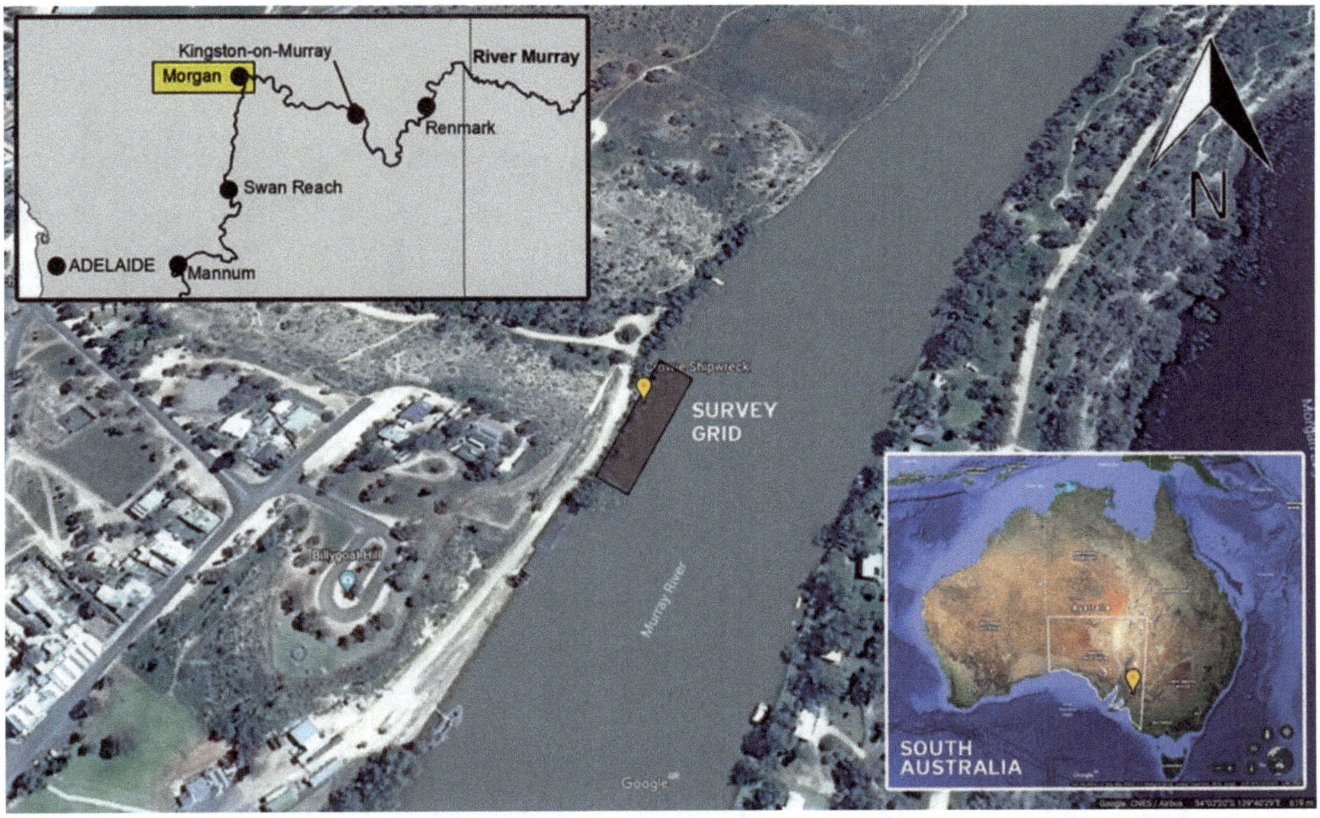

Fig. 11.1 Survey area (Background Image and Right Inset: Google Maps)

ums in underwater archaeology (i.e., Haydar et al. 2011; Liarokapis et al. 2017; Varinlioğlu 2011). This chapter summarizes the geophysical results relevant to the creation of a 3D model. Further details about the geophysical survey are available in Simyrdanis et al. (2018).

11.2 *Crowie*'s History, Context, Significance and Construction

11.2.1 History and Context

Launched on 9 November 1911, the river barge *Crowie* was the largest vessel in its class to operate on the Murray or Darling Rivers. Built by David Milne in the Goolwa shipyards for Captain George Arnold of Mannum, it was reported to measure 150 ft (45.7 m) in length, 30 ft (9 m) in beam, and 9 ft (2.7 m) in height, and was capable of carrying 700 tons, or 8000 bags of wheat (Anon 1911a, b, 1912a, b, 1913a, 1915, 1916, 1917, 1922a, 1950) (Fig. 11.2).

Crowie operated during the latter half of a booming trade era on the Murray and Darling Rivers, which began in the mid-late nineteenth century. The origins of the river trade were closely tied to the spread of pastoralism from Sydney to South Australia and the associated expansion of the wool industry (Kenderdine 1993). Prior to the establishment of river trade routes, wool produced on these pastoral properties had to be carried along barely formed tracks by bullock and dray which was relatively slow and expensive (Younger 1976). The river trade provided a more efficient means of transporting wool and provisioning of stations until the establishment of railways in the area.

The size of *Crowie* initially raised some concern, with one critic writing ‘the general opinion is inclined to question the serviceability of a barge so large’ (Anon 1911b). *Crowie* proved, however, able to successfully transport record-breaking cargo loads including 7200 bags of wheat in 1912 (Anon 1912a, b, c), 7500 bags of wheat in 1913 (Anon 1913b) and 2700 bales of wool in 1918 (Anon 1918). The largest ever consignment of flour (580 tons) shipped on the river was also carried by *Crowie* (Anon 1920). Other known cargo carried by *Crowie* included dried fruit (Anon 1912c, 4), red gum piles (Anon. 1925), stringybark piles (Anon 1927), chaff (Anon 1919), telegraph poles (Anon 1924), agricultural implements (Anon 1913b), cement (Anon 1922b), and steel plates (Anon 1939).

Crowie was also critically important during the freshwater famines on the Murray (Anon 1915, 1928). These events resulted from salt water incursions that occurred when sea water entered the river system via the Murray mouth, turning fresh water brackish. *Crowie* was deployed (as the largest barge available) to pump fresh water into its hull and trans-

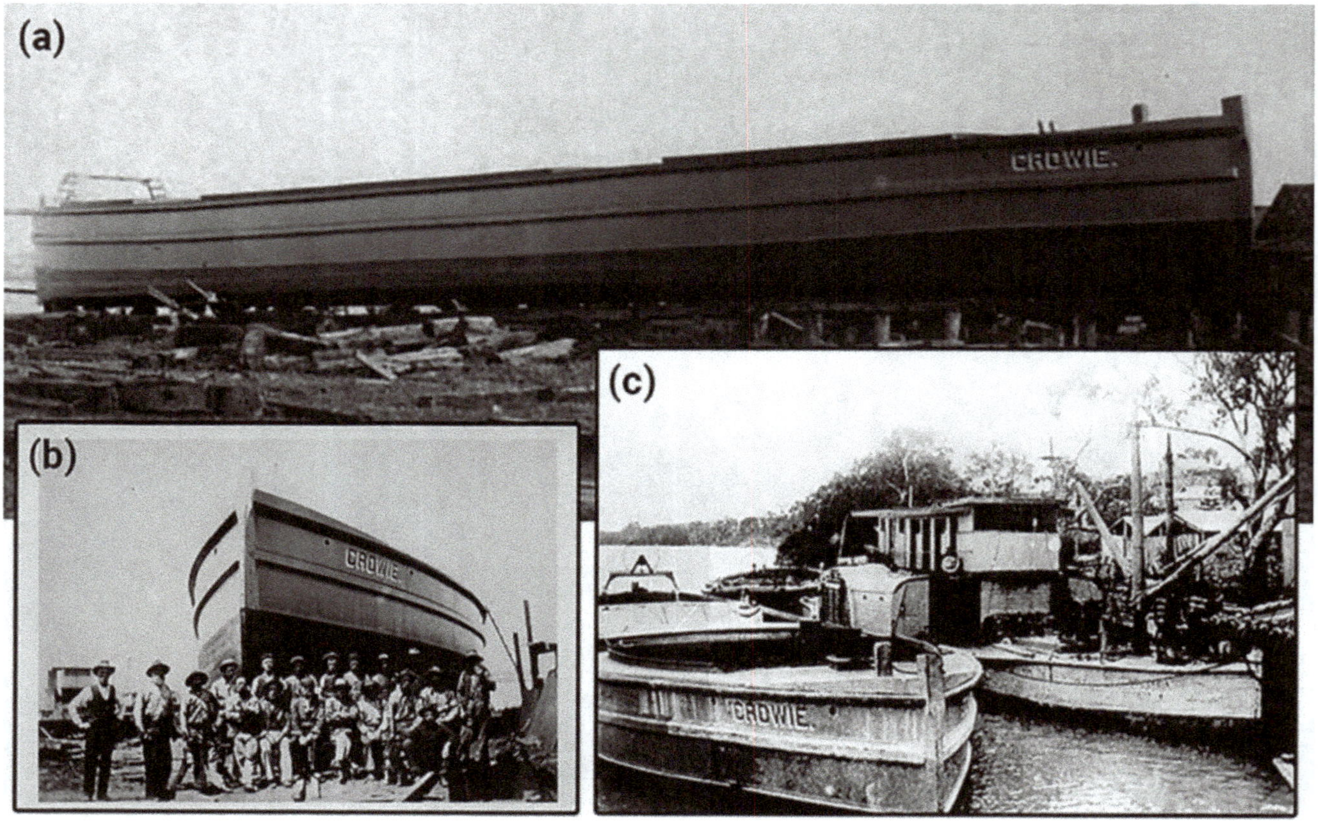

Fig. 11.2 (**a**) 'The barge *Crowie* on the stocks at Goolwa, built by J.G. Arnold and was the largest ever put on the Murray…', B6429, from the Goolwa Collection. (**b**) 'Murray River barge *Crowie*, built in 1911 at Goolwa', B12310, from the Murray River Collection. (**c**) 'P.S. *Wilcannia* and *Crowie* barge at Mannum (Godson number 257A/23)', PRG1258_1_709, from the Godson Collection. (Photographs courtesy of the State Library of South Australia)

port it to towns in need. *Crowie* was able to move approximately 600 tons each trip (Anon 1915).

The exact date of *Crowie*'s sinking is unknown. Historical records show *Crowie* appearing for sale on 11 April 1946, but by 1950 it had sunk (Anon 1950; Roberts et al. 2017). According to the Australian Heritage Database, as well as subsequent investigations by Roberts et al. (2017), *Crowie* is located approximately 100 m upstream from Morgan Wharf, and 10 m out from the western bank of the Murray River. The reasons for the sinking of *Crowie* are unknown. It is likely, however, *Crowie* was simply abandoned and, in the absence of any maintenance, eventually sank at its mooring.

11.2.2 Significance

Roberts et al. (2017) demonstrated that river vessels such as *Crowie* can contribute to the telling of more complex narratives relating to Indigenous riverscapes and cross-cultural entanglements. Their collaborative research, which incorporated historical data, oral histories and geophysical surveys, reminded us that the river trade took place within a riverscape that was and continues to be the 'country' of Aboriginal people (Roberts et al. 2017, 143). Such riverscapes were and are 'animated' spiritual worlds that intersect with people, the environment and material culture (such as river vessels) (after Bradley 1997, 177; Kearney 2009, 171–172). The river boat industry was also entangled with Aboriginal lives in other ways, often overlooked in contemporary histories, through the naming of vessels and the employment of Aboriginal people (Roberts et al. 2017). The naming of *Crowie* is a case in point as it is derived from the Ngarrindjeri (the Aboriginal language belonging 'to the people of the Lower Murray, Lakes and Coorong region of South Australia' (Gale and Sparrow 2010, 387)) word *krawi* which means 'big' and was hence appropriated for the barge (Anon 1911a; Nathan and Fang 2014, 51; Roberts et al. 2017, 136).

11.2.3 Construction

Crowie's dimensions and construction materials have been estimated from a range of sources including historical documents (although no known plans are extant) and sidescan sonar and multibeam surveys. Roberts et al. (2017) con-

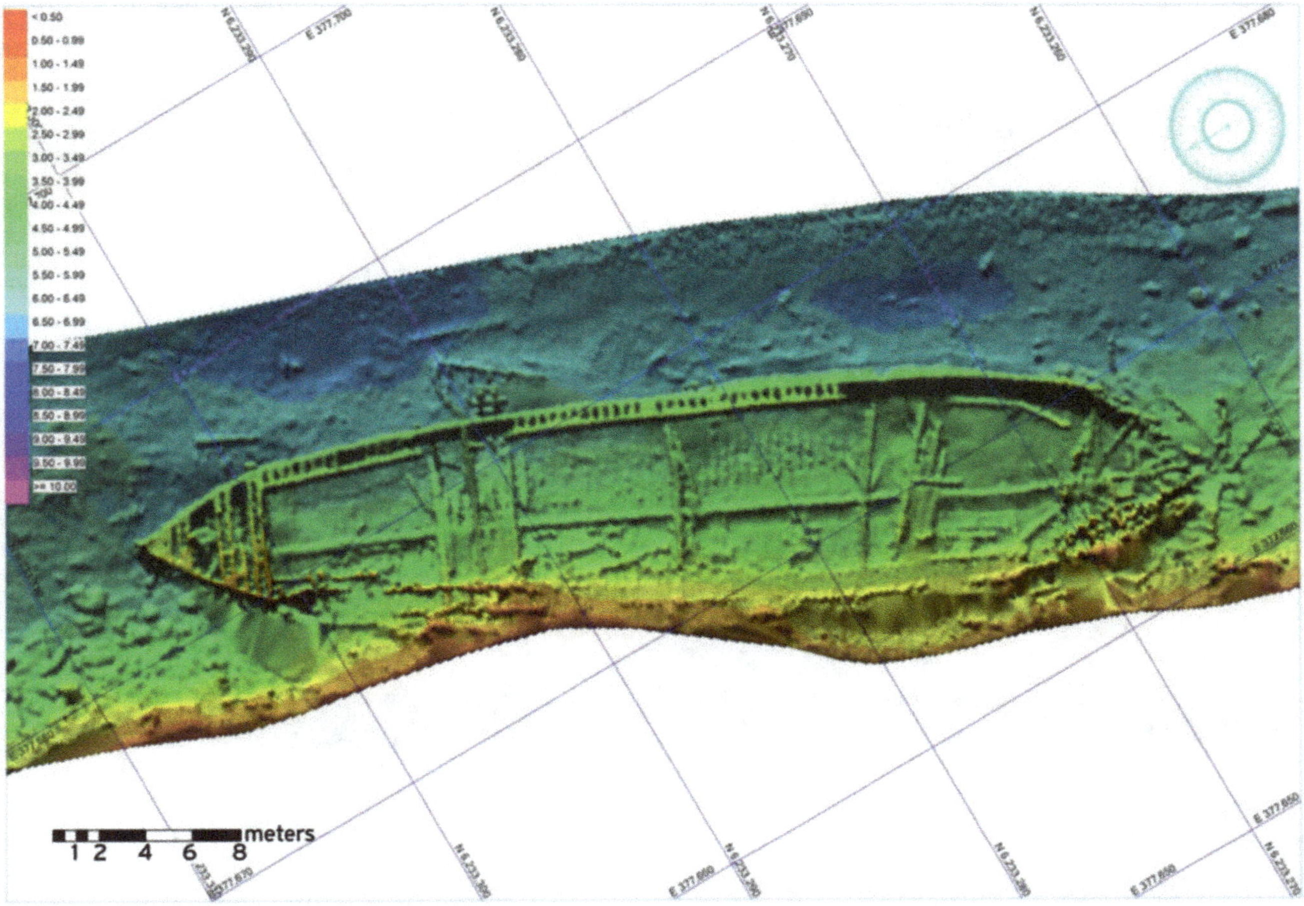

Fig. 11.3 Multibeam image of *Crowie* (19/3/2012) (G. Carpenter) in Roberts et al. (2017: 141)

ducted a sidescan sonar survey and analysed an earlier multibeam survey undertaken in 2012. The multibeam survey was undertaken by Gareth Carpenter, on 19 March 2012, on behalf of SA Water. Multibeam and sidescan data were consistent with the description from the Australian Heritage Database as well as the results from archival searches. The dimensions reported from Anon (1911a) indicated *Crowie* was about 45.7 m in length, 9 m in beam, and 2.7 m in height, while multibeam data suggested it measured 46 m in length and had a beam of 9 m. Multibeam data clearly highlighted the vessel's hull shape and construction features although it could not confirm whether the nine iron bulkheads listed in historical sources were in place (Roberts et al. 2017, 140). Features that were visible included remnants of the iron deck beams, the keelson, the angle-iron floors and the iron hatch coaming (Roberts et al. 2017, 140) (Fig. 11.3).

Sidescan data collected by Roberts et al. (2017, 140) was also able to highlight the key construction features, including the 'keelson, regularly spaced floors, shape of the bow, deck beams in the stern, partially preserved hatch coaming, remains of its bulkheads and its bow and stern section' (Fig. 11.4).

The construction of a number of barges, including *Crowie*, were undertaken in the Goolwa shipyards. The construction technique for *Crowie* cannot be confirmed through geophysical data, however the following description of typical bottom-based construction paraphrased from Roberts et al. (2017, 141), likely applied to *Crowie*. 'After the keel was laid, wooden bottom planking was assembled, followed by the insertion of angle-iron floors. Iron futtocks were then through-bolted onto the floors to erect the vessel's framework. The frame was planked up with wooden planking strakes below the waterline and with iron plating above the waterline—both fastened with rivets. A heavy timber keelson was then fastened on top of the floors with keel bolts. *Crowie* also had an iron-plate stern deck, as well as iron gussets and a barn-door rudder' (Roberts et al. 2017, 141). The bottom-based construction technique used to build *Crowie* meant that the largest area possible was left free in the barge for storage.

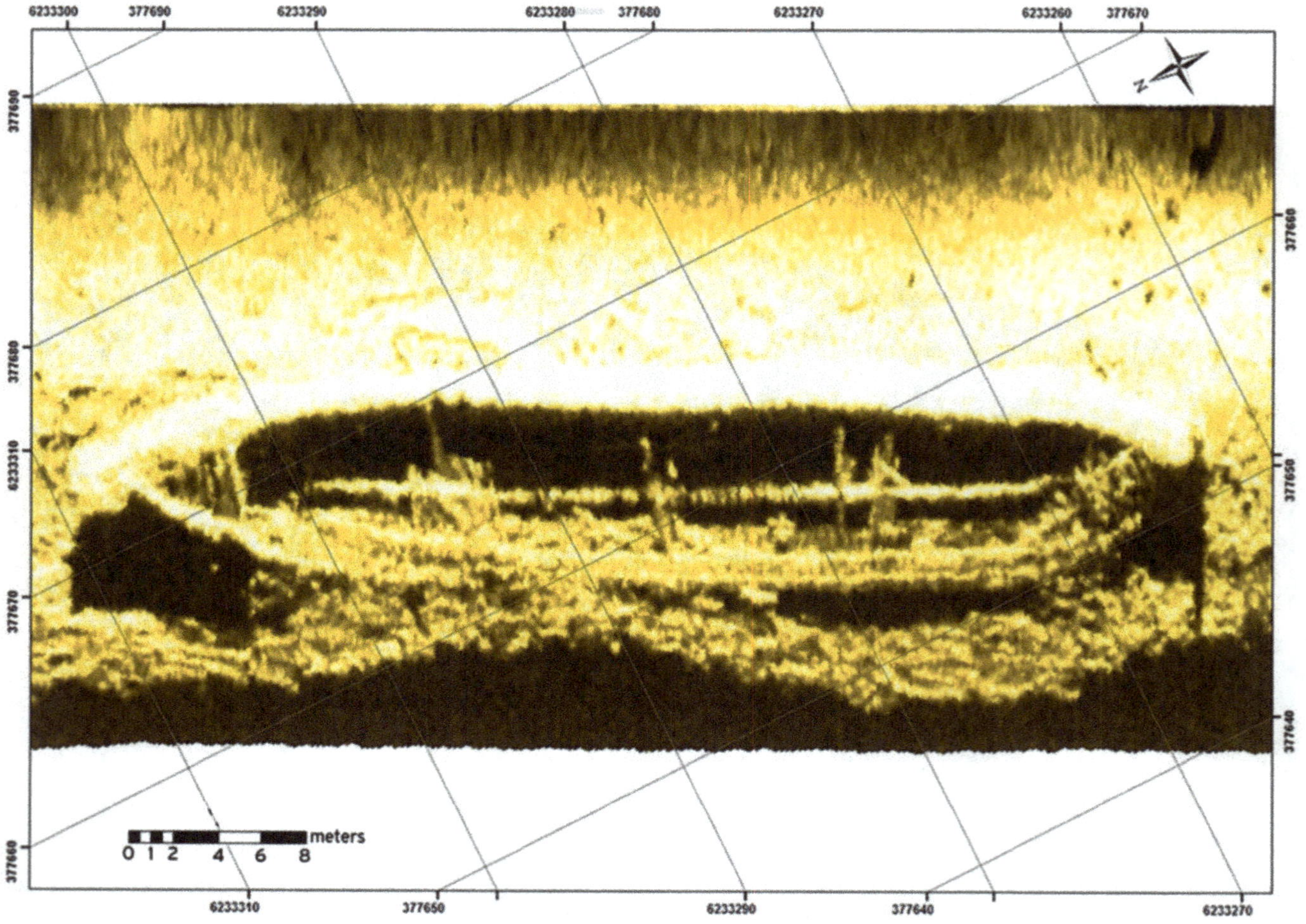

Fig. 11.4 Sidescan image of Crowie (3/5/2016) (Roberts et al. 2017: 142)

11.3 Geophysical Modelling

Previous research did not examine the portion of the vessel buried in sediment. Thus, whilst the length and beam measurements of *Crowie* were confirmed through multibeam and sidescan data, the depth of the extant vessel remained unknown, as well as the degree of preservation of the portion of the vessel buried in the riverbed. This project aimed to produce a complete 3D model of the wreck using geophysical data.

11.3.1 Electrical Resistivity Tomography (ERT)

ERT is a geophysical method used for archaeological prospection where a current is injected into the ground and the resulting electrical potential is measured at a variety of locations along a survey line. ERT can resolve buried archaeological and geological features with characteristic electrical signatures ('anomalies') that are easily distinguishable from the surrounding environment (Clark 1990). In archaeological investigations, electrical resistivity survey has most commonly been used for mapping of tumuli (burial mounds) (Tsourlos et al. 2014) and imaging buried archaeological features (Papadopoulos et al. 2011).

There has been an increasing trend towwards the use of ERT methods in marine and freshwater environments, particularly for geological mapping (Rucker et al. 2011) and the location of archaeological material (Passaro et al. 2009; Passaro 2010; Ranieri et al. 2010; Simyrdanis et al. 2015, 2016, 2018). Electrical resistivity can be deployed in aquatic environments with either floating or submerged sensors, as shown in Fig. 11.5. Orlando (2013) used numerical simulation modelling to estimate the resolution of these two configurations and demonstrated that floating cables result in poor images when the contrast between the resistivity of water and sediment layer is too small (resistivity ratio less than 1).

The application of ERT in submarine archaeology has been relatively uncommon to date. Ranieri et al. (2009, 11) used 3D geoelectrical data to map buried and submerged archaeological features including the ancient settlements at Nora (South Coast of Sardinia), which included Phoenician, Punic and Roman remains and the Roman town of Pollentia

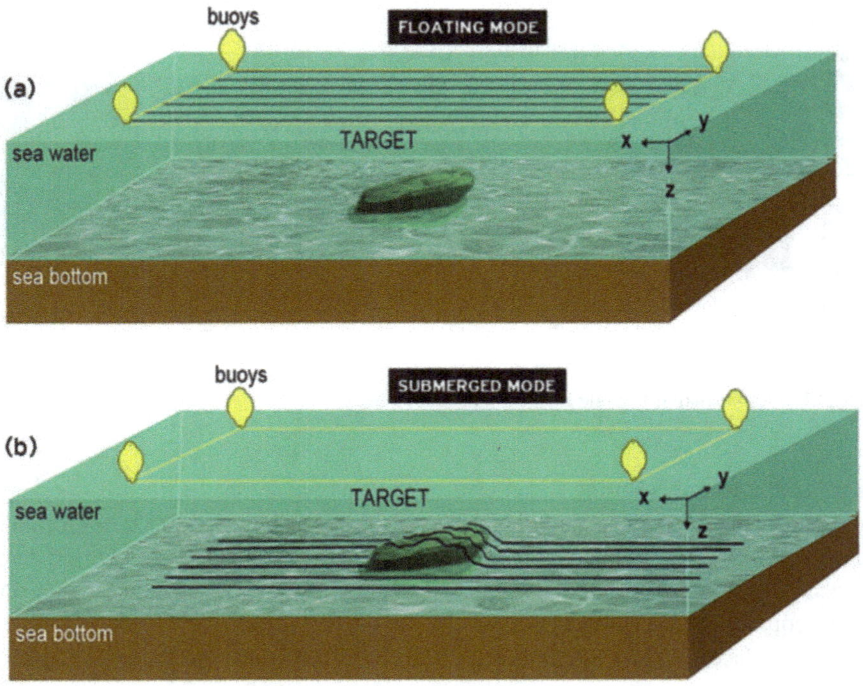

Fig. 11.5 Position of ERT cables in (**a**) floating or (**b**) submerged mode in a maritime environment. Red dots indicate sensors' position

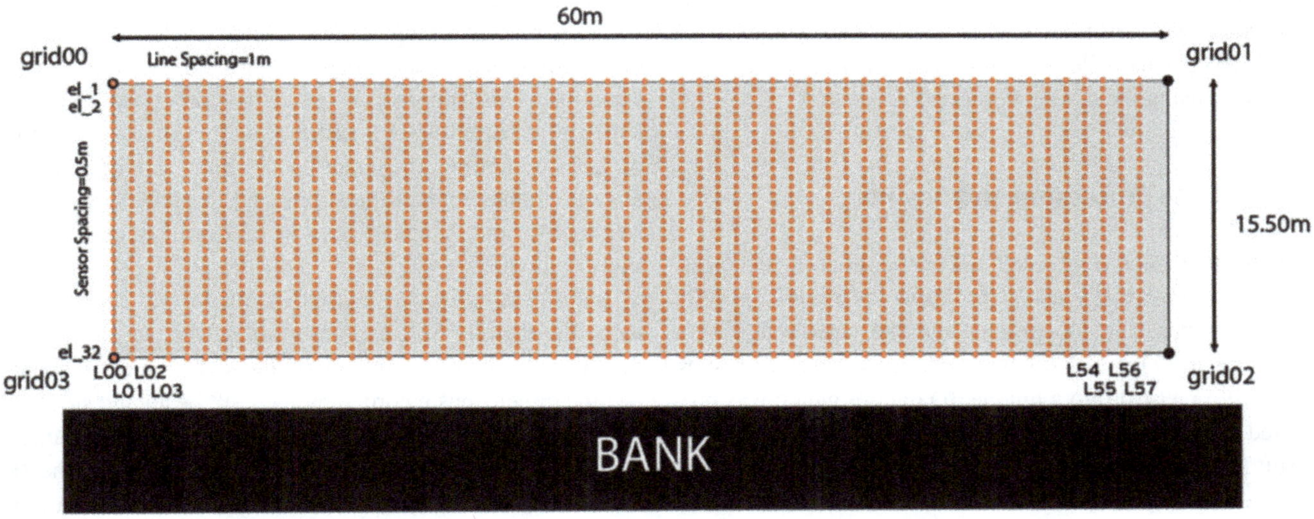

Fig. 11.6 Survey grid that was used for data acquisition. Red dots indicate the sensors' positions

(NE of the Isle of Majorca). A comprehensive feasibility study was also undertaken by Simyrdanis et al. (2015, 2016) who investigated the efficacy of ERT for reconstructing submerged archaeological material in shallow seawater environments. That research was undertaken at the Minoan archaeological site of Agioi Theodoroi in Crete, which contains a number of stone walls that were submerged due to recent tectonic activity. Passaro et al. (2009) and Passaro (2010) applied ERT to the investigation of a shipwreck at the Agropoli town of Salerno in Italy. The success of these studies indicates that ERT is an appropriate method for imaging conductive (metallic) objects and resistive (wooden) bodies in aquatic environments. This project represents the first time, however, that ERT has been used to map an entire shipwreck in 3D.

11.3.2 Data Acquisition and Modelling

ERT was applied at the *Crowie* site in order to reconstruct the shape of the buried portion of the barge. The survey grid was 60 m by 15.5 m with the long axis parallel to the riverbank as shown in Fig. 11.6. The four corners of the grid were established using heavy rocks as anchors. Floats were placed above each of these corner points and floating measuring tapes were then run between them such that they were taut

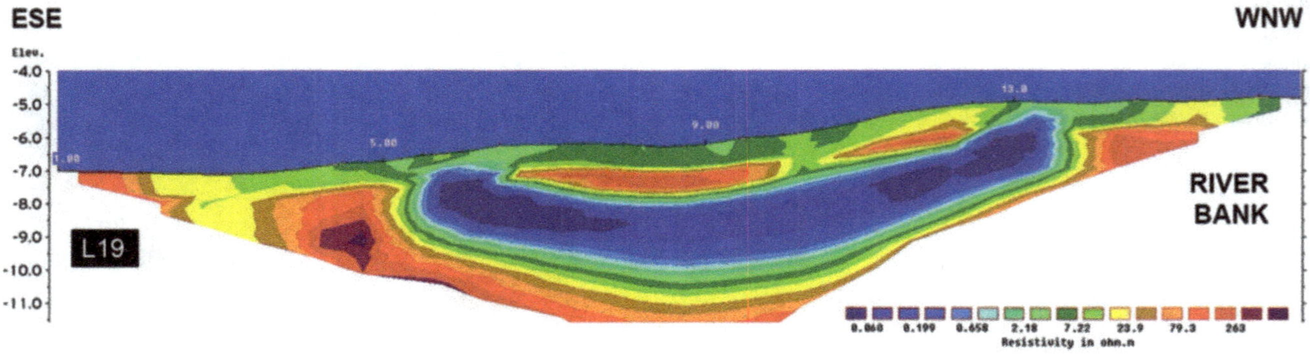

Fig. 11.7 2D resistivity profile image from Line 19

and unable to move during the course of the survey. These measuring tapes were used to guide the acquisition of 58 parallel lines oriented perpendicular to the bank and equally spaced (L = 1 m apart). The sensors (1856 electrodes in total) were equally spaced on each survey line (a = 0.5 m apart) and were placed on the water bottom (either on top of *Crowie* or directly on the surrounding river bottom). The depth to the water bottom was mapped throughout the survey area using a Leica Total Station and a prism on an extended staff. This instrument was positioned with reference to a number of static GPS points collected with a CHC 90+ GPS and post-processed using the AUSPOS service.

11.3.3 Data Processing and Results

Initially, the data from each line were filtered and post-processed individually using *Res2DInv* inversion algorithm software where the topography and the river water were incorporated (Fig. 11.7). Afterwards, data from all survey lines were merged into a unique 3D dataset which was processed with *Res2DInv* 3D inversion algorithm software. The result from the processing is a 3D visualization of the resistivity values in X, Y and Z orientations.

The resistivity values were exported into *Voxler* 3D representation software with each colour representing different resistivity values. Figure 11.8 demonstrates four different ways of visualising resistivity data within a 3D cube. In Fig. 11.8a the entire cube of resistivity values is shown with the water included, while the water is removed in Fig. 11.8b. In Fig. 11.8c the data are shown in a series of 2D 'slices' in a variety of orientations from within the resistivity model. In this case 3 slices (randomly chosen among many) have been presented, which correspond to the X, Y and Z orientations of the survey area. In Fig. 11.8d the data are plotted to show features with the same resistivity values as continuous surfaces. The approach demonstrated in Fig. 11.8d is ideal for mapping features with discrete resistivity values. The key material of interest on the Crowie was metal and so the ERT results were plotted with a low resistivity isovalue of 0.06 ohm.m. This was able to map the external boundaries of the ship (metallic parts), which can be clearly distinguished from the highly resistive background (sand and limestone sediments).

11.4 Visual Model

An alternative approach to creating a 3D model of a sunken vessel is by combining historic photos, measurements and descriptions from the literature to create a virtual reconstruction. This approach provides an important comparison to other forms of 3D modelling, such as photogrammetry or laser scanning, the results of which can be used to answer archaeological questions and to provide an effective tool for public engagement (i.e., Kormann et al. 2017; Plets et al. 2009).

In the case of *Crowie,* a visual 3D model was constructed, using the *Blender* 3D software suite, on the basis of photographs, published descriptions of the vessel's measurements and the dimensions recorded by the multibeam and sidescan sonar. Initially, a virtual box was created using the barge's maximum dimensions that acted as the outer limits of the 3D model. A virtual tube shape with the approximate form of the barge was then added. The dimensions of the vitual box and tube were informed by the measurements summarised in Roberts et al. (2017). Some detailed features, such as the name of the barge, internal division blocks and steering wheel base structure, were subsequently included based on historic photographs.

Having defined the form of the barge, *Octane Renderer* software was used to create a realistic texture for the exterior surface to enhance the visual appeal of the model. A semi-realistic appearence was used, utilizing the advantages of the 'Direct Lighting' kernel, which created a visually appealing, rather than photorealistic depiction, of this vessel (as shown in Figs. 11.9 and 11.10). A less stylized and more realistic model would be possible with better documented vessels but was unfeasible for *Crowie* given the limited number of photographs of the barge and their lack of colour.

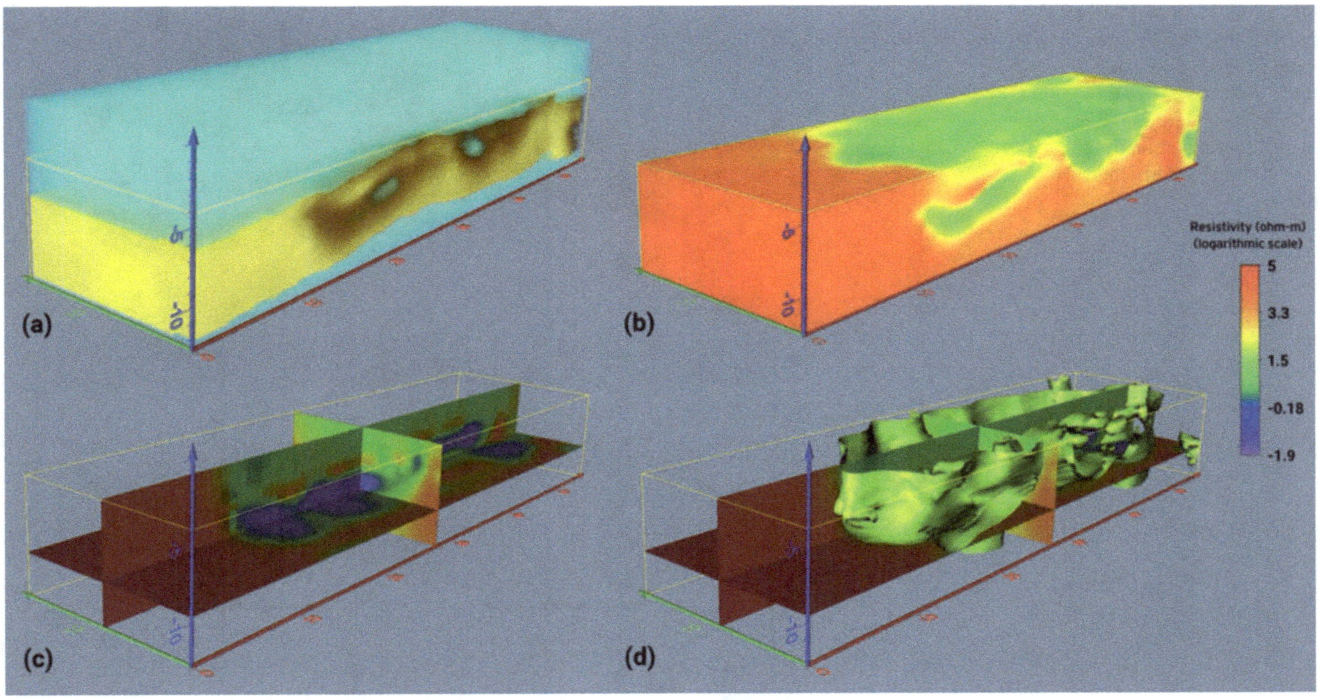

Fig. 11.8 Various representations of resistivity data collected with ERT method: (**a**) river water (light blue color) and bottom topography (light brown), (**b**) resistivity values distribution with 'volume' mode, (**c**) 2D profiles and (**d**) combined 2D profiles with isosurface mode (green color) depicting the metallic part of the barge

Once the visual model was created, it was transferred to the 3D visualizing and processing software *Meshlab* for verification. Indeed, a specific algorithm implemented in *Meshlab* (Corsini et al. 2009) allows for detailed comparison of digital model to photographs. The visual inspection of this alignment provided important clues on the morphology of the barge and helped improved the accuracy of the final model.

The model created based on historic information and sidescan/multibeam data was orientated in *Voxler* to match the ERT model allowing their dimensions and form to be compared as shown in Fig. 11.11. The results show an exceptional correspondence despite the models being generated from independent data sets. This suggests that both approaches are valid methodologies for creating 3D models of submerged or sub-surface vessels.

11.5 Discussion

The *Crowie* case study illustrates the relative advantages and disadvantages of two different 3D modelling methodologies for documenting archaeological materials which cannot be measured using conventional approaches. Clearly these methods cannot provide the same degree of spatial accuracy as is possible from survey techniques such as laser scanning or photogrammetry but are well suited to particular survey conditions, such as where the target is buried or in turbid or shallow water.

ERT was successful in the case of *Crowie* at imaging the parts of the wreck with a high degree of resistivity contrast from the surrounding materials (as shown in Figs. 11.8 and 11.11). In this case, the metal parts of the wreck (which have extremely low resistivity) were well resolved but the wooden features were much more ambiguous. An important advantage of ERT is that ferrous and non-ferrous metals do not have markedly different resistivity values and so ERT is unlike magnetometry in being able to image aluminium and other non-ferrous metals. ERT is also very suitable for shallow water contexts where sub-bottom profiling is problematic due to the abundance of 'ringing' from the sea floor reflector. A disadvantage of ERT is that it provides data with much lower resolution (0.5 m horizontal in this case) than would be possible from other methods. This resolution is governed by the minimum electrode spacing which is usually 0.5 m or 1 m, although it could be reduced for small survey areas. Another disadvantage of this method is that it requires a fixed survey grid and needs to be collected in static fashion, meaning it is much slower than other comparable methods.

The 3D digital model is visually appealing and easily recognisable as a cargo barge despite the image being stylized. While the image appears detailed, it is based on relatively sparse information and so the representation of the vessel's features is interpretive rather than accurate. In the context of public outreach, these (necessary) inaccuracies are trivial, however they may be more important for detailed research

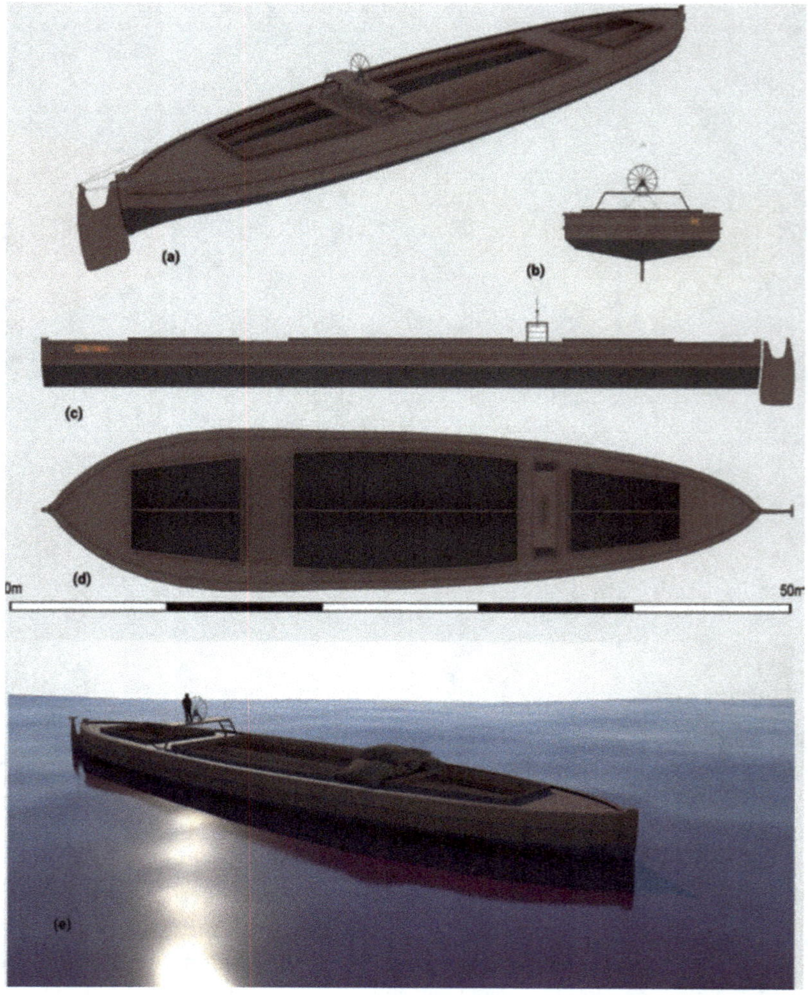

Fig. 11.9 3D representation of the *Crowie* barge from various perspectives (**a**) three-quarter, (**b**) front, (**c**) side, (**d**) top view and (**e**) a realistic presentation of the *Crowie* barge (3D Blender model)

Fig. 11.10 3D view of *Crowie* during the modeling procedure

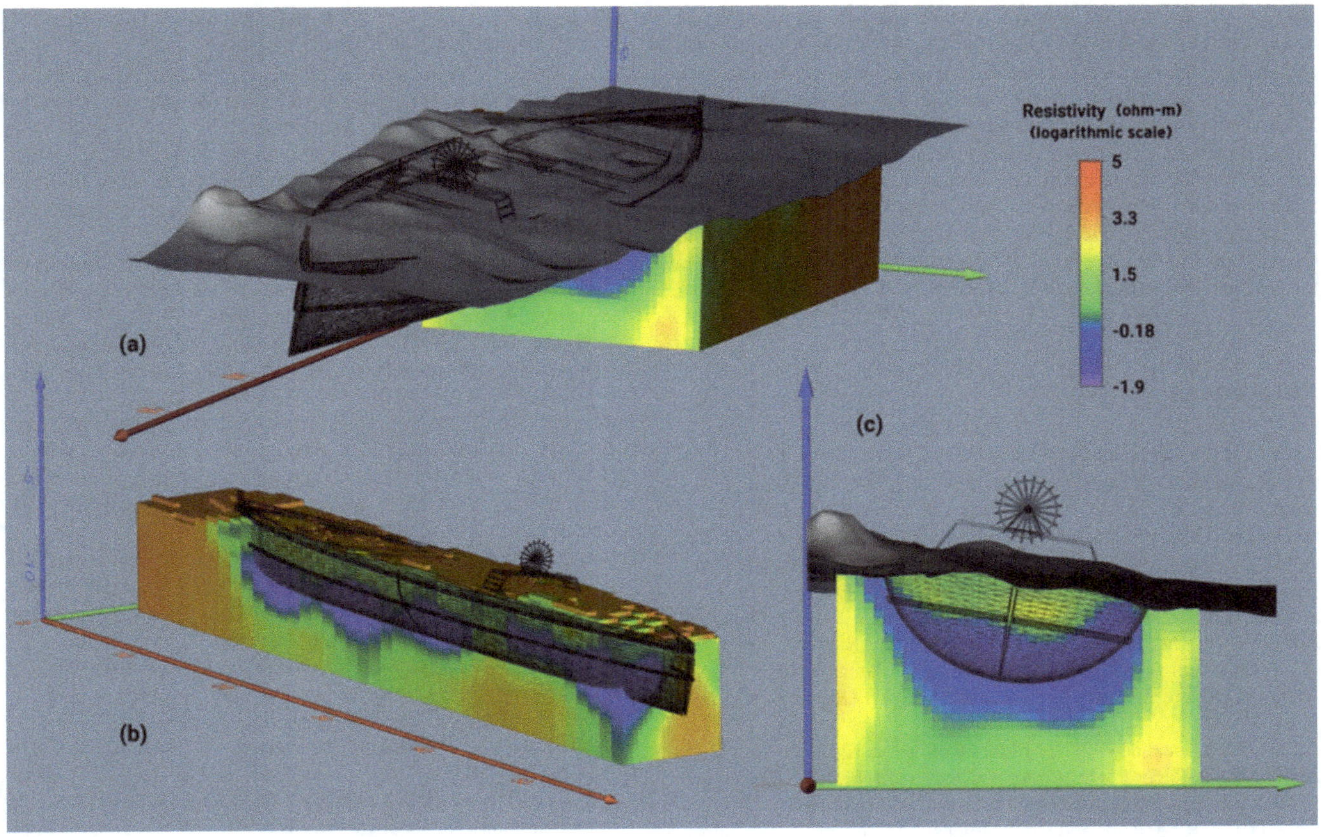

Fig. 11.11 Various views of electrical resistivity data (blue to red colours, representing respectively metallic parts to river sand) with the *Crowie* model (wireframe representation) submerged under the seabed (gray surface)

on shipbuilding. Due to the data sources, the image captures the form of the contemporary vessel when it is intact and not buried. In contrast, the 3D geophysical model based on the ERT data accurately represents the wreck in its current condition and provides a model that is much lower resolution, less visually appealing and more difficult to understand. The ERT survey also requires intensive fieldwork and specialized equipment. Nonetheless, it provides a quantitative image that is very useful for understanding the current condition of the vessel, particularly the sub-surface portion which is inaccessible to other, more conventionally applied, geophysical techniques.

11.6 Conclusions

The submerged and partially buried barge *Crowie* was used as a case study to test the applicability for a 3D reconstruction of shipwreck using both geophysical survey and historical research. The model created from ERT data provided an image of the current condition of the buried portion of the wreck while the model created from historic research combined with sidescan sonar and multibeam data provided a visually appealing 3D model with an excellent spatial correspondence with the ERT model. The final products, while different, are an evocative representation of a vessel that previously played an important role in the Murray River trade and which has been used to illustrate Aboriginal significance of riverscapes in the region. This study demonstrates that both geophysical and historical data can serve an important role in providing quantitative geometric information to constrain 3D models, particularly in low visibility conditions or when the target is buried. This project has also established that ERT is an effective geophysical method for maritime archaeology contexts, particularly in relation to shallow and turbid water environments.

Acknowledgements The overall project, including fieldwork activities, was funded by the 2017 Australia Awards-Endeavour Research Scholarships and Fellowships provided by the Australian Government and granted to Kleanthis Simyrdanis between July and December 2017. Ian Moffat is the recipient of an Australian Research Council Discovery Early Career Award (project number #DE160100703) funded by the Australian Government and a Commonwealth Rutherford Fellowship funded by the Commonwealth Scholarships Commision. That you to

Homerton College which hosted Ian Moffat as a Research Associate during the writing of this manuscript. Flinders University provided equipment and financial assistance needed for geophysical data acquisition. Thank you to Lisa and Barry from Morgan Waterfront Marina and ZZ Resistivity for their support of this research. Special thanks to Nikos Papadopoulos for his assistance during the fieldwork and data interpretation as well as to Lee Rippon, John Naumann, Celeste Jordan, Belinda Duke and Anika Johnstone who contributed to the field work for the project. We also acknowledge the River Murray and Mallee Aboriginal Corporation and the prior work of Roberts et al. (2017) which formed the basis for this methodological study.

References

Anon (1911a) The country. The Register, 10 November, p 3

Anon (1911b) Local news. Renmark Pioneer, 15 December, p 9

Anon (1912a) Wheat on the river. Chronicle, 24 February, p 16

Anon (1912b) Loxton prospects. Observer, 9 November, p 12

Anon (1912c) River matters. The Mount Barker Courier and Onkaparinga and Gumeracha Advertiser, 16 August, p 4

Anon (1913a) Wentworth on the Murray. Chronicle, 22 November, p 33

Anon (1913b) Activity at the wharf. The Mount Barker Courier and Onkaparinga and Gumeracha Advertiser, 24 January, p 4

Anon (1915) Fresh water famine on the Murray. Observer, 10 April, p 32

Anon (1916) Local news. The Mildura Cultivator, 12 July, p 10

Anon (1917) Record wheat cargo. The Mount Barker Courier and Onkaparinga and Gumeracha Advertiser, 23 March, p 4

Anon (1918) River lands and wool cargoes. The Register, 8 November, p 5

Anon (1919) River shipping. Murry Pioneer and Australian River Record, 28 November, p 6

Anon (1920) Flour mill's record output. The Mount Barker Courier and Onkaparinga and Gumeracha Advertiser, 23 April, p 1

Anon (1922a) Works on the river. Chronicle, 4 March, p 11

Anon (1922b) Barge stops on snag. Recorder, 2 December, p 2

Anon (1924) River shipping. Murray Pioneer and Australian River Record, 21 November, p 13

Anon (1925) Lock nine notes. Murray Pioneer and Australian River Record, 18 September, p 3

Anon (1927) Lock four notes. Murray Pioneer and Australian River Record, 29 April, p 4

Anon (1928) River shipping. Murray Pioneer and Australian River Record, 3 February, p 4

Anon (1939) Barge swept against lock. The Mail, 26 August, p 33

Anon (1950) Paddle steamers of the River Murray. Murray Pioneer, 14 September, p 8

Bradley JJ (1997) LI-ANTHAWIRRIYARRA, people of the sea: Yanyuwa relations with their maritime environment. Unpublished PhD dissertation, Northern Territory University

Clark A (1990) Seeing beneath the soil-prospecting methods in archaeology. B.T. Batsford Ltd, London

Corsini M, Dellepiane M, Ponchio F, Scopigno R (2009) Image-to-geometry registration: a mutual information method exploiting illumination-related geometric properties. Comput Graph Forum 28(7):1755–1764. https://doi.org/10.1111/j.1467-8659.2009.01552.x

Gale M, Sparrow S (2010) Bringing the language home: the Ngarrindjeri dictionary project. In: Hobson J (ed) Reawakening languages: theory and practice in the revitalisation of Australia's indigenous languages. Sydney University Press, Sydney, pp 387–401

Haydar M, Roussel D, Maïdi M, Otmane S, Mallem M (2011) Virtual and augmented reality for cultural computing and heritage: a case study of virtual exploration of underwater archaeological sites (preprint). Virtual Reality 15(4):311–327. https://doi.org/10.1007/s10055-010-0176-4

Kearney A (2009) Before the old people and still today: an ethnoarchaeology of Yanyuwa places and narratives of engagement. Australian Scholarly Publishing, North Melbourne

Kenderdine S (1993) Historic shipping on the River Murray: a guide to the terrestrial and submerged archaeological sites in South Australia. State Heritage Branch, Department of Environment and Land Management, Adelaide

Kormann M, Katsonopoulou D, Katsarou S, Lock G (2017) Methods for developing 3D visualizations of archaeological data: a case study of the early bronze age Helike Corridor House. STAR: Sci Technol Archaeol Res 3(2):478–489. Proceedings of the International Symposium on Archaeometry 2016 (Kalamata, Greece). https://doi.org/10.1080/20548923.2017.1372934

Liarokapis F, Kouřil P, Agrafiotis P, Demesticha S, Chmelík J, Skarlatos D (2017) 3D modelling and mapping for virtual exploration of underwater archaeology assets. Int Arch Photogramm, Remote Sens Spat Inf Sci Arch XLII-2/W3:425–431. https://doi.org/10.5194/isprs-archives-XLII-2-W3-425-2017

Nathan D, Fang M (2014) Re-imagining documentary linguistics as a revitalisation-driven practice. In: Jones M, Ogilvie S (eds) Keeping languages alive: documentation, pedagogy and revitalization. Cambridge University Press, Cambridge, pp 42–55

Orlando L (2013) Some considerations on electrical resistivity imaging for characterization of waterbed sediments. J Appl Geophys 95:77–89. https://doi.org/10.1016/j.jappgeo.2013.05.005

Papadopoulos NG, Tsourlos P, Papazachos C, Tsokas GN, Sarris A, Kim JH (2011) An algorithm for the fast 3D resistivity inversion of surface electrical resistivity data: application on imaging buried antiquities. Geophys Prospect 59(3):557–575. https://doi.org/10.1111/j.1365-2478.2010.00936.x

Passaro S (2010) Marine electrical resistivity tomography for shipwreck detection in very shallow water: a case study from Agropoli (Salerno, Southern Italy). J Archaeol Sci 37(8):1989–1998. https://doi.org/10.1016/j.jas.2010.03.004

Passaro S, Budillon F, Ruggieri S, Bilotti G, Cipriani M, Di Maio R, D'Isanto C, Giordano F, Leggieri C, Marsella E, Soldovieri MG (2009) Integrated geophysical investigation applied to the definition of buried and outcropping targets of archaeological relevance in very shallow water. Il Quaternario (Ital J Quat Sci) 22(1):33–38

Plets RMK, Dix JK, Adams JR, Bull JM, Henstock TJ, Gutowski M, Best AI (2009) The use of a high-resolution 3D Chirp sub-bottom profiler for the reconstruction of the shallow water archaeological site of the *Grace Dieu* (1439), River Hamble, UK. J Archaeol Sci 36(2):408–418. https://doi.org/10.1016/j.jas.2008.09.026

Ranieri G, Loddo F, Godio A, Stocco S, Cosentino PL, Capizzi P, Messina P, Savini A, Bruno V, Cau MA, Orfila M (2010) Reconstruction of archaeological features in a mediterranean coastal environment using noninvasive techniques, In: Frischer B, Webb Crawford J, Koller D (eds.) Making history interactive. Computer Applications and Quantitative Methods in Archaeology (CAA). Proceedings of the 37th International Conference, Williamsburg, Virginia, United States of America, March 22–26 (BAR International Series S2079). Archaeopress, Oxford, pp 330–337

Roberts A, Van Duivenvoorde W, Morrison M, Moffat I, Burke H, Kowlessar J, Naumann J with the River Murray and Mallee Aboriginal Corporation (2017) They call 'im *Crowie*': an investigation of the Aboriginal significance attributed to a wrecked River Murray barge in South Australia. Int J Naut Archaeol 46(1):132–148. https://doi.org/10.1111/1095-9270.12208

Rucker DF, Noonan GE, Greenwood WJ (2011) Electrical resistivity in support of geological mapping along the Panama Canal. Eng Geol 117(1-2):121–133. https://doi.org/10.1016/j.enggeo.2010.10.012

Simyrdanis K, Papadopoulos N, Kim J-H, Tsourlos P, Moffat I (2015) Archaeological investigations in the shallow seawater environment with electrical resistivity tomography. Near Surf Geophys 13(6):601–611. https://doi.org/10.3997/1873-0604.2015045

Simyrdanis K, Papadopoulos N, Cantoro G (2016) Shallow off-shore archaeological prospection with 3-D electrical resistivity tomography: the case of Olous (Modern Elounda), Greece. Remote Sens 8(11):897. https://doi.org/10.3390/rs8110897

Simyrdanis K, Moffat I, Papadopoulos N, Kowlessar J, Bailey M (2018) 3D mapping of the submerged *Crowie* barge using electrical resistivity tomography. Int J Geophys 2018:1–11. https://doi.org/10.1155/2018/6480565

Tsourlos P, Papadopoulos N, Yi M-J, Kim J-H, Tsokas G (2014) Comparison of measuring strategies for the 3-D electrical resistivity imaging of tumuli. J Appl Geophys 101:77–85. https://doi.org/10.1016/j.jappgeo.2013.11.003

Varinlioğlu G (2011) Data collection for a virtual museum on the underwater survey at Kaş, Turkey. Int J Naut Archaeol 40(1):182–188. https://doi.org/10.1111/j.1095-9270.2010.00304.x

Younger RM (1976) Australia's great river. Horizon Publishing, Swan Hill

HMS *Falmouth*: 3D Visualization of a First World War Shipwreck

12

Antony Firth, Jon Bedford, and David Andrews

Abstract

This chapter outlines an opportunistic yet innovative approach to developing a 3D visualization of HMS *Falmouth*, a Town Class light cruiser sunk during the First World War on the Yorkshire coast, England. The results of a multibeam echosounder survey of the seabed were combined with photogrammetry and laser scanning of the original builder's model of HMS *Falmouth*, which is in store in the collections of the Imperial War Museums (IWM). The visualization, made available via *Sketchfab*, helped to generate considerable public and media interest in an important heritage asset. This chapter also comments on the role of visualizations in engaging people for whom underwater archaeology is otherwise inaccessible, and considers the potential for visualizations to integrate research and prompt further investigation.

Keywords

Naval history · Historic shipwreck · Ship model · 3D modelling

12.1 Introduction

There are so many shipwrecks around the coast of the UK especially from the conflicts of the twentieth century that their heritage interest is overlooked both in archaeological terms and more widely amongst the public. The wealth of the UK's underwater cultural heritage receives a fraction of the attention and care directed to otherwise comparable heritage on land, perhaps only because it is out of sight and out of mind. Yet when shipwrecks are brought to light, public fascination is clearly apparent (Kenderdine 1998, 23). In such circumstances, the potential role and value of 3D visualization in engaging people's interest is self-evident.

This chapter outlines an opportunistic yet innovative approach to developing a 3D visualization of HMS *Falmouth*, sunk as a result of U-boat attacks in August 1916 and lying wrecked just off the coast of Yorkshire. The scope of the visualization was set by circumstances and the component methodologies were not new (Firth 2011; Menna et al. 2011), but the juxtaposition within an online, accessible visualization of the wreck and the ship together is relatively novel. The visualization successfully achieved its principal aim of bringing the story of HMS *Falmouth* to many more people than might have been the case otherwise. The visualization, however, also prompts a series of further considerations about visualizations as a focus for engagement, research and heritage management.

12.2 Background

HMS *Falmouth* is an accessible and well-known wreck site. The general depth of the seabed is 16 m below Chart Datum and the wreck is located about 8 km offshore, about 12 km from the nearest harbour at Bridlington. After it was sunk by U-boats in August 1916, salvage work by the Royal Navy commenced almost immediately (ADM 116/1508). Despite the relatively shallow depth the ship could not be recovered intact, but most of the main armament and some other fittings were removed. Although details are unavailable it seems that the wreck was cleared as a navigational hazard and targeted by commercial salvors in the interwar period. HMS *Falmouth* has been dived by recreational divers since the 1970s and further piecemeal salvage has

A shorter version of this chapter previously appeared in Issue 7 (Autumn 2017) of *Historic England Research*.

A. Firth (✉)
Fjordr Limited, Tisbury, Wiltshire, UK
e-mail: ajfirth@fjordr.com

J. Bedford · D. Andrews
Historic England, York, UK
e-mail: jon.bedford@historicengland.org.uk; david.andrews@historicengland.org.uk

J. K. McCarthy et al. (eds.), *3D Recording and Interpretation for Maritime Archaeology*, Coastal Research Library 31,
https://doi.org/10.1007/978-3-030-03635-5_12

Fig. 12.1 Contemporary postcard of the Town Class light cruiser HMS *Falmouth*

occurred—possibly on a large scale. Given this amount of disturbance, exacerbated by natural processes, HMS *Falmouth* has been described as a 'mangled wreck … fascinating for rummagers' (Divernet n.d.).

The basic details of HMS *Falmouth* and its loss are well known (Lyon 1977; Newbolt 1928) but the overall significance of the ship and the wreck have been overlooked. Fjordr Limited, a heritage consultancy, proposed a project to Historic England—the national heritage agency for England—to examine the significance of HMS *Falmouth* and to raise awareness, especially amongst the wider public who visit or live at the coast but are not aware of their heritage just offshore. The project linked the question of significance directly to the potential to develop greater social and economic benefits from maritime heritage, especially in struggling coastal communities. The project was commissioned by Historic England and resulted in a series of outputs, including a formal Statement of Significance (Firth 2016a) and a fold-out leaflet (Firth 2016b) that was distributed through Tourist Information Centres and local museums.

Both the Statement of Significance and the fold-out leaflet addressed the build, use and loss of HMS *Falmouth* as a complete 'ship biography.' This included setting HMS *Falmouth* in its context as a Town Class light cruiser, a class that embodied features of the Dreadnought revolution in a vessel with global reach. Members of this class were heavily used in the First World War, participating in many key engagements. HMS *Falmouth* was no exception. Stationed in the North Sea at Scapa Flow and then Rosyth on the Firth of Forth, *Falmouth* took part in numerous sweeps to intercept fishing vessels and merchant ships as well as in fleet actions, notably the First Battle of Heligoland Bight and the attempted interception of the German raid on Scarborough, Whitby and Hartlepool. At the Battle of Jutland, HMS *Falmouth* was the flagship of the Third Light Cruiser Squadron attached to the Battle Cruiser Fleet, engaging repeatedly with the German fleet (Fig. 12.1). *Falmouth* was again involved in a major operation by the Grand Fleet to intercept the German fleet on 19 August 1916 when the ship was torpedoed by a U-boat in the North Sea about 110 km east of Scarborough. Strenuous efforts were made to reach the safety of the Humber over the next 28 h, despite torpedo hits from a further U-boat, but the ship succumbed just offshore (Firth 2016a).

The century between 1850 and 1950 saw significant change in steam-powered iron and steel cargo ships and warships. Few larger vessels from this revolutionary period, however, survive in preservation as museum ships and, outside the former Royal Dockyards, little survives of the UK's shipbuilding heritage on land. It is a paradox that the principal survivors of Britain's mid-nineteenth- to mid-twentieth-century maritime heritage are those that were sunk and now lie on the seabed; almost all of the vessels that reached the ends of their careers still afloat were scrapped to leave no physical trace (Firth 2016c; Firth and Rowe 2016). The apparent absence of the material remains of this aspect of the UK's national story in this period—times that are otherwise well-represented by built heritage, industrial archaeology and all manner of preserved vehicles and aircraft—ought to be a concern not only for those who have a technical interest in ship construction or military history. The material heritage of all those communities engaged in shipbuilding, in seafaring and in maritime commerce and conflict more broadly has been erased except for those elements that are currently hidden by the waves. Furthermore, for most of the period since sinking this heritage has often been subject to damaging salvage and clearance activities, whilst its intrinsically unstable character suggests that it will suffer further degradation in coming decades.

This is true of all that HMS *Falmouth* represents. The civil shipbuilding yard of Beardmore's on the Clyde where *Falmouth* was built now lies under an industrial estate, a hospital and a hotel. *Falmouth* is the only known survivor of the important Town Class; all the others were scrapped—mostly in the 1920s and 1930s (Lyon 1977)—except for HMS *Nottingham* also sunk by U-boats in the same operation as *Falmouth* but whose wreck has yet to be found. Although there are others elsewhere—and the surviving HMS *Caroline* now receiving attention in preservation—*Falmouth* is also the only substantial wreck of a veteran of Jutland in England's territorial waters. Both in itself and as a representative of its class, this overlooked 'rummage' is a rare and significant part of our twentieth-century maritime heritage. Subsequent to the project, HMS *Falmouth* was designated under the *Protection of Military Remains Act* 1986; but this designation, administered by the Ministry of Defence, reflects the loss of service personnel and entails no proactive provision for future management of the wreck as a heritage site.

12.3 Origins of the 3D Visualization of HMS *Falmouth*

It is fortunate that 3D technologies for acquisition and visualization of underwater sites have become available just as the heritage value of nineteenth- and twentieth-century wrecks has begun to be recognized. Indeed, the ability to 'see' what survives on the seabed has probably contributed—with some key anniversaries—to recognising more recent wreck sites as heritage, even amongst archaeologists. Shipwrecks built from, or with major components made from, iron and steel tend to be extensive, complex and highly three-dimensional in the field, presenting difficulties to the conventional recording methods available to marine archaeologists up to the mid-1990s. The capability and increasingly mainstream availability of multibeam, photogrammetry and 3D visualization software have made the recording and interpretation of mid-nineteenth- to mid-twentieth-century shipwrecks possible in a way that could barely be contemplated just a few decades ago. Achieving a basic 2D survey would be the work of many hours underwater; the results of which are now far surpassed in minutes. The radical impact that this is having on our capacity to interpret underwater sites and to share them with a much wider public who need never approach the water, let alone go diving, will take time to fully appreciate as the technology sprints along. Although relevant to many types of site, for the archaeology of twentieth-century wrecks such as HMS *Falmouth* these are very exciting times.

Nonetheless, creating a 3D visualization of HMS *Falmouth* was not part of the original project. The possibility arose opportunistically and with only a short timeframe available before the results of the project were due to be launched. The first piece of luck was that the Maritime and Coastguard Agency (MCA)—the government agency responsible for navigational safety—was planning a high resolution multibeam survey off the Yorkshire coast as part of their Civil Hydrography Programme. Although not in the planned survey area, the MCA agreed to add a survey of HMS *Falmouth* to their contractor's programme for a modest contribution from Historic England. The survey was carried out by the MCA's survey contractor MMT aboard M/V SeaBeam using a Kongsberg EM2040D, which is a dual-head multibeam echosounder used for high resolution bathymetric surveying, and the survey was conducted in accordance with the MCA's standard specification for wreck investigation surveys (Fig. 12.2). The results were made available to the HMS *Falmouth* project in May 2016. The survey is excellent and captured many details of the wreck, showing that despite a great deal of degradation the wreck clearly retains a fair amount of overall coherence. Features such as the distinctive Y-shaped Yarrow boilers can be identified in the multibeam data and related to original drawings and to photographs taken by divers of the wreck on the seabed. Obtaining such a detailed survey was itself a major contribution to better understanding the survival and significance of the wreck.

The second piece of luck concerned the builder's model of HMS *Falmouth* (Fig. 12.3). As a result of collaboration with the National Museum of the Royal Navy over the Jutland 36 Hours exhibition (Firth 2016c), Fjordr became aware of a model of HMS *Falmouth* in the stored collection

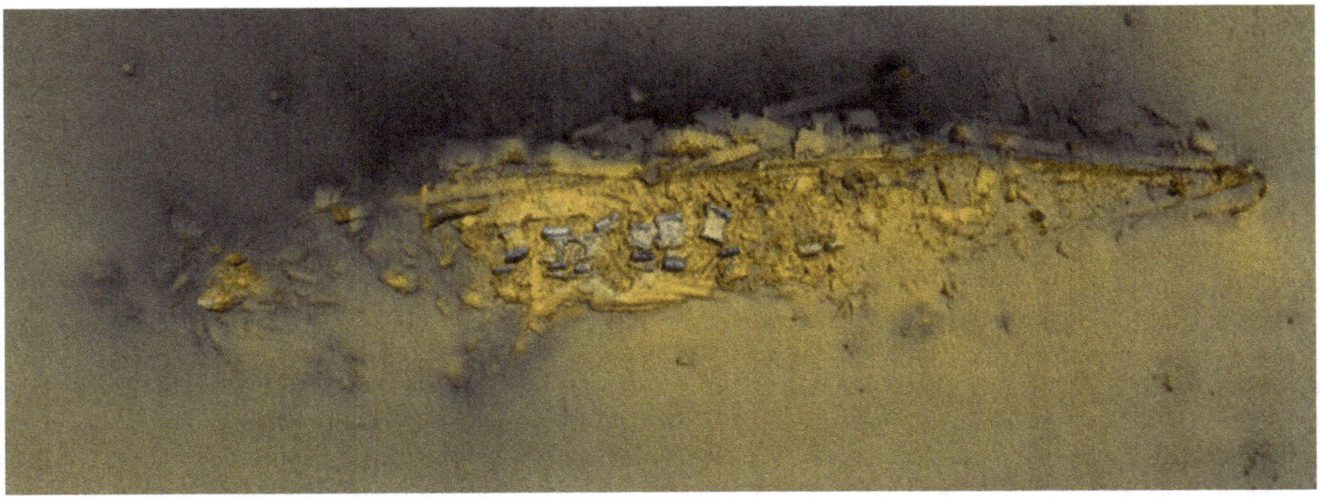

Fig. 12.2 Results of multibeam survey of HMS *Falmouth* by MMT for Civil Hydrography Programme. (Maritime and Coastguard Agency © Crown copyright)

Fig. 12.3 Photogrammetric survey of the builder's model of HMS *Falmouth*, carried out by Historic England Geospatial Imaging team at the Chatham store of the Imperial War Museums. Antony Firth, Fjordr

of the Imperial War Museums. Few details and no photographs of the model were available from IWM. Model maker John Haynes, however, had been commissioned to restore the model in 1979. Haynes kindly supplied photographs of the model before and after restoration. The model was a large-scale builder's model which is presumed to have been made by Beardmore's in 1910–1911. The model formed part of the King's ship model collection, which became part of the Imperial War Museums' collection when it was established in 1917. The model was damaged when the IWM was bombed during the Second World War and had deteriorated further. Haynes fully restored the model in superb detail for IWM but the model remained in store thereafter.

With the results of the multibeam, and knowing about the model, the idea came about to try and combine both into a single 3D visualization that juxtaposed the wreck with the original ship and could be made available to the public. The initial focus was on obtaining suitable access to the model from IWM, and taking initial advice on the feasibility of making a virtual model from John McCarthy, then at Wessex Archaeology in Edinburgh. Historic England's own Geospatial Imaging team came on-board to acquire data from the physical model and to develop the visualization. The timescale was limited because of the plan to publish and distribute a leaflet together with a media release in time for the centenary of *Falmouth*'s loss on 19–20 August 2016. The IWM kindly provided access to the model at their store at Chatham Historic Dockyard. In view of the technical innovation and the short timescale, the initial aspiration was to obtain still images that juxtaposed the wreck and the ship model. The hope, however, was also to create a full visualization that the public could access and explore. As the opportunity and the aspiration were opportunistic and unavoidably short-term, no attempt could be made to set the exercise in a broader context in terms of methodologies or outcomes; the authors relied on the practical experience of their craft rather than a formal design process.

12.4 Data Acquisition and Processing of the Ship Model

The Historic England Geospatial Imaging team visited Chatham Historic Dockyard on two days in mid-June 2016. The model was recorded using laser scanning and multi-image photogrammetry (Figs. 12.3 and 12.4).

The model was scanned using a Leica ScanStation P40 terrestrial scanner. The aim of the laser scanning was to provide control for the multi-image photogrammetry that was used to produce the finished model. The model was scanned from six positions with an average point spacing of 3 mm. The scans were registered together to form a point cloud using resection to common targets. The scanner did not cope very well with the rigging so there was a lot of noise in that area. There were, however, plenty of points on the hull that could be used for control. The final model was generated from 891 photographs taken with a Sony ILCE-7RM2 camera. The photographs were taken so that they overlapped with at least two others and in most cases, many more. They were taken from as many different angles as possible to achieve complete coverage of the ship. Each image is 40 Mp resulting in a 120 Mb TIF file, so the photography resulted in 104 GB of imagery.

The photography was processed using the multi-image photogrammetry software *RealityCapture* by *CapturingReality* of Slovakia. This software allows the integration of laser scan data. Even using a high-end workstation with 128 Mb RAM and specialist graphics card the processing took several days. Although the model was much less noisy than the laser scan data, it still required a lot of cleaning. In the end, the masts and associated rigging were removed from the final model.

The still images from the visualization were ready in time to include in the design of the fold-out leaflet. A preliminary animation was also prepared. It was decided, however, to concentrate on *Sketchfab* (2018) as a means of making the visualization available to the public, so that people could explore the visualization themselves rather than passively watching a fly-through. On *Sketchfab*, the visualization was accompanied by text with links to further information—including online versions of the fold-out

Fig. 12.4 Static image from the 3D visualisation of HMS *Falmouth*, as used in fold-out leaflet and press release. (Courtesy of Historic England. Maritime and Coastguard Agency © Crown copyright)

leaflet and Statement of Significance—and numbered annotations were added to the visualization itself to highlight features of the ship and the wreck, and to tell the story of the seven torpedoes that resulted in *Falmouth*'s loss.

12.5 Publication of the 3D Visualization

The 3D visualization of HMS *Falmouth* on *Sketchfab* was made public to coincide with a media release on the centenary of *Falmouth*'s loss (Fig. 12.4). The fold-out leaflet with an image from the visualization was distributed to museums and Tourist Information Centres in time for the centenary. A link to the *Sketchfab* visualization was included in Historic England's own web page on HMS *Falmouth* (Historic England n.d.). The process of creating the visualization provided the hook for Historic England's press release: 'Jutland Wreck Brought to Life' (Historic England 2016). The still image from the visualization featured in the extensive national and regional press coverage of HMS *Falmouth*'s centenary and the online versions of many newspapers embedded the *Sketchfab* visualization within their pages, adding to impact and connectivity (Table 12.1). The visualization also provided a striking image for posts on Twitter (hashtag #HMSFalmouth); each event relating to the sinking of HMS *Falmouth* was 'live tweeted' on its to-the-minute centenary, adding to overall impact.

Creating a 3D visualization from such a detailed physical model was very demanding, especially within the short timescale available from data acquisition (15–16 June) to printed output as a fold-out leaflet in time for the centenary (19–20 August). As a result, there are limitations to the visualization. Much detail of the rigging had to be cut out and the ship is decidedly 'blocky' when zoomed in to the ship model. Nonetheless, the visualization contributed very significantly to the overall objective of raising awareness of HMS *Falmouth*, especially as a hook for the media. The visualization was a 'staff pick' on *Sketchfab* and as of May 2018 it has had over 21,400 views. The visualization, undoubtedly, will continue to serve as an intriguing conduit for people to find their way to more detailed information about HMS *Falmouth*.

12.6 Development Potential of 3D Visualization for Further Research and Public Engagement

The opportunistic development of a 3D visualization of HMS *Falmouth* achieved its immediate purpose, generating a great deal of media interest and views on *Sketchfab* to mark the centenary of *Falmouth*'s loss. The visualization still holds great potential for further development, however, which may be relevant to how visualizations are considered more broadly amongst the tools available to archaeologists. In this regard, it is appropriate to consider the visualization on *Sketchfab* as a principal tool or output, rather than as popular outreach. A key benefit of visualizations on *Sketchfab* and other comparable platforms is their accessibility without recourse to the specialist software used in their preparation. They can be likened to the pdf of a journal article (especially as pdfs can now contain 3D content); easy to distribute and discuss, reflecting but not including the detailed data that underpins the conclusion. There is a need, of course, for methodological transparency, peer review and contestability. Such visualizations will undoubtedly improve in detail and capability as technology continues to progress. Even if current visualizations will seem rudimentary in only a few years, they can still be regarded as a reasoned foundation upon which further research, management options and public engagement can be based.

Arguably, the real foundation is the 3D survey data—acquired using multibeam, photogrammetry, laser scanning and so on—rather than the visualization. The presentation of 3D data in a way, however, that can be examined and explored directly by others—including those without specialist skills in the specific survey technologies—changes the context. Visualizations might be regarded as hypotheses reflecting the selectivity and choices made methodologically, but the result can still be tested and interrogated more than a 2D representation of the same data on a physical page. The accessibility of the platform adds to the openness of visualizations, which is important if discussions over

Table 12.1 Examples of press coverage of HMS *Falmouth*, August 2016

The Guardian	19 August 2016	https://www.theguardian.com/world/2016/aug/19/first-world-war-wreck-gets-virtual-restoration-off-coast-of-yorkshire. Accessed May 2018
The Telegraph	19 August 2016	http://www.telegraph.co.uk/science/2016/08/18/shipwreck-of-hms-falmouth-brought-back-to-life-on-100th-annivers/. Accessed May 2018
The Mirror	19 August 2016	http://www.mirror.co.uk/news/uk-news/digital-wizardry-brings-sunken-world-8660979. Accessed May 2018
The Yorkshire Post	19 August 2016	http://www.yorkshirepost.co.uk/news/analysis/first-world-war-wreck-off-yorkshire-s-coast-brought-back-to-life-1-8077498. Accessed May 2018
BBC News	20 August 2016	http://www.bbc.co.uk/news/uk-england-humber-37142617. Accessed May 2018

future management or physical accessibility are being raised with for example the fishing community or local divers.

One of our aspirations was to incorporate other sources into the visualization, especially the large-scale plans, profiles and sections of *Falmouth*'s sister-ship HMS *Weymouth* which are held by the National Maritime Museum. The capacity of visualizations to enable the public to access wrecks which are remote to most people is often commented upon (Kenderdine 1998, 17; Reunanen et al. 2015, 24.3). The same is also true of resources such as documents, drawings, photographs and models that are not intrinsically remote but which are not easily accessible because they are spread around various institutions, in store and/or require special handling. Enabling people to explore such sources virtually—juxtaposed with the ship as a 3D entity and with the remains on the seabed—is a prize worth pursuing.

As well as physical remoteness, the HMS *Falmouth* visualization might also be considered to counteract a form of conceptual remoteness also. Shipwreck data are not always easy to 'read.' As Adams notes, 'even to experienced eyes the relationship of many wrecks to the complete entity they once were is often far from clear' (Adams 2013, 94). One of the benefits of multibeam is that images are more readily understood by lay people than other forms of remote sensing, at least in the case of relatively intact wrecks, but HMS *Falmouth* is perhaps more typical of the many wrecks that are already quite degraded. Using 3D visualizations to juxtapose wrecks with the ships they once were, as we have done in this instance, is therefore a means of making shipwrecks less remote conceptually as well as physically. Even this simple juxtaposition suggests that the remains of HMS *Falmouth* are more complete and coherent than the history of clearance and salvage might suggest—especially in the buried portion of the lower hull. The impression is at odds with the earlier perception of the wreck as a scrapyard and a good rummage. Hence, conceptual access is not just about reaching audiences: making an association with the ship could help elevate the physical material from random wreckage to meaningful heritage, nudging behaviours towards maritime archaeology both on site and in wider society.

The visualization has also helped in identifying the original position of photographs of crew taken aboard *Falmouth* (Figs. 12.5 and 12.6), which it might be possible to include in the visualization in future. This has two important aspects. First, it helps in placing people back aboard the ship, to present it as a human, lived-in space, even aboard a warship that was in the thick of the action at times. Although it is possible to place crew photographs based on ship drawings or models, the ability to obtain the same viewpoint as the camera and recreate the same immediate landscape of the people in the historic image provides—literally—a new perspective on the vessel (Fig. 12.5). This is especially valuable insofar as the human dimension of ships in use tends to be obscured by

Fig. 12.5 Top: Photograph of HMS *Falmouth*'s officers, probably taken in summer 1916 around the time of the Battle of Jutland. (Courtesy John McDonald). Bottom: The 3D visualisation of the area corresponding to the officers' photograph. (Maritime and Coastguard Agency © Crown copyright)

a focus on their design and construction, or on the circumstances of loss. As Adams notes, visualization can help ensure that 'a ship as a thing cannot be separated from the people who conceived, designed, built, used and either lost or disposed of it' (Adams 2013, 94). Thinking especially about models such as that of HMS *Falmouth*, which exist in museum collections and stores in profusion, the comments of Cooper et al. (2018, 17) are apposite:

> An abiding challenge for the presentation of watercraft in a dry and static museum gallery is the fact that boats and ships are, in their intended applications, dynamic structures in ever changing aquatic environment. Digital modelling … enables museums to overcome the stasis of the museum object … and engage visitors with the lived experience of vessels …

In this respect, the attempt to place crew photographs aboard *Falmouth* using the visualization resonated strongly with a line from the diary of a young gunnery officer,

Fig. 12.6 Top: Photograph of one of Falmouth's Y-shaped Yarrow boilers, taken by local diver Mike Radley. (Mike Radley). Bottom: Yarrow Boiler Screenshot. (Maritime and Coastguard Agency © Crown copyright)

Arnold Pears, aboard *Falmouth* at Jutland and at the time of its loss (LIDDLE/WW1/RNMN/235):

> I have no heart to write … the loss of that ship, the symbol to me of my home, my work, my play, my life, my companion in danger, hits me too hard …

The second important aspect of crew photographs is that the visualization has formed part of a project that has become a focus for members of the public to contribute their own stories, associations and sources to HMS *Falmouth*. This was not a key objective but is a somewhat unintentional (though very welcome) consequence of social media. Photographs and documents relating to the crew have some to light from privately held, personal archives that have been handed down or acquired in conjunction with, for example, medal collecting as a hobby. The personal connections to *Falmouth*'s crew add considerably to the significance of the vessel as well as being a further source of primary evidence. It cannot be claimed that visualization of HMS *Falmouth* was an exercise in participation and co-production, as encouraged by Jones et al. (2018). HMS *Falmouth* does, however, at least point in the direction that such a project might take in the marine sphere. Visualization could prompt and provide a focus for the public in researching their own connections to ships like HMS *Falmouth* and major themes such as the First World War at sea. This would help to unlock the huge potential of historic material held privately in families and communities, enabling the public to contribute their own knowledge and associations and thereby shape their own maritime histories.

Further opportunities for community-based co-production are provided by the scope to combine the visualization with still photographs and videos of the wreck taken by divers (Fig. 12.6), enabling components of the wreck to be identified and observations to be made on survival and condition. Although the current visualization does not incorporate the ship drawings and diver photographs, it has been used alongside such sources to better understand what survives of the machinery spaces of HMS *Falmouth*, for example. The visualization could also be used as a focus for future fieldwork by volunteers, perhaps adding detail to the current visualization with localized 3D models derived from underwater photography. Together with the supporting fold-out booklet, the visualization can already help recreational divers to better understand the wreck and the relationship of its features to the original ship. There is potential also for the visualization to provide the basis for a 'virtual dive trail' that will enable the non-diving public to visit the wreck.

Remarkably, HMS *Falmouth*'s Armed Steam Cutter—the principal ship's boat, which provided an important element of the ship's capability as a light cruiser—has survived to the present day, despite being abandoned at the time of loss. The cutter was recovered as salvage by a fishing boat (ADM 116/1508) and, after a career of its own, has been acquired by Portsmouth Naval Base Historic Trust. The Trust intends to restore the cutter to steam as part of its Memorial Fleet of small boats. The cutter is prominent on the builder's model and has been highlighted on *Sketchfab*, underlining the tremendous range of hitherto disparate historical and archaeological sources that are being re-connected and integrated by the visualization.

12.7 Conclusions

3D visualization of HMS *Falmouth* has contributed substantively to the project's objective of raising awareness of the significance of a known but undervalued wreck, making the story of the ship accessible to a much wider audience.

The project has successfully combined specialist survey data from a degraded wreck with a builder's model that has been hidden from the public for at least 70 years. In a sense, this was an entirely satisfactory end point, but it should also be apparent that 3D visualizations such as that of HMS *Falmouth* are not only a product to be measured in terms of the number of views they receive. Rather, the visualization of HMS *Falmouth* is part of a process that will extend well into the future, alongside the wreck and the many other archive sources—physical and documentary; held publicly and privately—that relate to *Falmouth*'s story. The visualization has a special place because of its capacity to help integrate so much disparate and often inaccessible material: a seed dangling in 3D space around which the overlooked history of HMS *Falmouth* can, in time, crystallize. Importantly, the visualization exists in a public space and engagement with communities, whether locally on the Yorkshire coast or globally across the web, will continue to add many facets to *Falmouth*'s story. Looking beyond this particular wreck, it is hoped that the project points the way to far greater use of 3D visualization to bring to life underwater cultural heritage, representing and reconnecting the full range of evidence upon which we draw. The capacity to juxtapose survey data and ship models is especially worth pursuing, to mobilize plentiful but underused resources to shine a light on the UK's mid-nineteenth- to mid-twentieth-century maritime history. Finally, the 3D juxtaposition of wreck and ship reminds us to see shipwrecks as inhabited places; to focus on people as well as their technology.

Acknowledgements This chapter has been prepared by Antony Firth of Fjordr Limited with contributions from Jon Bedford and David Andrews of Historic England, who carried out the photogrammetry and laser scanning of the ship model and the combined 3D visualization. The HMS *Falmouth* project as a whole was funded by Historic England; our thanks go to Wayne Cocroft and Gareth Watkins. Access to the builder's model of HMS *Falmouth* was provided by the Imperial War Museums courtesy of Maria Rollo. John Haynes, model maker and restorer, kindly provided information and photographs that helped in identifying the builder's model. Thanks are also due to John McCarthy for initial advice on the feasibility of scanning the builder's model. The multibeam survey was commissioned, with the support of Historic England, by the Maritime and Coastguard Agency (MCA) from their survey contractor MMT; Robert Kinnear and Paula English of MCA and Martin Godfrey at MMT all have our gratitude. Information on the significance of HMS *Falmouth*'s cutter, and access to the boat itself, has been provided by Portsmouth Historic Naval Base Trust courtesy of Peter Goodship and Diggory Rose. The background research on HMS *Falmouth* has drawn on archives held by a wide range of institutions and private individuals, whose kind assistance is gratefully acknowledged. Attributions for images are noted in the figure captions. We are grateful also to the editors and an anonymous reviewer for comments which strengthened the paper in review.

References

Adams JR (2013) Experiencing shipwrecks and the primacy of vision. In: Adams JR, Ronnby J (eds) Interpreting shipwrecks: maritime archaeological approaches. The Highfield Press, Southampton, pp 85–96

ADM (116/1508) HMS *Falmouth* salvage of guns etc. The National Archives

Cooper JP, Wetherelt A, Zazzaro C, Eyre M (2018) From Boatyard to museum: 3D laser scanning and digital modelling of the Qatar Museums watercraft collection, Doha, Qatar. Int J Naut Archaeol 47(2):419–442. https://doi.org/10.1111/1095-9270.12298

Divernet (n.d.) 100 Best UK wreck dives: the HMS *Falmouth*—78. http://www.divernet.com/100-best-uk-wreck-dives/p301908-the-hms-falmouth-78.html. Accessed May 2018

Firth A (2011) Marine geophysics: integrated approaches to sensing the seabed. In: Cowley D (ed) Remote sensing for archaeological heritage management, EAC Occasional Paper No, 5. Europae Archaeologia Consilium, Brussels, pp 129–140

Firth A (2016a) Wreck of HMS *Falmouth*, off Bridlington: statement of significance with supporting narrative. Unpublished document for Historic England. https://content.historicengland.org.uk/content/docs/wreck-hms-falmouth-statement-of-significance.pdf. Accessed Sept 2017

Firth A (2016b) The wreck of HMS *Falmouth*: First World War 'Town Class' light cruiser and Jutland veteran sunk by U-boats off the Yorkshire coast on 19–20 August 1916. Leaflet prepared on behalf of Historic England. http://content.historicengland.org.uk/content/docs/hms-falmouth-broadsheet.pdf. Accessed May 2018

Firth A (2016c) Jutland: an archaeological perspective. In: Sheldon M (ed) 36 hours: Jutland 1916—the battle that won the war. National Museum of the Royal Navy, Portsmouth

Firth A, Rowe P (2016) The national importance of cargo vessels: Tees pilot. Unpublished report for Historic England. Project Number HE 7051, Fjordr Ref: 16261. Fjordr Limited, Tisbury. https://historicengland.org.uk/images-books/publications/national-importance-cargo-vessels-teespilot/national-importance-of-cargo-vessels-tees-pilot-report/. Accessed Dec 2018

Historic England (2016) Secrets of the deep—Jutland Wreck brought to life. https://historicengland.org.uk/whats-new/news/jutland-wreck-brought-to-life. Accessed May 2018

Historic England (n.d.) HMS *Falmouth*: a veteran of the Battle of Jutland. https://historicengland.org.uk/whats-new/first-world-war-home-front/what-we-already-know/sea/hms-falmouth/. Accessed May 2018

Jones S, Jeffrey S, Maxwell M, Hale A, Jones C (2018) 3D heritage visualisation and the negotiation of authenticity: the ACCORD project. Int J Herit Stud 24(4). https://doi.org/10.1080/13527258.2017.1378905

Kenderdine S (1998) Sailing on the silicon sea: the design of a virtual maritime museum. Arch Mus Inform 12(1):17–38

LIDDLE/WW1/RNMN/235 (PEARS, S ARNOLD) Leeds University Library, Liddle Collection

Lyon D (1977) The first town class 1908–31, Parts I–III. Warship 1977:1–3, 48–58, 54–61, 46–51

Menna F, Nocerino E, Scamardella A (2011) Reverse engineering and 3D modelling for digital documentation of maritime heritage. Int Arch Photogramm, Remote Sens Spat Inf

Sci XXXVIII(5/W16):245–252. https://doi.org/10.5194/isprsarchives-XXXVIII-5-W16-245-2011

Newbolt H (1928) Naval operations: history of the Great War based on official documents, vol IV. Naval & Military Press, Uckfield, Reprint, 2003

Reunanen M, Díaz L, Horttana T (2015) A holistic user-centered approach to immersive digital cultural heritage installations: case *Vrouw Maria*. J Comput Cult Herit 7(4):24.1–24.16. https://doi.org/10.1145/2637485

Sketchfab (2018) https://sketchfab.com. Accessed 23 July 2018

13

Beacon Virtua: A Virtual Reality Simulation Detailing the Recent and Shipwreck History of Beacon Island, Western Australia

Andrew Woods, Nick Oliver, Paul Bourke, Jeremy Green, and Alistair Paterson

Abstract

Beacon Virtua is a project to document and virtually preserve a historically significant offshore island as a virtual reality experience. In 1629, survivors of the wreck of VOC ship *Batavia* took refuge on Beacon Island, Western Australia, followed by a mutiny and massacre. In the 1950s the island became the base of a successful fishing industry, and in 1963 human remains from *Batavia* were located. The fishing community has recently been moved off the island to protect and preserve the site and allow a thorough archaeological investigation of the island. Beacon Virtua exposes users to the history of both the shipwreck survivors and the fishing community. The project uses the virtual environment development software Unity to present a simulation of the island, with 3D models of buildings and jetties, photogrammetric 3D reconstructions of graves and other features, 360° photographic panoramas, and information on the history of the island. The experience has been made available on a wide range of different platforms including via a web-page, as part of an exhibition, and on head mounted displays (VR headsets). This chapter discusses the features included in Beacon Virtua, the storytelling techniques used in the simulation, the challenges encountered and solutions used during the project.

Keywords

3D visualization · Beacon Virtua · Maritime landscape · Storytelling · Unity

13.1 Introduction

On the morning of 4 June 1629, the VOC (Dutch East India Company) ship *Batavia* struck Morning Reef off Beacon Island—1 of about 100 islands in the Houtman Abrolhos Archipelago, off the coast of Western Australia. Of the 322 people on board *Batavia* when it grounded, only 122 would eventually reach the vessel's intended destination of *Batavia* (modern-day Jakarta) (Roeper 2002, 220–221). Two-hundred people died at this remote location, either by drowning, due to illness or injury, or by murder or execution (Roeper 2002, 220–221). Survivors from the wreck initially mainly took refuge on Beacon Island. The senior merchant Francisco Pelsaert took a ship's boat and yawl, with 48 people on-board to look for water and supplies, but soon decided to head to *Batavia* which they finally reached a month later (Roeper 2002, 69). Left in charge of the survivors was junior merchant Jeronimus Cornelisz who progressively oversaw the murder of 125 individuals, initially under the pretence of limited provisions, but later for no obvious reason (Drake-Brockman 2006, 130; Roeper 2002, 85). The story of the *Batavia* period of Beacon Island's history is a real-world 'Horrible History' and has been written about widely (Dash 2003; Drake-Brockman 1956, 2006; Edwards 1966; FitzSimons 2011; Pelsaert 1647; Roeper 2002). Beacon Virtua is a virtual reality (VR) simulation of Beacon Island which explores ways of exposing visitors to five periods of Beacon Island's history: (a) the wrecking of *Batavia* in 1629 and the subsequent experience of the crew and passengers on the island, (b) the fishing history of the island from the 1950s through to around 2010, (c) the discovery of shallow graves and the *Batavia* wreck site and subsequent excavation in the 1960s and 1970s (Green 1989) (d) the recent history with the

A. Woods (✉) · N. Oliver
HIVE, Curtin University, Perth, WA, Australia
e-mail: A.Woods@curtin.edu.au

P. Bourke · A. Paterson
Social Sciences, University of Western Australia (M257), Perth, WA, Australia
e-mail: paul.bourke@uwa.edu.au; alistair.paterson@uwa.edu.au

J. Green
Department of Maritime Archaeology, WA Museum, Fremantle, WA, Australia
e-mail: jeremy.green@museum.wa.gov.au

J. K. McCarthy et al. (eds.), *3D Recording and Interpretation for Maritime Archaeology*, Coastal Research Library 31,
https://doi.org/10.1007/978-3-030-03635-5_13

removal of the fishing shacks in 2014 and an ongoing archaeology program, and (e) the future of the island as a location for heritage tourism.

Beacon Virtua also allows an exploration of how a virtual environment simulation can be used as a digital preservation tool. Beacon Virtua shows the island as it was in 2013 when it housed a small fishing community, with the island dotted with small buildings (colloquially referred to as 'shacks'), jetties and other structures (see Fig. 13.1). In 2014 all the shacks and jetties were removed which allowed a more detailed examination of the archaeology of the island to take place (Department of Fisheries 2014). The fishing history of the island is an important phase in the life of the island so there was a desire to document the physical embodiment of that activity before it was removed.

Visualization and simulation are an important tool to enable the interpretation and representation of cultural heritage particularly maritime (Adams 2013). The goals of Beacon Virtua were to provide an accurate digital record of the island and also to provide a way for people to experience the island in an accessible manner. The simulation is based on actual data over an artist's impression, and as far as possible the simulation is intended to be used by anyone regardless of familiarity with game like experiences or computers in general. Beacon Virtua has been targeted at multiple platforms to assess the practicality of using different types of computer systems for displaying virtual environments and also to make the simulation available to a larger audience.

Virtual reality has a long history of development, but it is in roughly the last 5 years that the evolution of a number of key technologies has reached an important stage to allow highly realistic experiences to be presented to users. Virtual reality experiences can be delivered to users via a range of different techniques including head-mounted displays, large-scale immersive displays and also more conventional displays; however, it is usually head-mounted displays that the public associates with VR. Virtual reality techniques offer an important ability to immerse a user in a fictitious or remote location. It has been shown that the use of immersive technologies to present cultural heritage experiences results in better understanding and improved retention (Jacobson 2013). This project aimed to explore the impact, experience and usability of VR technologies to communicate the story of Beacon Island.

The original development of Beacon Virtua is discussed in Bourke and Green (2016). This chapter will focus on development of Beacon Virtua as a storytelling tool—the adaption of the simulation to expose users to the rich history of the island and doing so across multiple platforms. We will also discuss its features and how the user interacts with them and the challenges encountered in creating the simulation. The methods used to capture the original assets for the simulation are discussed in Bourke and Green (2016).

Fig. 13.1 Aerial view of Beacon Island from the Beacon Virtua simulation showing the jetties and shacks

13.2 Simulation

The level of visual realism in a virtual reality simulation can vary considerably—options can include: more or less photo-realistic, impressionistic or even schematic (Denard 2009). The choice of level of photo-realism can be driven by a range of factors including audience type (e.g. expert or non-expert) and purpose of the simulation (e.g. archaeological interpretation or public education) (Frankland 2012). The level of photo-realism used for Beacon Virtua was chosen on a more pragmatic level. In this particular project there was a desire to make the simulation as photo-real as possible, however this had to be metered with the effort involved in gaining photo-realistic qualities for the various aspects of the simulation, and the load that this may place on the computer platform used to run the simulation.

The level of detail that can be implemented in a full 3D virtual environment is dependent on a range of factors including: the time/resource available to create the simulated environment, the tools and techniques used to document the location, and the computing power available to render the environment in real-time. The resultant realism and accuracy of the simulation will also depend upon the actual geometric complexity of the real-life environment—a very simple environment can be simulated very accurately with limited resources and effort, but a complex environment may be impossible to accurately depict even with unlimited resources.

3D modelling and virtual reality simulation have been used on a wide range of maritime archaeology related projects. Examples include the Pianosa Island wreck site in Italy which has used a wide range of emerging techniques to document, map, model and visualize the wreck site including traditional photogrammetry (for point-by-point measurements), 3D manual modelling, early experiments with photogrammetric 3D reconstruction, along with GIS and VRML for visualization (Drap et al. 2007, 2008; Haydar et al. 2008). Others, such as the Mazatos shipwreck in Cyprus, which started excavation in 2007 using manual 3D modelling and visualization in a large CAVE display (Katsouri et al. 2015) and the *Vrouw Maria* shipwreck in Finland which has used multibeam sonar, laser scanning of ship models, manual 3D modelling, and VR simulation using Unity (Reunanen et al. 2015). Also notable, the *Le Boullongne* virtual sailing ship simulation has been developed from the ship's plans and unlike the others listed above, is not a wreck site simulation (Barreau et al. 2015).

Beacon Virtua is different from these examples in that the focus is on the island, and it uses the island as the platform to expose users to the story of the ship and her crew, and other aspects of the history of the island. All of the content in Beacon Virtua is presented on or in relation to the island. The user can access all of this content by exploring the island, similar to how one might walk around a museum to see exhibits. At the time of writing, and to our knowledge, this is the only VR simulation of an entire 'maritime landscape' used to educate the public about the cultural heritage of the location. Rather than just a series of individual wrecks in 3D, this project presents an entire environment and although it does not show the actual wreck of *Batavia* in the simulation now, there are plans to extend the simulation to include the offshore site in one seamless environment in the future.

13.2.1 Guided Tour

One of the challenges in Beacon Virtua was how to turn a fully explorable island simulation into a guided tour to tell the story of Beacon Island's history. The reason for guiding users was that there are many uninteresting or insignificant locations on the island and hence a user might not find the key locations of the island and discover important aspects of the story without assistance. Any guidance mechanism that was implemented had to be compatible with all of the different platforms on which the simulation would be deployed. There were also limitations of file-size and modes of interaction on the various platforms. Different deployments might also have different requirements - such as the exhibition version that is discussed later. The mechanism that was implemented to guide the user was a series of footprints which indicate a preferred path around the island. The path was chosen to guide the user to important locations on the island that in turn could be used to expose the user to the various aspects of the island's history. Along the path are a series of floating information panels containing text information about the history of Beacon Island or some aspect about the particular location they are seeing. The user is also able to freely explore the island if they wish. The simulation starts at the end of the main jetty looking towards the island – giving the user the impression of having just arrived at the island by boat (Fig. 13.2). The walk along the jetty provides time for the user to become accustomed to the controls and receive some guidance about the simulation. The guided path continues around the island through some of the shacks, past some of the shallow graves and ends at the southern tip of the island which is the closest point of the island to the *Batavia* wreck site, which is a nice point to complete the guided tour aspect of the visit.

13.2.2 Technical Features

Beacon Virtua has been built in the 'Unity' virtual environment development platform. Unity is a versatile piece of

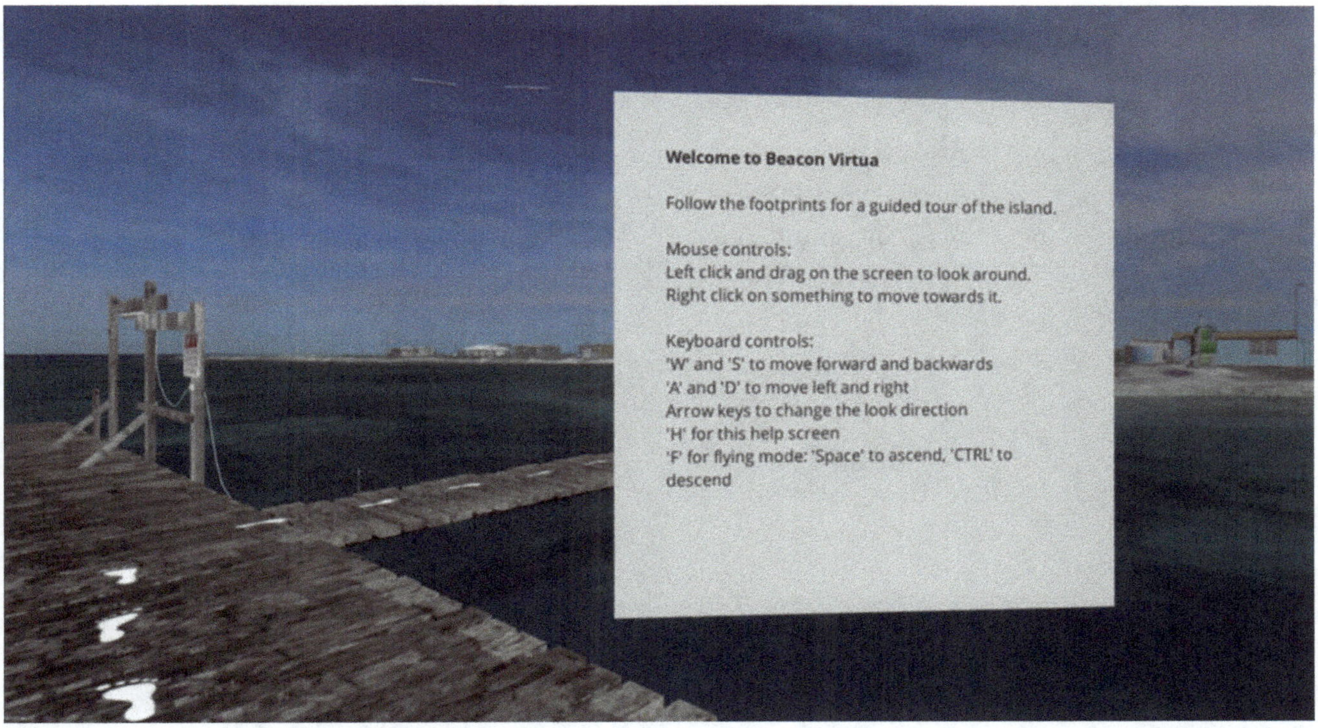

Fig. 13.2 The Beacon Virtua simulation commences at the end of the main jetty

software intended for making many different kinds of applications, including games and simulations (see also Benjamin et al., Chap. 14, this volume). It provides an editor capable of building standalone applications that can be distributed to multiple end users without them needing the Unity editor or the source project. In the editor, virtual environments can be constructed out of 3D models and other assets, and systems can be created to allow the user to navigate the environment and interact with features. Unity can build the same project to a wide range of different platforms (the target hardware and operating system software) which makes it very versatile. Beacon Virtua uses a mixture of techniques to achieve an advanced level of realism appropriate for the task of telling the story of Beacon Island within various technical constraints (Bourke and Green 2016).

13.2.2.1 Island and Ocean

The ground layer of Beacon Island is reproduced as an aerial image draped over a Digital Elevation Model (DEM) of the island. The surrounding ocean is simulated using a selection of different water simulation techniques and algorithms depending upon the compute capacity of each target platform—which in turn have different levels of realism. Several nearby islands that play an important part of the story of *Batavia* have also been experimentally implemented but have not been included in the first museum release of the simulation.

13.2.2.2 Buildings and Jetties

The buildings and jetties have been recreated as 3D models made by an artist from measurements and photographs taken of the originals on the island. The interiors of the buildings have only been recreated in shape, with blank grey walls (Fig. 13.3). However, the interiors of each room of each shack were captured photographically with series of 360° panorama photo bubbles. These bubbles record the inside of each structure in a level of detail that it is not feasible to recreate geometrically.

13.2.2.3 Graves and Coral Features

Three sets of graves, a coral cairn and a coral shelter have been captured as digital 3D models using photogrammetric 3D reconstruction (P3DR) techniques and specifically using the software *Agisoft Photoscan/Metashape*. P3DR combines the use of traditional photogrammetric methods with advanced image processing techniques to generate detailed digital 3D models of real-world objects from a series of photographs of that object. P3DR is good at generating digital 3D models of complex objects that would be difficult to do any other way, however in this project it has been done sparingly because the models can often be extremely detailed and can place a high computational demand on the simulation (Cox 2017). Generating models that have a 'low poly count' but remain visually realistic can be very difficult. The use of P3DR to create 3D models of maritime archaeology related items has exploded in recent years and is particularly suitable

Fig. 13.3 One of the shacks on the island illustrating the internal photo bubble

for underwater items due to limited visibility and an inability to capture a single image of a large object using other techniques (Woods 2016a). The detailed digital 3D models are inserted into the Unity scene amongst the other project assets (Fig. 13.4).

13.2.2.4 360° Photo Bubbles

Some details of the island were unable to be accurately reproduced as 3D models, such as vegetation, and hence a series of over one hundred 360° panorama photo bubbles were captured across the island. In the simulation, the user is able to navigate and essentially 'pop their head inside a bubble' to see a photographically accurate depiction of the island from that particular point. In order to prevent the large number of photo bubbles polluting the view of the landscape, the photo bubbles only fade into view when the user comes close to them. This is not the first time that photo bubbles have been used in a VR environment – other examples include 'Eye of Nagaur' from 2008 (Bourke 2009; Shaw 2010) and 'Mawson's Hut' from 2010 (Morse 2010). Photo bubbles are a very economical way of increasing the realism of the simulation. The photo bubbles in Beacon Virtua are monoscopic (2D), although there are ways of capturing them as stereoscopic bubble pairs (Gurrieri and Dubois 2013).

13.2.2.5 Information Panels

As mentioned earlier, a series information panels have been inserted into the environment to provide instructions and tell the story of Beacon Island. The information panels are made to clearly stand out from the environment, so they are not mistaken for real features of the island. The panels are programmed to fade out when the user is far away so that they do not obscure the environment when they are not needed. Sherman and Craig (2002) discuss four types of VR interfaces: (i) in the world, (ii) in the hand, (iii) in front of the view (HUD), and (iv) on the panel. In Beacon Virtua, the information panels we used fit the last category, information and/or controls on a 3D panel in the world. This has proved to be an effective solution for the requirements of Beacon Virtua.

13.2.2.6 Text Menu

In addition to the information panels, a text menu is also used. This menu can be brought up by clicking a button or pressing a key and provides an overview of the controls as well as a set of options allowing the user to access more features such as jumping to a particular point of the island.

13.2.2.7 Audio

Ambient sounds around the island were recorded while on site—capturing sounds such as the birds, waves and wind. The different audio recordings have been located at their corresponding locations on the island, and as such provide a dynamic soundscape as the user moves around.

Fig. 13.4 The coral cairn as is it reproduced as a digital 3D model in Beacon Virtua

13.2.2.8 Birds

Beacon Island is home to a variety of birdlife, which have been incorporated into the simulation. The birds are represented in the simulation as white, winged objects which fly above the island in random patterns based on a home point. The birds will randomly pick a destination to fly to—the choice of destination is weighted so that the bird will favour destinations close to its home point. The algorithm provides both a pattern of birds gathering around a focal point as well as individuals that fly off on their own. The bird home points have been positioned over each of the wharves. The destinations are capped between certain altitudes so that the birds do not fly through buildings or other scenery but remain in view. Each bird has an audio source set to play recordings of the island's birds at random intervals, allowing them to squawk as they fly overhead. As the birds are flying above the island they can only be seen from a distance, allowing them to create the impression of birds without using detailed models or animations. The implementation of the birds in the simulation contributes to the dynamic nature of the simulation.

13.2.2.9 *Batavia* Marker

The *Batavia* was wrecked on Morning Reef about 1.6 km from the shores of Beacon Island. To indicate the presence of the shipwreck in the simulation a marker was placed floating above the shipwreck location (Fig. 13.5). Clicking on the marker brings up a text menu providing a brief explanation of the marker. At this stage the user cannot explore or navigate to the wreck site but it is something that we are working towards. In another part of the project, digital 3D models of the *Batavia* wreck site have been developed from roughly 3500 underwater photographs taken of the wreck in the 1970s using the P3DR technique described earlier (Woods 2016b). The team is working towards including these 3D models of the wreck site in the simulation in a future revision of the simulation.

13.3 Target Platforms

To date Beacon Virtua has been built for 14 different target platforms: Desktop Windows and Mac, WebGL/Web Browser, multiple Head Mounted Displays (Google VR Cardboard iPhone and Android, HTC Vive, Oculus Rift, Gear VR), four types of large-scale immersive displays such as those in the HIVE (high resolution tiled, Cylinder, Dome, and Wedge/CAVE), a touch-screen exhibition version, and videos (regular widescreen, and 360° 3D) for uploading to YouTube or other streaming service. The flexibility of Unity to export the same project to multiple platforms saves a considerable amount of development time and maximizes the potential target audience. Different platforms have different capabilities and different interaction modalities which will be discussed specifically in the following sections.

Fig. 13.5 Looking towards the *Batavia* wreck site from Beacon Island in Beacon Virtua

13.3.1 Desktop

The original or development version of Beacon Virtua is targeted at desktop or laptop computers running the Windows or Mac operating systems. This version has all of the high-level features and content available. This version is deployed as a downloadable application about 1 GB in size. The only sacrifice in quality for the PC version is that the panoramic bubbles are 4K resolution rather than 8K, since the use of 8K textures triples the application file size to 3 GB.

13.3.2 WebGL

The most accessible version of Beacon Virtua uses Unity's WebGL player that allows the simulation to run within a desktop web browser—Firefox, Chrome, Internet Explorer and Edge were all tested. This version is built into a webpage, which is hosted normally as part of a website. When the user navigates to the web page, their browser begins running Beacon Virtua in a similar way to view any web based content. This system requires the entire application to be downloaded into the web browser when the user navigates to the page. As such, the total size of the application is limited by the amount of time the user will wait for it to download, and how much memory their browser can allocate to run it. In our testing we found that the application needed to be kept within around 25 MB in size.

Unity heavily compresses the assets when it makes the WebGL build of the application, but to meet this much smaller build size some assets either had to be removed entirely or use lower quality copies. The 3D models captured using P3DR contributed significantly to the file size and hence the 3D models of graves and cairns were removed and replaced with flat images. Additionally, textures were downsized, and all but a small sample of the panoramic photo bubbles were removed.

The massive drop in application size from 1 GB to around 20 MB is possible because the majority of the original application's size is due to the texture resolution. Removing bubbles and downsizing other textures is relatively easy, though there is a noticeable decrease in the visual quality of the simulation.

13.3.2.1 Head Mounted Displays

Unity provides native support for several head mounted display VR systems. Unity can be configured to build applications that will run with minimal programming effort. However, the design of the simulation needs to be adjusted to account for the differences in user experience and supported user input devices on the various HMD platforms. The high-end HMD version is built off the PC version and employs an Xbox controller that operates relative to where the user is looking. This version has not been publicly released at this stage.

Google Cardboard is an entry level head mounted display which uses a smartphone to display stereoscopic content. The phone is mounted in a low-cost holder—often made from cardboard, hence the name—which when worn, uses lens to show each of the viewer's eyes only half the screen. An image for the left eye is rendered on one half of the smartphone screen and an image for the right eye is rendered on the other half, allowing the user to see stereoscopic 3D depth.

Unity can build for Cardboard with a plugin provided by Google. The Cardboard version of Beacon Virtua is also constrained by its file size. Users must download it, and their phone must be able to fit the application in memory. The project team targeted less than 100 MB for the Cardboard version. To reach this size limit the textures were downsized and bubbles were reduced as was done with the *WebGL* version. Cardboard applications can be built for iPhone and Android phones. Separate builds must be made for iOS and Android, and there are different stores and approval processes for distributing the application to end users.

The Samsung Gear VR is another system for turning an Android smartphone into a head mounted display. Like the Oculus Rift, the Gear VR is directly supported by Unity and needs no plugins. As the Gear VR is quite similar to the Google Cardboard it could use effectively the same version of Beacon Virtua; however, the Gear VR version can assume a higher level of computer performance and hence can offer a higher graphics performance level than the Google Cardboard version.

13.3.2.2 Large-Scale Immersive Displays

The Curtin HIVE visualization facility at Curtin University features four immersive large-screen displays in one facility (Woods 2016b). Each of the displays has unique capabilities. The Tiled display is a 24 mega pixel media wall made up of 12 full-HD 55″ LCD panels. The Cylinder display is a 3 m high 180° wide screen with an 8 m diameter which can operated in stereoscopic 3D. The Wedge display has two rear-projected flat screens mounted at 90°, each with a 3.7 m diagonal which can also operate in stereoscopic 3D—similar to two panels of a CAVE display. The Dome display is a 4-m diameter half-dome oriented vertically which fills the user's full primary and peripheral field of view. These displays are typical of the types of large-scale immersive displays available in visualization facilities around the world. Beacon Virtua has been customized to run on all four of the large screens in the HIVE.

The principal HIVE version of Beacon Virtua is for the HIVE Cylinder display. MiddleVR is used to run Unity applications in stereoscopic 3D on the Cylinder. To account for the curved surface of the Cylinder, MiddleVR is configured to use 12 stereoscopic cameras around the cylinder to render the environment, each one drawing to a small vertical strip section of the screen. When the application is run in stereoscopic 3D each camera needs to be duplicated so that there is one camera for each eye for a total of 24 cameras.

The Dome version of Beacon Virtua uses a special camera model that pre-distorts the fisheye image to account for the optical properties of the display. The pre-distortion ensures that the final image on the curved dome surface appears correct to the viewer. The HIVE systems can optionally use the SpaceMouse six degrees of freedom controller from 3Dconnexion for user navigation input. All HIVE display versions have unique executables but run the same content as the PC version of Beacon Virtua, except with 8K bubbles and adjusted information text to explain the different controls.

13.3.2.3 Exhibition Version

Towards the completion of this edition of Beacon Virtua there was an opportunity to showcase the simulation in a major exhibition; however, it was thought necessary to implement a special exhibition version. Although Beacon Virtua works well with a head mounted display, this configuration would require a full-time attendant which was not possible to resource for an exhibition which would run for almost 6 months—hence, it was decided to use a flat-screen display to present the simulation in this instance.

Some extra thought had to go into how the exhibition version would ideally operate – particularly the input method. There was concern that physical controllers such as a keyboard and mouse, a joystick, a SpaceMouse, or a gaming controller would be too confusing for a general audience across a wide range of ages. Touch-screen devices are very common these days so it was decided to adapt the simulation to run with a touch-screen interface. In the exhibition, Beacon Virtua was run on a gaming laptop and presented on a 55″ LCD touch screen mounted on a wall. Controls were adapted to work with the touch screen, so that look direction could be changed by swiping the screen and the player could move to a new location by tapping the screen in the direction in which they wish to move.

A number of additional features were added to this version to make it more suitable for the exhibition environment. Firstly it was important that the screen did not remain static between users, perhaps stuck looking at a blank wall, hence an automatic path follow mode was added which would commence after a predetermined period of no user input. Once the timeout is reached, the simulation navigates to the closest point on the guided pathway, and then proceeds along the pathway whilst pausing at all of the information panels and items of interest. When a new user approaches the display, they would see an interesting and enticing simulation. Users could just continue to watch the simulation as it auto-navigates around the island or touch the screen and take control. In order to encourage users to take control, a photo-realistic hand is animated to rise from the bottom of the screen and make a touch action at regular intervals. Once taking control, the user can choose to continue from their current location or start at the beginning. As this version did not need to be downloaded, the 8K bubble textures were used.

13.3.2.4 Videos

Two special plug-ins were used to export video sequences of the user auto-navigating through the entire Beacon Virtua experience—one in standard widescreen aspect in 4 resolution, and another in 360° stereoscopic 3D producing an over-under equirectangular video file. Both of these video sequences have been uploaded to YouTube to provide potential users with a quick way to preview the content of Beacon Virtua.

13.4 Multiple Target Platforms

A significant set of challenges were posed by Beacon Virtua's deployment to multiple platforms. Different platforms have their own technical and design considerations. Different display types require different cameras to render them—for example, the Google Cardboard setup requires a camera for each eye configured to render to the phone screen properly, while the HIVE Cylinder requires a special VR manager object. Different display systems support different user interface devices and/or modalities. Different systems also have different graphics compute capacity and some content may limit the frame rate of the simulation more than is acceptable. File size is an important consideration for versions that will be downloaded, possibly restricting how much content can be included in the simulation.

In *Unity*, content is arranged in files called scenes. A scene is like a room where things can be placed. At first multiple scenes were created for different versions of Beacon; this made sense when there were just two versions (*WebGL* and desktop) but it became difficult to manage as the number of versions increased and as testing refined which content was used in which version. A system was needed that could share content between versions that had similar capability and requirements, while also allowing customized content to be used for platforms that have different requirements. This was achieved using *Unity*'s Additive Load feature. Normally when a scene is loaded the previous scene is deleted. When an Additive Load is used, the new scene is added to the current one. Content was split between scenes, for example the island and buildings went in one scene while the panoramic bubbles went in another. A version is made by creating an entry scene, this scene includes the control method needed for that version, a UI that works for that display type, and a camera suitable for rendering to that display type. The entry scene also includes a list of all the content scenes needed by that version. When the application is run, the content scenes are loaded additively, adding the island, bubbles, information panels and so on to the entry scene.

This allows two different versions to share content scenes when they have content in common. When the content is updated, changes are inherited by both versions because they are loading the same content. Where content needs to be different, an alternate content scene can be made with the different content and the two versions can load two different scenes for that section of content.

13.5 Navigation

An ongoing challenge of Beacon Virtua's development was how to allow the user to explore the island. The navigation system is intended to invoke the sense of walking. This required moving over the ground, following the level of the ground, and not letting the user walk through walls, as that is both impossible in real life and visually jarring.

Unity includes its own character controller as a generic solution for letting the user navigate an environment, however this was found to not work well on Beacon's uneven surfaces and small building interiors. A new player controller was developed which used collision detection to move over the ground and bounce off walls, allowing the user to move around the island like they were walking.

During the early development of Beacon Virtua it was observed that users often struggled with this particular navigation system. In particular, moving around the small, narrow interiors of buildings proved challenging. The problem was that the system would take the user's input as a direction to move, and then if the collision detection encountered any objects it would either stop or bounce off. This meant that the user would have to select the exact direction that was clear of obstacles to avoid them. Another issue was that this navigation method required a control scheme capable of entering any direction the user wanted to move. This meant that the simulation could not be deployed to the exhibition touch screen or Cardboard without an overhaul of the navigation system, as these versions had limited input methods.

A new system was therefore devised which would take the user's input and treat it as a destination to reach, moving around any obstacles if needed. The new system makes use of Unity's pathfinding capability. Pathfinding (in games) is typically a system used by computer control characters to navigate the environment. It works by creating a map of all the areas in the environment a character could reach, and at runtime, characters use an algorithm to find the optimal path to their destination according to that map. Unity has its own system for implementing pathfinding, providing tools for building this map, and telling agents (the user in our case) where to move to. A navigation map was built for the scenes of Beacon Island, and a system was created to pass the user's input to their player as a destination to reach. The pathfinding then moves the player to the destination automatically, stepping around obstacles as it moves.

13.6 Dynamic Text

The information panels around the island present a lot of text to the user. Each information panel is an object floating in a Unity scene. Text on the panels could be updated in the editor. This text needed to be written, proof-read, sent around the team members for approval, and updated over the development of the project. The text also needed to change across the different versions, as it would refer to the control scheme used and content that may or may not be present.

At first the information panels were treated like the other content and used the additive scene load to load in different sets of panels depending on which version was used. This approach was problematic as panels often only had minor differences, such as an added sentence, between versions. A lot of scenes were added with very similar panels, which exaggerated another problem. Every time a change was made, the text had to be copied from the master text document to each of the information panels. The master document also had to include notes about how panels changed across versions. This was found to be a very labour intensive and potentially error prone technique to update the panels.

The solution was to create a system which read the text from a master file at runtime, and used commands embedded in the file text to filter what content was shown depending on what version was being played. This system uses a third-party plugin for Unity called Ink (2016). Ink is a mark-up language made by the game company Inkle for use in computer game versions of 'Choose your own Adventure' books. Text is written in an ink file, and at runtime Unity can read through this file and display text according to the command logic included in the text file. Text can be wrapped in an 'if' statement, so it will only display if that condition is true. In the entry scene, we then have a list of variables we set to true or false to determine what version of the text is displayed.

Integrating Ink also provided access to a system for offering choices to the user which was used for Beacon Virtua's main menu system. Options are displayed asking the user if they would like to go back to the start or jump to a particular part of the island. When a response is selected the simulation receives an instruction to move the user or similar.

13.7 3D User Interface

A minor challenge, though a common one in VR development, is that the user interface (UI) must be viewable in 3D. In 2D game like experiences, the UI can be drawn on top of everything the user sees, because the user sees everything on a flat surface. With a 3D experience, the user can see depth so an interface has to appear at a certain depth. This in turn means the UI elements should be drawn at a distance the user can comfortably see them, but at the same time appear in a way that they do not interfere with objects in the world, such as appearing behind or through walls.

The UI was built to appear as a 3D object in the world, and a script was written to position and orient the UI in relation to the user's location and facing direction. To avoid clipping through other objects, when the menu is opened all the other objects in the world are made invisible. With the menu displayed the user is unable to move, so there is no need for them to be able to see the environment.

13.8 Discussion

The museum version of Beacon Virtua was launched to the public on 11 October 2016. Beacon Virtua is made available via the WA Museum's website via www.museum.wa.gov.au/BeaconVirtua. The web page provides an explanation of Beacon Virtua followed by four ways to experience the simulation: (1) A video preview available via YouTube, (2) The *WebGL* version which provides a simplified version of the simulation, (3) a Google Cardboard (Google VR) version which will run on iPhone and Android smartphones, and (4) the full desktop version which can run on Mac and Windows computers. Allied pages provide information about Beacon Island, the *Batavia* and the ARC Linkage Roaring 40s project.

The exhibition version of Beacon Virtua was displayed at the *Travellers and Traders in the Indian Ocean World* exhibition at the WA Maritime Museum in Fremantle and officially launched by the King and Queen of The Netherlands. The exhibition ran over the period 31 October 2016–23 April 2017 (Figs. 13.6 and 13.7). The simulation was run on a laptop plugged into a large 55-in. touch screen LCD. This configuration provided an effective way for visitors to experience Beacon Virtua, with visitors able to walk up and explore for as long as it held their interest. Promotional postcards located beside the display provided visitors with information about the downloadable versions of Beacon Virtua (Fig. 13.8). One idea was that visitors could experience the HMD version of Beacon Virtua by using their smartphone in combination with a cheap Google Cardboard viewer. The simulation proved to be highly reliable over the length of the exhibition

Fig. 13.6 Beacon Virtua in situ at the Travellers and Traders exhibition during a curator tour. (Natali Pearson, University of Sydney)

Fig. 13.7 Beacon Virtua being interacted with by a user at the Travellers and Traders exhibition

with only two restarts necessary—both due to power failures.

While Beacon Virtua has not undergone formal user experience testing, the team has generally observed positive reactions from people who have been shown the simulation. The simulation's development has mostly concentrated on getting content into the simulation, presenting it in a way that gives a clear sense of the island's layout and allowing users to explore it easily. The best way to use this kind of simulation to inform users about the subject is another consideration. Unity Analytics has been used to collect limited user activity information about the use of Beacon Virtua. Unity Analytics was in beta during the development of Beacon Virtua and its results have not been validated as yet so the

Fig. 13.8 Beacon Virtua promotional postcard

analytics results will not be commented on here. As of June 2018, there was a reported 335 installs of Beacon Virtua across the Android and iOS App Stores.

13.9 Future Work and Conclusions

Formal user testing is something that the team would like to progress and some work is being conducted in that direction currently. One of the most often requested additions to Beacon Virtua is a simulation of the wreck site, which is not currently included except for a floating marker which indicates the location of the wreck site. At the *Travellers and Traders* exhibition visitors could be seen touching the wreck marker indicating their desire to visit that location. It is worth noting that in parallel with the development work on Beacon Virtua there has been another project supported by the ARC Linkage Roaring 40s project focussing on developing digital 3D models of the *Batavia* wreck timbers as they existed on the seafloor before excavation. A selection of digital 3D models have been developed from approximately 3500 underwater photographs taken during the 1970s and processed using P3DR. Plans are currently in progress to include the *Batavia* wreck site in a future upgrade of Beacon Virtua.

The *Batavia* wreck site today is characterized by a large patch of sand, from where the wreck timbers were removed, surrounded by reef. A PhD student also supported by the Roaring 40s project has created a digital 3D model of the current site using P3DR techniques (McAllister 2018). This 3D model could also be integrated into Beacon Virtua to provide a representation of the wreck site as it is today. Further work could also include the inclusion of a CAD model of the *Batavia* into the simulation based on knowledge of the timbers and general design of the ship itself (Van Duivenvoorde 2015, 2005).

Virtual reality technologies are currently undergoing rapid development and improvements—both in terms of display and computer hardware, but also in terms of software support and capability. These advancements in turn are spurring significant growth in the development of applications which exploit these technologies. The development of Beacon Virtua has served as an exploration of the use of virtual reality technologies to act as both a digital record of the place and as a way for people to virtually experience the site.

Beacon Virtua uses a rich mix of asset types to implement the simulation of Beacon Island. The challenge and conflict is the ability, or otherwise, to preserve high quality assets while at the same time providing an interactive experience. The level of detail required in a simulation so it can act as a digital record will depend upon the expectations of the individual user and the purpose for which the simulation is created. In some user's eyes the current level of detail will be insufficient for their requirements—but to add more detail could sacrifice the ability to run the simulation in real-time and hence to provide a realistic interactive visualization. In our view there needs to be much better support for automated level of detail allowing progressively more detailed mesh and textures. In its current form Beacon Virtua has provided a good example of how a remote and significant cultural heritage site can be delivered as an interactive experience to a general audience. The techniques used to provide structure to the guided tour is fairly simplistic but seems to work well. We look forward to conducting formal user experience testing on the simulation to further optimise its operation.

Acknowledgements This project has been funded by the ARC Linkage project 'Shipwrecks of the roaring forties: a maritime archaeological reassessment of some of Australia's earliest shipwrecks' (project LP130100137). Initial project work was supported by a Your Community Heritage Grant awarded in 2012 to the Department of Maritime Archaeology at the WA Museum in conjunction with the Department of Fisheries, The University of Western Australia and Flinders University. We wish to thank the Curtin HIVE team including Joshua Hollick and Jesse Helliwell for their contributions to the project. Further project funding for the onward development of Beacon Virtua has recently been received from the Australian Government Department of Environment and Energy, Protecting National Historic Sites scheme.

References

Adams JR (2013) Experiencing shipwrecks and the primacy of vision. In: Adams JR, Ronnby J (eds) Interpreting shipwrecks: maritime archaeological approaches. The Highfield Press, Southampton, pp 85–96

Ariese C (2012) Databases of the people aboard the VOC ships Batavia (1629) & Zeewijk (1727) - An analysis of the potential for finding the Dutch castaways' human remains in Australia. Special Publication No. 16, Australian National Centre of Excellence for Maritime Archaeology Report—Department of Maritime Archaeology, Western Australian Museum, No. 298. Online: http://museum.wa.gov.au/maritime-archaeology-db/maritime-reports/databasespeople-aboard-voc-ships-batavia-1629-zeewijk-1727. Date Accessed 5 Sept 2018

Barreau J-B, Nouviale F, Gaugne R, Bernard Y, Linares S, Gouranton V (2015) An immersive virtual sailing on the 18th-century ship Le Boullongne. Presence 24(3):24

Bourke P (2009) EON—Eye of Nagaur. http://paulbourke.net/exhibitions/EON/. Accessed 15 July 2018

Bourke P, Green J (2016) Keeping it real: creating and acquiring assets for virtual environments. Comput Games J 5(1–2):7–22. https://doi.org/10.1007/s40869-016-0018-z

Cox G (2017) Working on the invincible: the deal with photogrammetry. Artasmedia blog. https://artasmedia.com/2017/03/26/working-with-the-invincible-photogrammetry/. Accessed 14 Sept 2017

Dash M (2003) Batavia's graveyard: the true story of the mad heretic who led history's bloodiest mutiny, 2nd edn. Phoenix, London

Denard H (2009) The London charter for the computer-based visualisation of cultural heritage (version 2.1). http://www.londoncharter.org/. Accessed 4 Sept 2018

Department of Fisheries (2014) Restoration under way at Abrolhos Batavia site. http://www.fish.wa.gov.au/About-Us/Media-releases/Pages/_archive/Restoration-under-way-at-Abrolhos-Batavia-site.aspx. Accessed 15 July 2018

Drake-Brockman H (1956) The reports of Francisco Pelsaert. West Aust Hist Soc 5(2):1–18

Drake-Brockman H (2006) Voyage to disaster: the life of Francisco Pelsaert (translated by ED Drok). University of Western Australia Press, Nedlands

Drap P, Seinturier J, Scaradozzi D, Gambogi P, Long L, Gauch F (2007) Photogrammetry for virtual exploration of underwater archaeological sites. Proceedings of the 21st international symposium of CIPA, Athens, Greece, 1–6 October 2007. 6 pp

Drap P, Seinturier J, Conte G, Caiti A, Scaradozzi D, Zanoli SM, Gambogi P (2008) Underwater cartography for archaeology in the VENUS project. Geomatica 62(4):419–427

Edwards H (1966) Islands of angry ghosts. Hodder & Stoughton, London

FitzSimons P (2011) Batavia: betrayal, shipwreck, murder, sexual slavery, courage: a spine-chilling chapter in Australian history. William Heinemann, Sydney

Frankland T (2012) A CG artist's impression: depicting virtual reconstructions using non-photorealistic rendering techniques. In: Chrysanthi A, Murrieta Flores P, Papadopoulos C (eds) Thinking beyond the tool: archaeological computing and the interpretive process, BAR International Series 2344. Archaeopress, Oxford, pp 24–39

Green J (1989) The loss of the Verenigde Oostindische Compagnie retourschip BATAVIA, Western Australia 1629. In: Hands AR, Walker DR (eds) BAR International Series 489. B.A.R, Oxford

Gurrieri L, Dubois E (2013) Acquisition of omnidirectional stereoscopic images and videos of dynamic scenes: a review. J Electron Imaging 22(3):030902. https://doi.org/10.1117/1.JEI.22.3.030902

Haydar M, Maidi M, Roussel D, Drap P, Bale K, Chapman P (2008) Virtual exploration of underwater archaeological sites: visualization and interaction in mixed reality environments. In: Ashley M, Hermon S, Proenca A, Rodriguez-Echavarria K (eds) Proceedings of VAST: international symposium on virtual reality, archaeology and intelligent cultural heritage. The Eurographics Association, pp 141–148. https://doi.org/10.2312/VAST/VAST08/141-148

Ink (2016) https://www.inklestudios.com/ink/. Accessed 4 Sept 2018

Jacobson J (2013) Digital dome versus desktop display: learning outcome assessments by domain experts. Int J Virtual Personal Learn Environ 4(3):51–65 (Article 4)

Katsouri I, Tzanavari A, Herakleous K, Poullis C (2015) Visualizing and assessing hypotheses for marine archaeology in a VR CAVE environment. J Comput Cult Herit 8:1–18

McAllister M (2018) 'Seeing is Believing': investigating the influence of photogrammetric digital 3D modelling of underwater shipwreck sites on archaeological interpretation. Unpublished PhD dissertation, University of Western Australia

Morse P (2010) Heritage visualisation on iPhone. http://www.petermorse.com.au/2010/04/heritage-visualisation-on-iphone/. Accessed 15 July 2018

Pelsaert EF (1647) Ongeluckige voyagie, van't schip Batavia, nae de Oost-Indien: gebleven op de Abrolhos van Frederick Houtman, op de hooghte van 28 1/3-graet, by zuyden de linie Aequinoctiael. Uytgevaren onder den E. Francoys Pelsert. Jan Jansz, Amsterdam

Reunanen M, Díaz L, Horttana T (2015) A holistic user-centered approach to immersive digital cultural heritage installations. J Comput Cult Herit 7:1–16

Roeper V (2002) De schipbreuk van de *Batavia*, 1629. Walburg Pers, Zutphen

Shaw G (2010) Eye of Nagaur. https://www.jeffreyshawcompendium.com/portfolio/eye-of-nagaur/. Accessed 15 July 2018

Sherman WR, Craig AB (2002) Understanding virtual reality: interface, application, and design. Elsevier, New York

Van Duivenvoorde W (2005) Capturing curves and timber with a laser scanner: digital imaging of *Batavia*. INA Q 32(3):3–6

Van Duivenvoorde W (2015) Dutch East India Company shipbuilding: the archaeological study of Batavia and other seventeenth-century VOC ships. Texas A&M University Press, College Station

Woods AJ (2016a) An underwater odyssey: the mission to image the wrecks. In: McCarthy M (ed) From great depths: the wrecks of HMAS Sydney II and HSK Kormoran. UWA Publishing and WA Museum, Perth, pp 255–277

Woods AJ (2016b) Curtin HIVE—hub for immersive visualisation and eResearch. Presentation at stereoscopic displays and applications conference, electronic imaging symposium, San Francisco, California, USA, 15–17 January 2016. https://youtu.be/iSLESjF-ZuXg. Accessed 5 Sept 2018

Integrating Aerial and Underwater Data for Archaeology: Digital Maritime Landscapes in 3D

14

Jonathan Benjamin, John McCarthy, Chelsea Wiseman, Shane Bevin, Jarrad Kowlessar, Peter Moe Astrup, John Naumann, and Jorg Hacker

Abstract

Archaeologists have aspired to a seamless integration of terrestrial and marine survey since maritime archaeology began to emerge as a distinct sub-discipline. This chapter will review and discuss how 3D technology is changing the way that archaeologists work, blurring the boundaries between different technologies and different environments. Special attention is paid to the integration of data obtained from aerial and underwater methods. Maritime archaeology is undergoing an explosion of site recording methods and techniques which improve survey, excavation and interpretation, as well as management and conservation of material culture, protected sites, and cultural landscapes. An appraisal of methods and interpretive tools is therefore necessary as well as a consideration of how theoretical concepts of maritime landscapes are finding new expressions in practice. A thematic focus is placed on integrating land and sea through case studies of maritime archaeological sites and material which range chronologically from the recent past to several thousand years before present.

J. Benjamin (✉) · J. McCarthy · C. Wiseman · S. Bevin
J. Kowlessar · J. Naumann
Flinders University, Adelaide, SA, Australia
e-mail: jonathan.benjamin@flinders.edu.au; john.mccarthy@flinders.edu.au; chelsea.wiseman@flinders.edu.au; shane.bevin@flinders.edu.au; jarrad.kowlessar@flinders.edu.au; john.naumann@flinders.edu.au

P. M. Astrup
Moesgaard Museum, Aarhus, Denmark
e-mail: pma@moesgaardmuseum.dk

J. Hacker
Flinders University, Adelaide, SA, Australia

Airborne Research Australia (ARA), Parafield, SA, Australia
e-mail: jmh@flinders.edu.au; jorg.hacker@airborneresearch.org.au

Keywords

Underwater archaeology · Aerial archaeology · Digital archaeology · Maritime cultural landscapes · Archaeological theory

14.1 Introduction

The discipline of archaeology is currently undergoing a step-change in site recording methods and techniques that have improved and enhanced scientific archaeological survey, excavation and interpretation. This can be described as a shift from a reliance on separate technologies in parallel to the use of converged and integrated technologies and a shift from 2D methods to 3D methods. This trend was identified by Wheatly and Gillings (2002, 216–217) at an early stage and examples include engagement with the technology and analytical techniques (Spring and Peters 2014), theory (Garstki 2017), public engagement (Tait et al. 2015), illustration (Morgan and Wright 2018) and archiving (Austin et al. 2009). While it is now widely recognised that photogrammetry supplements and enhances, rather than replaces, existing techniques, it has nevertheless drawn digital 3D recording firmly into the mainstream of archaeological practice. Furthermore, it has arguably pushed spatial recording through a watershed such that the gaps between pre-existing survey technologies, usually separate, have begun to be bridged. While maritime archaeology projects have employed a wide array of survey techniques, dissemination has typically been presented sequentially in archaeological reports, or perhaps overlaid in a 2D format. As software packages increase their capabilities, they increasingly overlap. Coupled with a general rise in computing power, archaeologists increasingly find that high resolution 3D survey datasets from separate sources can be spatially combined without having to flatten them into 2D beforehand.

The drivers behind a spatial approach to archaeological research include a wide array of software and hardware

J. K. McCarthy et al. (eds.), *3D Recording and Interpretation for Maritime Archaeology*, Coastal Research Library 31,
https://doi.org/10.1007/978-3-030-03635-5_14

developments. Perhaps the two most important factors are the near universal adoption of multi-image photogrammetric techniques and of unmanned aerial vehicles (UAVs) by archaeologists, with both technologies enhancing each other (Colomina and Molina 2014). The dramatic rise in the application of photogrammetry for archaeology generally (Doneus et al. 2011; McCarthy 2014; Remondino 2011) has been facilitated by recent advances in software algorithms that allow semi-automated 3D reconstruction, which is a technology that is equally applicable to land, intertidal and, with a somewhat higher degree of technical preparation, underwater sites (McCarthy and Benjamin 2014). As a result, a much larger proportion of archaeologists have begun to develop 3D modelling software skills to make the most of the recording process. The desire to fully exploit these photogrammetric datasets naturally draws researchers to attempt to bring in other spatial datasets and to attempt a synthesis of 3D information within a fully 3D analytical environment. To some extent, this process of 3D data integration in archaeology has been underway for some time, most notably through the advent of laser scanning. However, photogrammetry has moved beyond the realm of the specialist into the skillsets of the maritime archaeology generalist as it is much more easily accessible and versatile, without requiring comparatively complex hardware. At the time of writing, it is likely that the vast majority of underwater archaeological divers will have been exposed to the technique at some point. This is a dramatic change from just five years ago when only very few would have heard of the technique. At the same time, the rising ubiquity of drones or UAVs has also facilitated capture of data for 3D models of entire landscapes on spatial scales accessible to such aerial platforms. This is discussed in more detail below, but regardless of which driver, it is clear that maritime archaeologists are increasingly reliant on single-environment multi-source spatial workflows. These lead to greater contextualization of the maritime archaeological resource. In particular, photogrammetry, which can operate at a variety of scales from the smallest artefact to the various larger scales that can be defined as a 'landscape' (e.g. local, area, regional, etc.), lends itself to a multi-scalar survey approach (Olson et al. 2013).

Alongside these benefits, the changing practice in maritime archaeology means that theoretical issues surrounding use of digital technologies must be considered. In the broader discussions of archaeology, these discussions are well underway. Huggett et al. (2018, 44) describe the 'grand challenges' facing digital archaeology as 'fundamental (addressing theory and practice); innovative (not simply adopting concepts and techniques from other fields); revolutionary (potential for paradigm change, creating new technological competencies and ways of knowing); inspiring (engaging across the sector and beyond); measurable (with intermediate goals to gauge progress and achievement, at the same time allowing for the possibility of failure); and co-operative (involving more than just an individual researcher or team, and crossing national and potentially disciplinary boundaries).' Huggett et al. also point out that advances in digital archaeology are not about the development of technologies by archaeologists per se, but rather, how archaeologists decide to adopt and apply digital technologies and the resulting impacts on the field.

With these concepts in mind, this chapter will review and discuss how 3D technology is changing the way maritime archaeologists work. We examine how boundaries between archaeology and environments are blurred and interpreted through emerging technology and how this allows theoretical concepts in maritime archaeology to find practical expression, specifically the concept of the Maritime Cultural Landscape (Westerdahl 1992). The wider theoretical reconsideration includes a focus on spatial relationships of material cultural within a landscape. The case studies included were chosen to illustrate the wide-ranging impacts of digital workflows across temporal and national boundaries in an integrated field of maritime and underwater archaeology. The chapter discusses methods, equipment and results of case studies illustrating contemporary practice, before reviewing the current state-of-the art in digital maritime archaeology. Emphasis is placed on the data and analyses that allow for the integration of land and sea, which make up the 'seamless' cultural landscape. Case studies range chronologically from historical periods to those sites originally formed several millennia before present. Special attention is paid to the integration of data obtained from aerial and underwater methods as these two sub-specialisations within archaeological recording are not often combined, with research groups most frequently publishing output in distinct specialist publications (with some notable exceptions, e.g. Firth 2011). This chapter will show how this gap can be bridged, integrating the aerial and underwater datasets through increasingly complex case studies that range from a simple intertidal survey, to a fully integrated maritime landscape above and below the waterline.

14.2 Maritime Archaeological Theory and Integrated Cultural Landscapes

Westerdahl (1992, 5) first introduced the concept of Maritime Cultural Landscapes (MCL) in his seminal article and explained that 'during the maritime archaeological survey… the need arose for a scientific term for the unity of remnants of maritime culture on land as well as underwater.' In the ensuing three decades after Westerdahl's influential 1992 paper the concept has become 'a dominant research area within North European maritime archaeology' (Tuddenham 2010, 6). The theoretical reaches have expanded beyond

Northern Europe, entering the global discussion. Westerdahl himself describes the maritime cultural landscape as a physical and cognitive place 'where the boats were built, the materials, the area where the wood was taken, and finally the vessel itself and its parts' (Westerdahl 2008, 19). This eventually expanded to include an all-encompassing conceptual definition of 'maritimity', not only in landscape but in the more conceptual 'mindscape.' This idea was welcomed at the time and provided a new theoretical framework in a discipline where such frameworks were in short supply. As a result, Flatman (2011, 311) states that 'archaeology has witnessed an unprecedented and consistent array of work either directly on or indirectly contributing to this field of analysis by both "terrestrial" and "maritime" archaeologists.' The theory remains in fashion and appears in the title of numerous recent books and articles by maritime archaeologists (Delgado et al. 2016; Ford 2018a; Harris 2017).

Despite this ongoing popularity, difficulties in the way MCL theory is applied are apparent. Stewart (2011, vii) felt it necessary to say of MCL theory that, 'we are still at an early stage, feeling our way through, but unsure of exactly how to go about it, or what we hope to accomplish.' Likewise, Ford (2018b, 198) states 'MCL supports varying perspectives. Multiple theoretical perspectives can be pursued under the MCL aegis; cultural ecology to phenomenology and Marxism to practice theory can all be explored within an MCL framework. Importantly, MCL also takes in a management perspective, allowing us to organize and manage cultural resources. It is a broad church.' The definitions of the term are certainly broad, and this has created unintended consequences that are arguably to the detriment of maritime archaeology. While increasing in breadth, without a stronger definition, its value as a concept has eroded over time. This is because the body of discussion of MCL exists as a spectrum, nebulous and expanding like its own universe, born from Westerdahl's original survey of the Swedish coasts in the 1980s. The reason may be, as Flatman (2011, 326–327) notes, that the theory is attractive to students and to researchers because it is grounded in the tangible physical world. The type of spatial relationships at the core of MCL make it one of the more accessible theoretical frameworks with which most people can identify, even for newcomers to maritime archaeological theory. Unfortunately, the very accessibility and the openness of MCL to different interpretations has, over time, led to an increasing vagueness of definition and lack of theoretical direction.

Those who have sought to apply the theory in a practical Cultural Heritage Management context, have run into very difficult to answer questions, not least of which is how far from the shore we should consider an MCL to extend (Ford 2018b, 199–201) and what is and what is not 'maritime' within the anthropological or archaeological discourse (see Gately and Benjamin 2018). Many of the publications referencing the theory, even as part of their title, are little more than catalogues of the maritime culture and history of a region rather than attempts to develop the theory into something more complex or to explore cognitive relationships with material culture past and present, as well as the more prosaic physical environment aspects. This is perhaps because maritime archaeologists are generally conscious of the need to engage with archaeological theory and have tended to rely on MCL as a convenient seat filler which does not interfere with the scope of their particular study. After all there is probably no archaeological site to which MCL theory could not be applied in some form. The use of MCL in this way contributes little to the broader discipline, which is already overly techno-particularist and sorely needs long-term and genuine engagement with well-defined, useful, theory. In some cases, MCL has been treated as though it were a concept which could be tested to see if it existed in a given area, but as Rönnby (2007, 80–81) notes, 'the challenge is probably not so much to prove that a special environment plays a role in the formation of societies and cultures. It is evident that it does.'

If technology itself is neutral, can technological advancements be used to focus the development of archaeological theory? In an appraisal of the state of MCL studies just prior to the boom in photogrammetry algorithms that have so greatly impacted archaeology, Stewart (2011) identified two major challenges for the future of MCL studies. Firstly, he notes the need to bring cultural interpretation into maritime landscape surveys. Secondly, he notes the technical challenges of past landscape reconstruction. In the case of the first challenge, as we have noted above, MCL theory has been attractive and accessible due to its spatial nature but this has been to a large extent a subjective mental exercise. As a result, so plentiful are the interpretative possibilities for a past maritime landscape, that the collective body of research has been unable to make much progress. Choices are made by the archaeologists on which sites to focus within a landscape and at which scale they are to be surveyed. This is often invisible in the final products. For the second challenge, Stewart states that landscape reconstruction 'requires geographical, geological, or geoarchaeological experts, but the specialists in these fields do not generally have expertise in maritime culture. By the same token, specialists in maritime life—nautical archaeologists, maritime historians, and maritime ethnologists—do not typically have expertise in reconstructing landscapes' (Stewart 2011, viii). In the years since these two challenges were identified by Stewart, the rise of much more user-friendly software and the relative ubiquity of powerful hardware means that maritime archaeological practitioners and even maritime archaeology students may have at least basic skills of landscape reconstruction within their grasp. In many cases discussions of MCL have been bound by a lack of materiality but new technologies allowing comprehensive landscape and seascape survey increasingly allows for a practical implementation of MCL theory in praxis.

These emerging techniques offer a higher level of detail than previously available to most field archaeologists. Rather, there is a newfound, and reasonably accessible means for many archaeologists to capture data to reconstruct a physical landscape digitally and in 3D. This provides a previously unattainable baseline for developing and testing theories related to landscape-material-culture interconnection and context, spatial relationship and setting which leads to the broader opportunities to review more anthropological questions as related to cognitive and social significance. We consider the MCL as a social construct based on cumulative characteristics of the physical environment and all of the natural resources and challenges they offer to societies past and present. Rönnby's point is worth heeding: (2007, 81): 'The challenge is instead to show how people within a very special maritime milieu and linked to 'maritime durees' within these surroundings have nevertheless constructed their social situation so differently….' While mapping in 3D alone might not resolve these issues, the added value of these outputs provides a newfound baseline to study maritime cultures. The physical environment impacts the cognitive landscape and must be considered, along with archaeological material, as a starting point, with consideration for change over time. A review of physical landscape over time begins with a data-driven exercise of material culture to interpret and understand the physical environment, exploited and culturally modified by past cultures. The development of technological approaches allows for a deliberate re-emphasis on the material as a data source, evaluated in detail and spatially, in a physical and virtual space.

One of the greatest advantages of a 3D survey-driven approach for maritime archaeology is the potential to digitally 'dissolve' the boundary between terrestrial and underwater environments. The modern coastline gives the archaeologist in the field a misleading impression of the contemporary physical landscape of an archaeological site, obscuring and distorting the lost landscapes of the past. In his introduction of MCL, Westerdahl (1992:6) recognised the importance of a survey strategy that spanned from land to sea so that 'a vision of the total topography of the waterfront area is applied, features on nearby land being as important as depth curves underwater.' Only through a survey strategy that seeks to capture the topography across the modern land, intertidal and submerged, can we achieve an informed approximation of the contemporary landscape context of the site and therefore study the cultural relationship between past people, materiality, land and sea. Considering as examples two of the most widely studied types of maritime archaeological sites, shipwrecks and submerged prehistoric sites, Westerdahl's statement is certainly true for submerged prehistoric sites. For shipwrecks, Westerdahl recognised that they could be challenging to include in the maritime landscape concept as they lacked, 'at first sight any obvious relationship to their immediate surroundings' (Westerdahl 1992: 6) but there are many links, including the relationship between the coastal topography and the wrecking event, post-wrecking site formation processes and interactions between the wreck and onshore communities from the time of wrecking onwards. For this reason, the comprehensive 3D survey approach has the potential to be equally relevant to all maritime archaeological sites.

14.3 Aerial Archaeology

Airborne observation and recording techniques for archaeology have been practiced for several decades and are discussed in technical breadth and depth in various publications (e.g. Bewley and Rączkowski 2002; Cowley 2011; Cowley et al. 2018; Duel 1969; Riley 1982). While the work undertaken from the air during the earlier twentieth century focused primarily on oblique photography or on vertical orthographic stereophotography, recently the focus has shifted to photogrammetry capable of dense 3D capture of land surfaces (e.g., Remondino 2011). This has also been true of coastal maritime archaeology (e.g., Benjamin et al. 2014; Cowley et al. 2012). Beyond photogrammetry, Lidar has also grown in popularity thanks to wider availability of data and more powerful computers. Bathymetric Lidar has also had an impact on coastal and shallow water archaeology (e.g., Doneus et al. 2013) and there have been studies exploiting publicly available bathymetric Lidar data for palaeolandscape reconstruction (Bicket et al. 2017). These methods for aerial recording have so far focused on 'manned' aircraft using traditional survey and data processing and interpretation methods.

Data acquisition at a local scale or site scale has been increasingly acquired by UAV. This popularity of UAV-based technologies prompted the dedication of a special issue of the Society for American Archaeology's newsletter (SAA 2016) to showcase the potential for archaeological sites and landscapes to be recorded by UAV. Here, we consider a shift in most photographic data from manned aircraft to UAVs. The advent of the small, consumer grade UAV means that any archaeological unit, department individual researcher or student can now take low altitude aerial photographs in a small scale, affordable manner. This latter point is perhaps the single greatest step-change for mapping sites located on dry land, since more sites and landscapes are now easily documented from above, which otherwise would not have been recorded in this way. We are nonetheless careful not to mistake this increase in the deployment of UAVs as an outright replacement for other methods. In terms of manned aircraft, there are now solutions available based on modern comparatively low-cost small motorglider and/or ultralight aircraft equipped with multitudes of sensors, including the most powerful full waveform-resolving small footprint lidar systems, high resolution cameras and more (Hacker et al. 2018a, b). Combined with new survey strategies and processing workflows, such solutions are delivering very detailed data for landscapes on local (~1 km) to regional scales (i.e. up to hundreds of kilometres). This is especially useful if larger datasets are employed for context, and particularly where the broader geomorphological context is relevant. Consequently, the needs of any given survey, and not a pre-determined choice of aerial technologies, should dictate the right tool for the job (Fig. 14.1) (Neininger and Hacker 2011; Cowley et al. 2018). This is evident in the

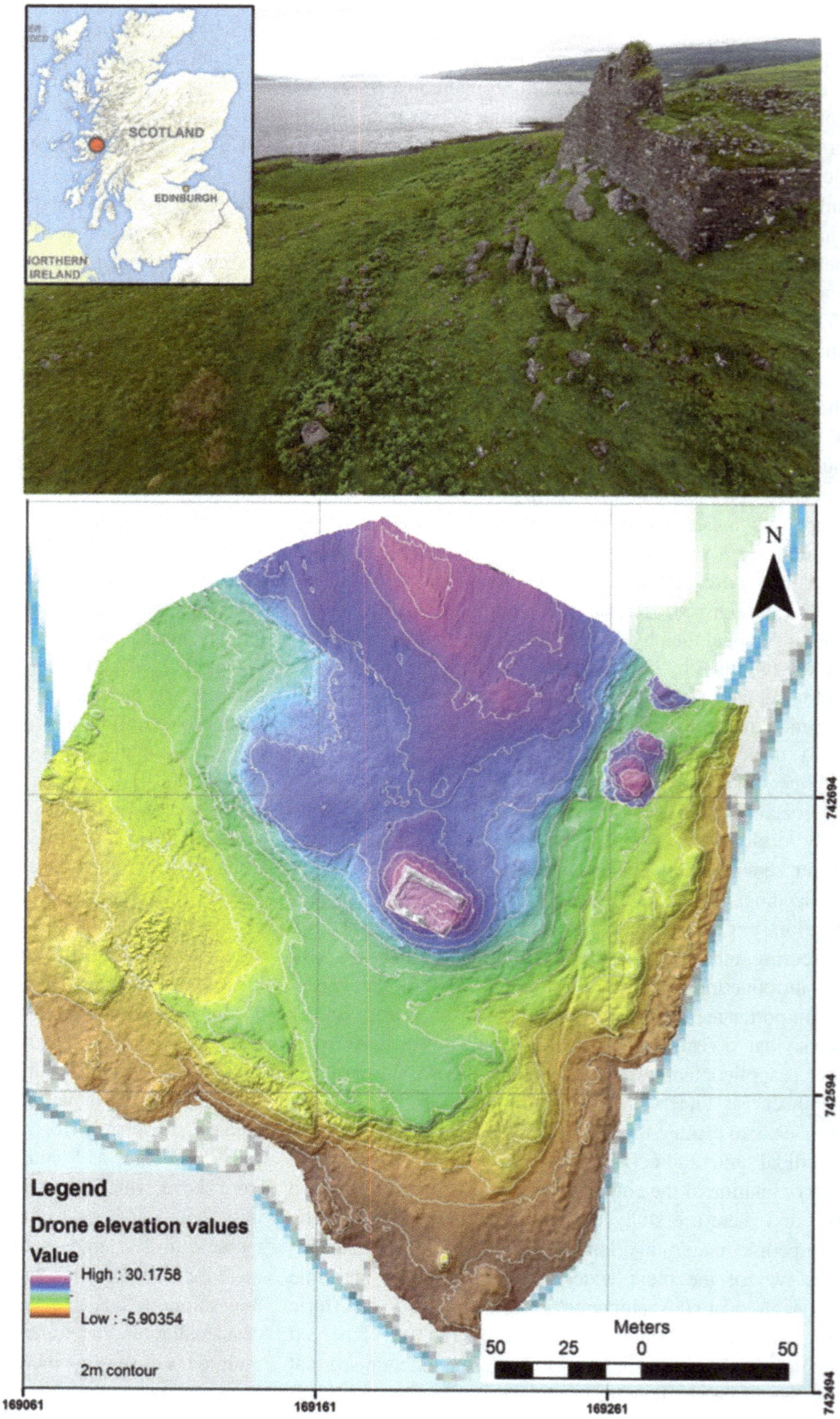

Fig. 14.1 Historic building recording through photogrammetry (Taken from the air. (Ardtornish Castle, Scotland). Figure contains Ordnance Survey OpenData, Crown Copyright 2018)

case studies presented herein which rely on data collected through a combination of UAVs and manned aircraft as both have their own strengths and weaknesses and their own place in coastal and marine landscape survey.

The availability of consumer-grade UAVs within the last decade allowed for the widespread adoption of aerial photography roughly in parallel with the advances in underwater digital photography, both of which provided raw data to feed into the newly available consumer-grade photogrammetry pipelines. Most notable in the low-cost range of UAV are those developed by Chinese UAV company DJI (with over 70% of the global market share as of 2017) (Chen and Lynch Ogan 2017, 57). Use of both multi-rotor and fixed wing UAVs (Remondino et al. 2011) have added to the suite of methods and tools for aerial recording. In many situations, drones have replaced the diverse and traditionally comparatively costly array of aerial photography techniques previously in use by archaeologists. Cowley et al. (2018) remind, however, that this is not necessarily true for all situations and the method of aerial survey should always be decided by the archaeological purpose. A typical example may be a study requiring identification and initial mapping of remote and/or larger areas not easily accessible to UAV-based surveys where the above mentioned small manned aircraft have considerable advantages, even in terms cost-efficiency. There are also a multitude of other methods on offer, the cost of which has to be carefully analysed, such as scaffolding towers, extendable photo poles, kites and balloons which can be combined with photogrammetry.

14.4 Technical Challenges: Shallow Water and Intertidal Zones

Working within the intertidal zone has always been a challenge for maritime archaeologists for both excavation and remote sensing survey. While sonar techniques have been the mainstay of marine remote sensing survey and aerial lidar or image-based techniques have been foremost terrestrially, the difficulty of applying remote sensing techniques to the shallow and intertidal zone resulted in the so-called 'white ribbon', a zone extending from the minimum depths for sonar and geophysical survey up to the coastline (Kotilainen and Kaskela 2017).

The recent rise in sophisticated photogrammetry algorithms, commonly available software, combined with the increase in computer processing power have had a strong impact on coastal and shallow water archaeology (e.g. Kreij et al. 2018; McCarthy and Benjamin 2018). Recording for 3D has become an important part of the maritime archaeological process, both technically and experientially, as 3D recording has the potential to bridge the gap between terrestrial and underwater archaeology. In this respect photogrammetry is similar to bathymetric lidar (Bicket et al. 2017; Doneus et al. 2013; Kotilainen and Kaskela 2017) although the latter technique has impacted the discipline less due to a much higher relative cost. Underwater photogrammetry, where the camera is submerged, is difficult in shallower environments, due both to the increased effect of caustics and to the forced proximity of the camera to its subject. The effect of caustics can be reduced to some extent by waiting for diffuse natural lighting due to oblique sunlight or cloud cover. For distance, the surveyor can use a very wide-angle lens with a very tight line spacing. One solution can be applied in areas with a large tidal range, where surveyors can increase the distance between camera and subject by waiting for high tide. The same approach can be used for fully intertidal sites, where terrestrial survey techniques can be applied at low tide while underwater techniques may be applied to the same location at high tide.

There have been a number of studies that have addressed the technical challenges of aerial air-to-water photogrammetry, sometimes referred to as photo bathymetry (Maas 2015, 18141). A growing number of studies that have reviewed methods for overcoming issues, including sun glitter (Mount 2005), refraction (Georgopoulos and Agrafiotis 2012) and wave-related distortion (Chirayath and Earle 2016). As with underwater photogrammetry, the value of shallow water survey is strongly correlated to water clarity, but it is also particularly limited by factors which are either absent from terrestrial and fully underwater photogrammetry or which have a much less-pronounced effect on the quality of results. These factors include the Fresnel reflection of water and glitter from sunshine as well as ray distortion caused by both wave action and by refraction of light as it passes from air into water. The effects of these factors can be limited by taking aerial photographs as near to vertical as possible to both reduce the effects of refraction and constrain the remaining refractive distortion to radial, use of circular polarising filters to cut down reflections and by shooting when the water surface is as calm as possible. This is the approach taken in the case studies presented below. In ideal conditions for coastal photography by UAV for photogrammetry, it is possible to capture 3D surfaces through the water for the first few metres, such that shallow bathymetry and features are expressed in the resulting 3D models. Finally, it is worth noting an interesting recent study which has combined under and above-water scans of floating vessels (Menna et al. 2015), dissolving this barrier in a similar way through a different method.

Considerable new developments have also taken place for using bathymetric lidar in the intertidal and shallow water zone, down to about 12 m water depth. This involves two full-size small footprint waveform-resolving lidars on an above-mentioned small traditional aircraft flown in innovative multi-overpass flight patterns together with sophisticated data processing and interpretation workflows (Hacker and Pfennigbauer 2017).

14.5 Underwater Photogrammetry

The various chapters found throughout this volume illustrate the widespread and ubiquitous nature of underwater photogrammetry, and how it has revolutionised maritime archaeology in a short period. The development of these technologies has generally not been driven by maritime archaeologists and has a much larger literature in technical publications from various disciplines. However, the adoption of the technology by maritime archaeologists as it developed is of direct relevance. Contemporary to some of the early work in aerial archaeology, photogrammetric methods in underwater archaeological site recording were implemented as early as the 1960s (Bass 1966). The subsequent decades saw incremental advancement in the adaptation of photogrammetric 3D recording for underwater archaeology, with pioneers (see Drap et al. 2003; Green and Gainsford 2003) pushing the boundaries of available software. From 2006 onward, publications appeared utilising multi-image photogrammetry for small archaeological objects underwater using highly technical workflows. Henderson et al. (2013) marked a step-change in photogrammetric site recording under water by analysing much larger numbers of images on a site-level scale by Autonomous Underwater Vehicle (AUV), in the shallow waters of Greece. Up to this point most of the adoption of multi-image approaches had relied on expensive workflows that were not practical for the wider archaeological community. From 2013, however, this began to change. McCarthy and Benjamin (2014) demonstrated a low-cost approach that was affordable for most maritime archaeology practitioners and was also effective, at the Drumbeg historic shipwreck site in NW Scotland (see also McCarthy et al. 2015). Yamafune et al. (2016) showcased new approaches to reconstruct large wooden wrecks in 3D using photogrammetric data as a baseline. Liarokapis et al. (2017) used underwater photogrammetry to map a fourth-century shipwreck in Cyprus and to create an immersive virtual reality experience for the site. In short, since 2009, maritime archaeology has developed the ability to move beyond a half century of reliance upon manual diver recording (with some assistance from sonar techniques) for complex underwater archaeological sites to being able to capture full shipwrecks and submerged settlements within only a few dives. While this has augmented rather than replaced existing techniques, it has significantly reduced the amount of time required to record a site, which is critical when considering the constraints imposed by dive times and working under water (which can often be a small fraction of what could be achieved on land). Thus, time is of the essence for the underwater archaeologist and new technological approaches have brought in considerable savings in time, cost and risk, while at the same time providing a much richer and more objective record of the archaeological sites.

14.6 Digital Maritime Landscapes in 3D: Case Studies

Combining the techniques referred to in the preceding sections allows us to dissolve the line dividing land and sea. The archaeological landscape can be composed of terrestrial, intertidal and marine environments. Thus, the integration of datasets can form a 'seamless' cultural landscape in 3D. This can significantly enhance interpretation and research as well as public outreach. In the following examples, aerial data were collected by traditional aircraft as well as by remote pilot operation. Underwater data were collected through in-water photography techniques, designed for photogrammetric modelling, while snorkelling, SCUBA diving and by boat. These case studies are presented with an increasing degree of complexity, from a straightforward intertidal zone site survey to a fully integrated terrestrial, intertidal and marine digital 3D landscape. This progressive demonstration provides a gradual introduction to the applied methods. They also become more intricate with each example in an effort to demonstrate how the techniques can build upon one another with increasing complexity, integrating more and more data from different sensors.

14.6.1 The Intertidal Zone

Two intertidal surveys from Scotland demonstrate the effectiveness of aerial documentation undertaken in the intertidal zone. Intertidal archaeology represents the transition zone, sometimes underwater and sometimes exposed. It can be difficult to record due to the dynamic environment and constantly changing conditions. It can also be dangerous, in some cases, where tides move quickly, and surface conditions are muddy. Surveyors must take extra care, or indeed cancel a survey plan that involves too much risk. This is precisely where intertidal archaeology benefits from aerial approaches to site and landscape survey (Cowley et al. 2012).

The remains of a small wooden boat at Ardno on the shores of Loch Fyne, Scotland was found with only the lower part of the hull intact (Fig. 14.2). The wooden carvel remains were obscured under intertidal seaweed and after a full manual and photogrammetric survey of the site, further research unearthed a probable identification of the vessel and early photographs

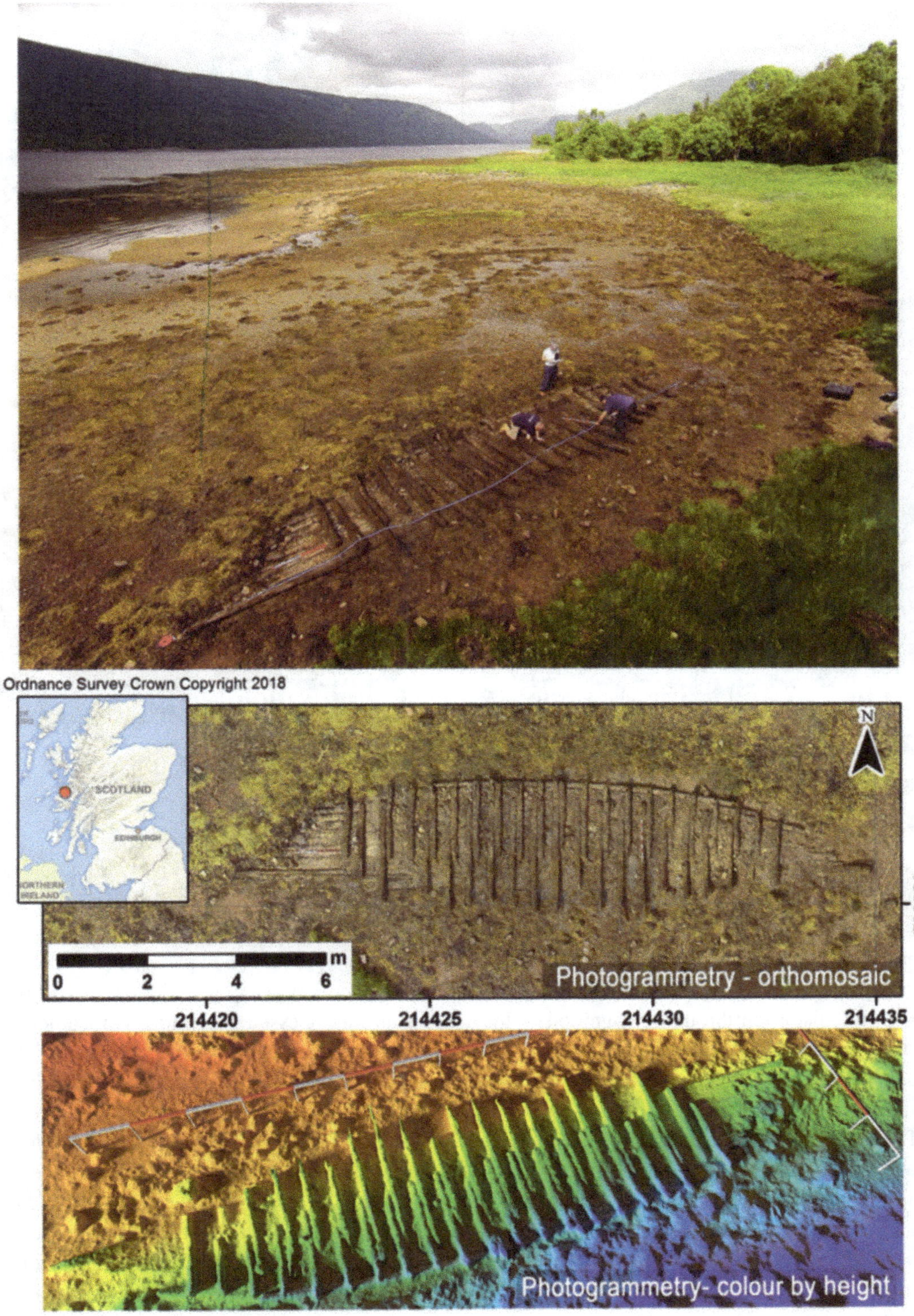

Fig. 14.2 The Ardno historic boat remains were reported to the SAMPHIRE project team (McCarthy and Benjamin 2018) and recorded through both traditional nautical archaeological techniques and aerial-based rapid recording methods

dating to around 1910 (McCarthy et al. 2015:C27–C29; McCarthy and Benjamin 2018:13–14). The Ardno hull represents a simple recording technique with a very low flying (below the tree level) UAV to record the previously unknown vessel. A DJI Phantom V2+ (v.3) was deployed with its built-in 12 mpxl camera. The camera itself is a fisheye with 115-degree, distorted view. This can be partially corrected in software such as Adobe Lightroom (through a standard lens

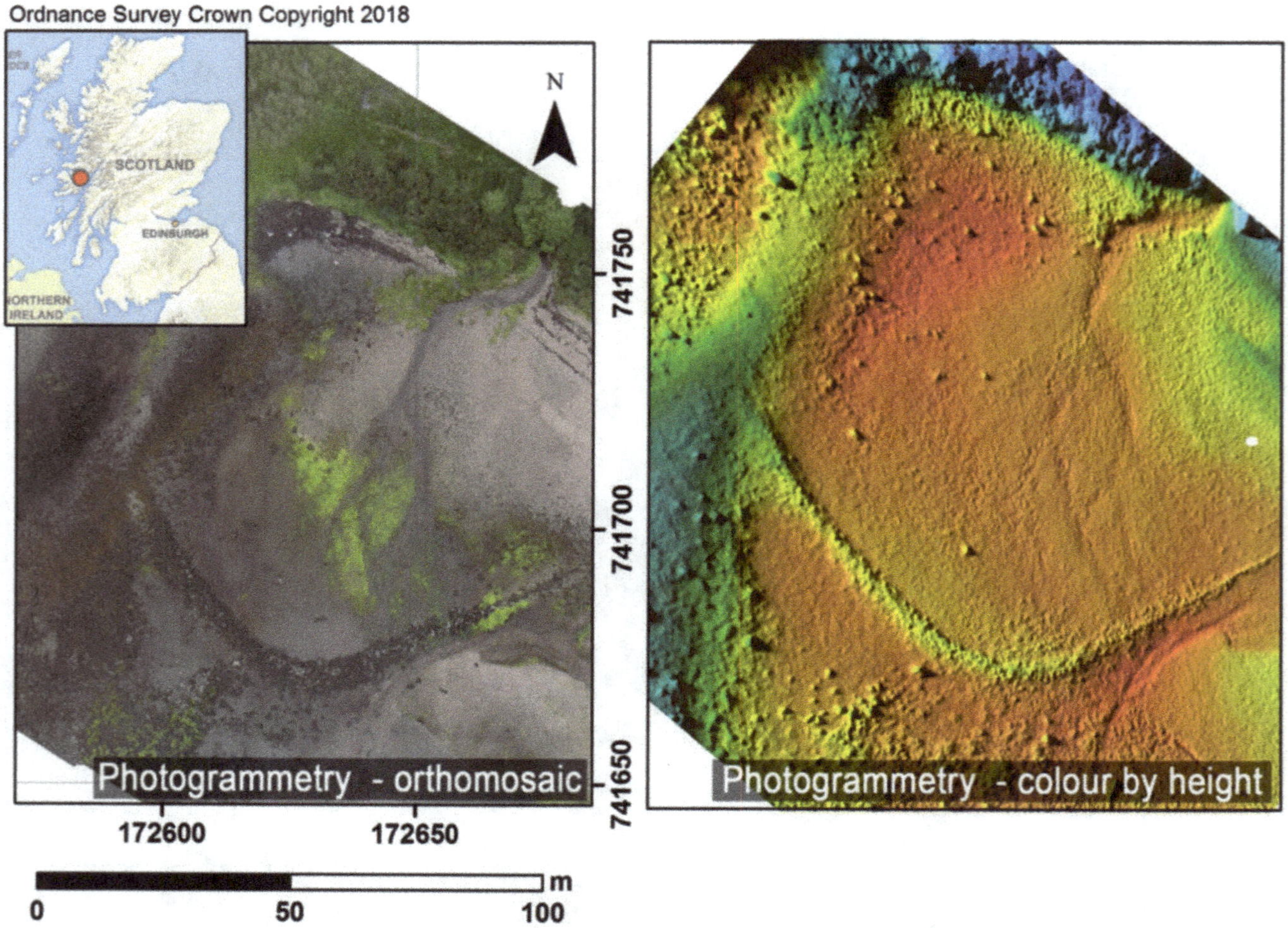

Fig. 14.3 The historic fish trap at Inninmore, sound of Mull, recorded by UAV deployed from a small boat as a platform for rapid survey of remote or inaccessible locations

profile) or, for 3D photogrammetric recording, Agisoft Photoscan/Metashape contains an algorithm that can account for the majority of the distortion through recognition of the camera in the image file metadata. The results at the Ardno survey, undertaken in less than a half hour, demonstrate use of a small, consumer-grade UAV flown at very low altitude. Both 3D and 2D images are useful to consider the site within the landscape, and for reconstructive purposes. A similar process was undertaken for the fishtrap at Inninmore, which was recorded from a UAV deployed from a vessel (Fig. 14.3). The use of a vessel-based platform for UAV operations brings another option for rapid recording of sites which are situated in locations that are otherwise difficult to access.

In contrast, two intertidal Mesolithic sites in NW Scotland are selected to demonstrate landscape-scale recording of Scotland's intertidal (or 'partially submerged') prehistory (Fig. 14.4). The lithic processing site at Lub Dub Aird (Loch Torridon) and the peat deposit at Clachan Harbour (Isle of Raasay), represent the only confirmed intertidal prehistoric sites in Scotland which have been impacted by postglacial sea-level rise (Hardy et al. 2015) though they both remained partially preserved until the time of their discovery (Bailey et al. 2019)[1]. At both sites, traditional aircraft (Cessna 172) and DSLR cameras were used to survey the contemporary landscape in 3D. The 3D capture of landscapes is particularly useful for interpretation of past landscapes and land surfaces where sea-level rise impacted the setting of the site itself. The lithics found at these sites represent Mesolithic deposits (Ballin et al. 2010; Hardy et al. 2015). Sea level during the Scottish Mesolithic was approximately between 3 and 15 m lower than today's mean sea level (MSL). This has an interpretive importance, not only for contextualising the otherwise ephemeral lithic scatters in today's landscape, but also for interpreting their geographic setting past and present.

[1] 'Recently media reports appearing just before this volume went to press have mentioned some important, but as yet unpublished, discoveries at Benbecula.'

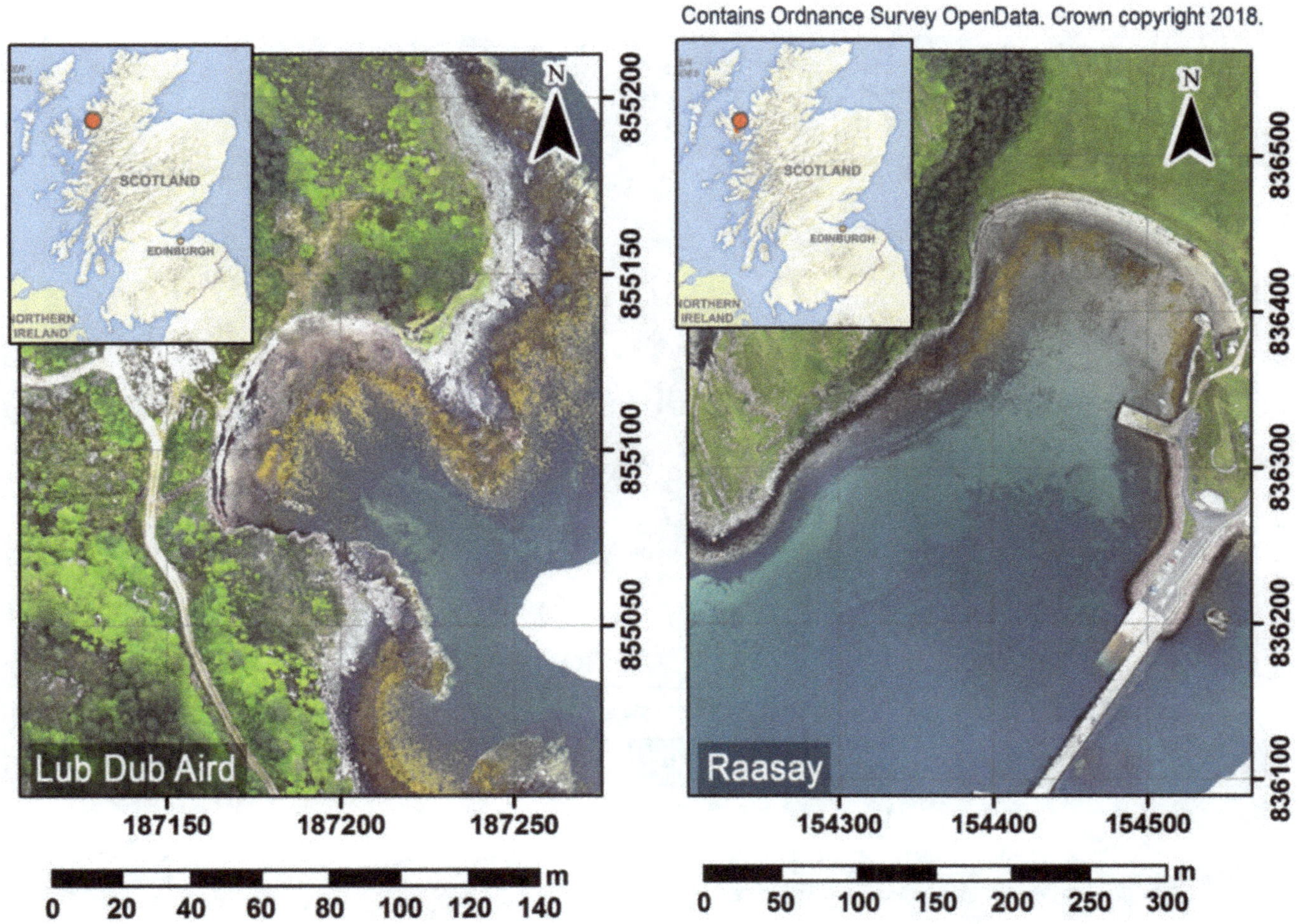

Fig. 14.4 Lub Dub Aird and Clachan Harbour landscape settings in 3D. These two locations represent the only two excavated find spots where Mesolithic deposits have been confirmed in the intertidal zone. The landscapes were recorded by traditional aircraft and camera equipment in order to capture the sites' settings. The past landscapes can be inferred by comparing sea-level models and adjusting the digital elevation accordingly, in line with the depth:age ratio of the Mesolithic period in western Scotland and taking into consideration location-scale isostatic uplift

14.6.2 Nearshore Historic Shipwrecks

Two historic shipwrecks in south-eastern Australia are presented as case studies. The recording of these two sites demonstrate different methods which can be used to integrate a fully underwater site within its surrounding environment. In these cases, the remains of the vessels represent historic wrecking events in the nineteenth century and the sites themselves were both partially salvaged by local inhabitants after the wrecking event took place. *Star of Greece* was the place of a loss of significant life, where 18 sailors perished near Port Willunga, South Australia (Ash 2007). *Leven Lass* wrecked in Victoria in comparable conditions, near shore on an exposed reef (Roberts et al. 2015) and both sites are protected by Australian law through *The Historic Shipwrecks Act* (1976).

The *Star of Greece* shipwreck is a well-known maritime archaeological site and tragedy in South Australia. Built in Belfast, Northern Ireland, by shipbuilders Harland & Wolff in 1868, it sank in bad weather, bound from Adelaide to London in July 1888. The site itself is approximately 2–5 m deep, located less than 200 m from shore. It was recorded by Ash (2007) requiring more than 30 dives to plan the site manually. Returning to the site in 2015, a single aerial photographic excursion and several attempts to record the site photographically were made; the latter proved difficult due to the size of the vessel and as such a complete site plan in high-definition 3D remains to be completed. The first attempt to record the site through underwater photogrammetry was undertaken by SCUBA divers, who managed to record only 15% of the vessel in two dives, although over 2000 photos were taken. Further attempts were made to record the vessel by boat-based photogrammetric techniques in 2017 and 2018. The resulting survey established a near-complete 3D site record through data acquired by a pole-mounted housed, stereo camera system (Sony RX100 M3) (Fig. 14.5).

Fig. 14.5 The *Star of Greece* shipwreck was recorded using boat-based photogrammetry based on over 6000 images to produce a 3D record. It is shown here as a 2D orthophoto alongside an aerial image of the wreck site. Resources like Google Earth now provide a simple way of providing wider context of archaeological sites

Fig. 14.6 (**a**) Location of *Leven Lass*, overlaid on aerial photogrammetry, showing landscape context; (**b**) Orientation map showing the location of the *Leven Lass* wreck site; (**c**) *Leven Lass* wreck site including survey points and trench locations; (**d**) Oblique view of photogrammetric survey of Trench A, looking northwest; (**e**) Oblique view of photogrammetric survey of Little Rookery Beach landscape, looking east; (**f**) Oblique view of photogrammetric survey of Trench B, looking southeast

In 1984, the remains of an unidentified wooden vessel were recorded by Heritage Victoria in approximately 4 m of water off the northern coast of Phillip Island, Australia. Flinders University-Heritage Victoria field investigations from 2012 to 2017 identified the shipwreck site as the nineteenth-century Clyde-built (Scottish) brig, *Leven Lass*. The site is used as a teaching location, where graduate students have been able to learn shallow-water 2D and 3D recording methods. In addition to underwater photogrammetric recording by simple housed compact camera (Fig. 14.6), aerial based photogrammetric data were acquired by UAV (DJI Phantom 2V+) at this site. The integration of the aerial and underwater data showcases the shipwreck, in its landscape context and with visible reef in immediate proximity. The vessel itself remains reasonably well-preserved, despite several early attempts to salvage the site followed by decades of artefact removal by local snorkelers and divers. A 2D orthophoto was produced to compare alongside student drawings of Trench A, excavated in 2015 (Fig. 14.6d), and Trench B excavated in 2016 (Fig. 14.6e).

14.6.3 Deep Time and the Integrated Maritime Landscape

The case studies illustrate the recent and rapid integration of digital 3D recording techniques. They demonstrate the site-scale, showcasing three historic (post-medieval or post-contact) sites and two local-scale archaeological landscapes of intertidal Mesolithic sites in Scotland and provide varying levels of data integration and archaeological setting. Here we reiterate that the spatial data and interpretation of 3D data differ when considering drowned palaeolandscapes and the cultural deposits therein/thereon. Historic shipwreck sites may be studied for their relationship with the surrounding terrain in which they sank, or ruined in a final punctuated event, however their origin (construction place) is not often their final resting place. This differs from the prehistoric sites inundated during postglacial sea-level rise. A submerged settlement provides an opportunity to apply digital techniques, to study site formation, preservation processes and, in a practical sense, future management. 3D data is particularly useful for underwater cultural heritage of drowned landscapes, within modern shorelines, intertidal zones and coastal (terrestrial) environments because the water level represents a somewhat arbitrary segregation that historically introduced research bias and certainly presents a modern, technical challenge.

The coastline of Denmark is surely one of the best places in the world to study impacts of climate change and sea-level rise on past peoples (e.g. Fischer 1995; Uldum et al. 2017). A Mesolithic site located on the island Hjarnø (Astrup et al. 2019; Skriver et al. 2017) represents an outstanding opportunity to integrate multi-scalar 3D recording techniques and present a complex site in high resolution. This is important because submerged material from Denmark has been recorded for several decades using traditional archaeological recording methods. However, the dissemination of the research has suffered historically from a lack of detail due to generalised illustrations and poor-quality underwater photography. The ability to recreate a landscape from aerial and underwater means has led to a high quality, scalable, photo-realistic resource that is more informative and therefore of higher value to international scholarship and museum visitors alike.

The landscape of Hjarnø (Figs. 14.7 and 14.8) was analysed using a terrestrial Lidar dataset with a resolution of 0.4 m, alongside a bathymetry dataset with a resolution of 50 m (interpolated data based on one point per 50 m). Both datasets are readily available to the public. Given the substantial variation in resolution between the two datasets, GIS-based interpolation allowed for the combination of the bathymetry and topography. The bathymetry and topography datasets were merged into a singular data layer, which was then interpolated to a DEM using the kriging method at a resolution of 5 m. Contour lines of 5 m were generated from the elevation model to be used in the creation of a TIN surface to then visualise this region in 3D. A variety of photogrammetric surveys were then made of the areas of archaeological interest, including UAV photography for photogrammetry of the terrestrial and intertidal areas undertaken at low tide, underwater snorkel and diver-based photogrammetric survey of excavated trenches and artefacts, surface photogrammetry of the entire seabed around the midden. These datasets were then merged into a single 3D environment in GIS and in the 3D modelling software Blender, with floating surveys georeferenced using the total station dGPS survey. Further work is ongoing to integrate the results of coring and excavation both at the site itself and in the wider landscape (Astrup et al. 2019).

14.7 3D GIS

Geographic Information Systems handle complex spatial information however this information is typically represented in a simple 2D format with 3D data and offer archaeologists a platform and collection of tools with which to manage complex spatial data sets. Despite a visual disconnect between the data and reality these types of displays have been widely accepted and utilized by archaeologists. While commercial 3D GIS packages have existed for some time, these are typically expensive products and designed for spe-

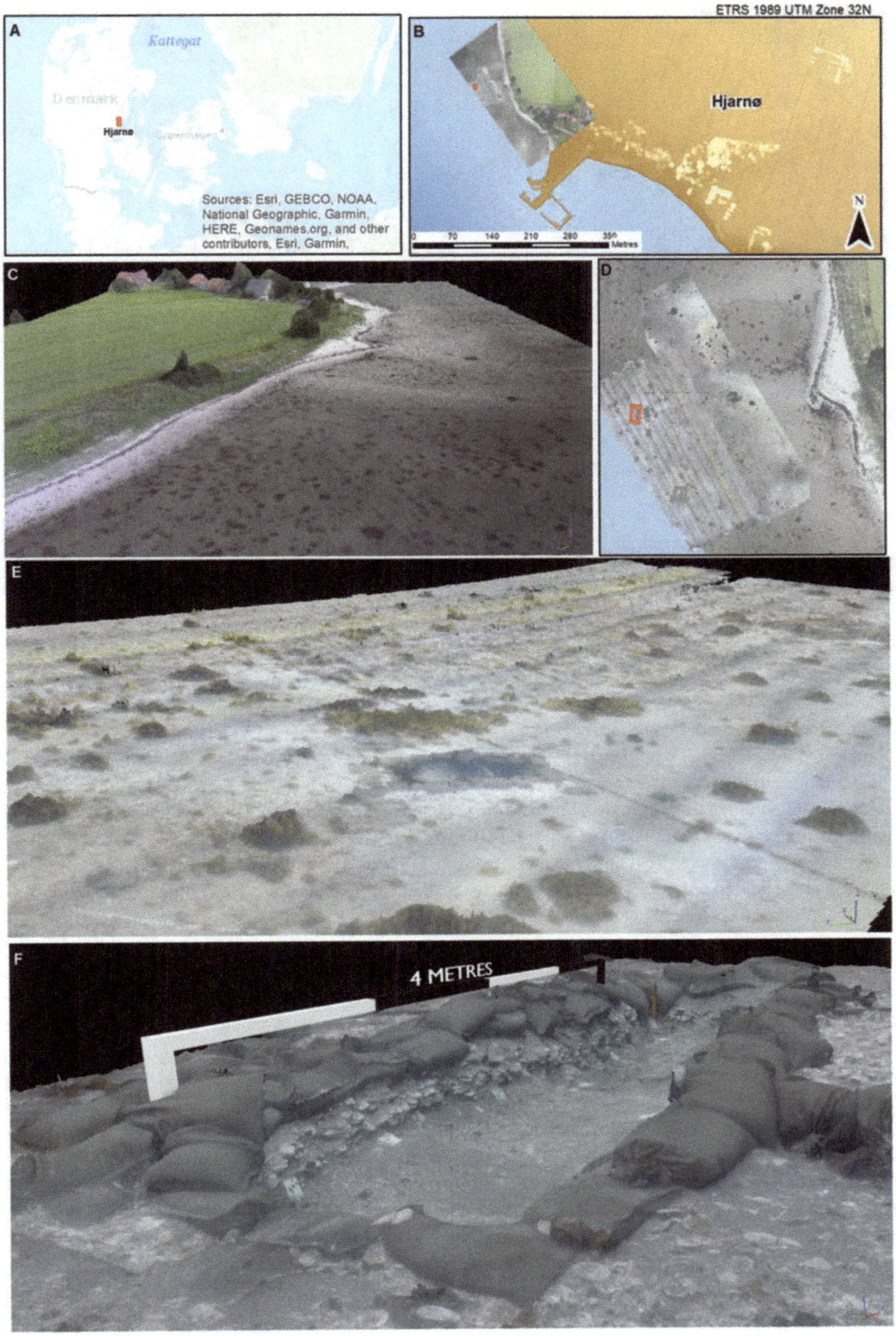

Fig. 14.7 (**a**) Orientation map showing the location of Hjarnø; (**b**) Site and landscape context, including aerial photogrammetric datasets, Lidar topography, and bathymetry. Location of excavated trench shown in red. Bathymetry dataset sourced from the Danish Geodata Agency. Contains data from the Data Security and Efficiency Board, DHM/Surface (0.4 m grid); (**c**) Oblique view of aerial photogrammetry and coastal environment, looking southeast; (**d**) Locations of the aerial and underwater photogrammetric surveys; (**e**) Array photogrammetry survey showing excavated trench and context underwater; (**f**) Oblique view of excavated trench, facing north

Fig. 14.8 A multi-dataset 3D working environment of the coastal prehistoric Hjarnø site and landscape, showing several layers of aerial and underwater photogrammetry, integrated with lidar data

cialised tasks in specific disciplines. (Wheatley and Gillings 2002: 241–242). Standalone general GIS packages only very recently allowed the integration of complex 3D models. As a result of the limitations of GIS packages previous attempts at 3D GIS for archaeology turned to custom developed, project specific platforms (e.g. Nebiker 2002; Wüst et al. 2004).

Within the last two decades development of 3D GIS platforms with the goal of integrating and visualizing complex 3D geometries and other data sources have been developed for the purpose of archaeology. An early example of pioneering work is the DILAS (Digital Landscape Server) platform (Nebiker 2002; Wüst et al. 2004). Another example of a custom platform for 3D GIS is the MayaArch3D project (Agugiaro et al. 2011a, b; Richards-Rissetto et al. 2012). Their project's approach to archaeological data organisation, visualization and analysis creates an interpretive window into the experiential aspects of past human behaviour as well as empirical analysis material culture (Agugiaro et al. 2011a). A considerable drawback of these purpose-built 3D GIS systems is that they require a specialized workflow using custom-coded software and complex database management system architecture. The project-specific, design makes them difficult to apply beyond their purpose-built archaeological research (Dell'Unto et al. 2015). In the context of such individual developments archaeologists called for a more general approach to 3D GIS, which led to a re-examining of the most modern features of commercial GIS packages (Dell'Unto et al. 2015; Ford 2004). A number of modern commercially available GIS systems exist that offer some form of true 3D data integration. These systems are often built with a variety of features offering different data management and analysis potentials.

Richards-Risetto (2017) recently asked 'What can GIS + 3D mean for landscape archaeology?' drawing attention to the historical tension and critique mainly by post-processual archaeologists to the application of GIS. Richards-Risetto points to the issues related to predictive modelling based on environmental data, the now ever-present debate around environmental determinism versus socio/cultural variables for locating, discovering and interpreting archaeological sites. The application of 3D GIS has an enormous potential for the study of maritime landscapes, bringing together 3D data into a realistically visualized geospatial environment and allowing analytical questions to be asked, empirically as well as providing rich visualizations of the data that allow more nuanced interpretations. Verhoeven (2017, 1021) points to the limited need for realistic appearance by the broader GIS community, which has historically limited the advancement for GIS packages to offer 3D visualizations natively. 3D modelling/animation software can fulfil many of the functions of 3D GIS but is generally limited to Euclidean space and incapable of dealing with geographic coordinate systems. There is currently a gap between these two technologies, but this is a clear example of convergence underway. Wheatley and Gillings recognized this in 2002 when they wrote that 'it seems likely that the convergence of spatial technologies will continue beyond what we currently understand as GIS—in fact this is one of the principal reasons that we favour the term 'spatial technologies' over the more defined (and contested) 'Geographic Information Systems' (Wheatley and Gillings 2002, 216). Once the technical computing capacity can routinely support large scale 3D GIS platforms, archaeologists are on track to become some of the earliest adopters of this technology.

14.8 Digital 'Realities'

Archaeologists are also adopting gaming engines as a form of data visualization and enhancement, which can be integrated into GIS (Richards-Rissetto 2017) or used for augmented reality (Eve 2014, 2017; Watterson 2015). The creation of interactive virtual spaces has become more accessible as the tools required evolve to meet the demands of a variety of users and applications. Watterson (2015, 20), points out that 'all image-making within archaeology involves implicit assumptions and explicit choices, but the context and technique behind the creation of these images and the ways they are consumed often obscure this process.' While digital tools and techniques are often in focus, the need to re-affirm the underlying rationale for how data are collected, adapted and presented should be consistently reviewed. Ewes (2014) recently discusses 'an opportunity to merge the real world with virtual elements of relevance to the past, including 3D models, soundscapes, smellscapes and other immersive data.' And suggests 'sophisticated desk-based GIS analyses can be experienced directly within the field and combined with a body-centred exploration of the landscape, creating an *embodied* GIS.'

The implementation of captured 3D data into navigable 3D spaces is now achievable within a reasonable budget and a team with the necessary specialist skills. The technological developments, which enable digital media experts and archaeologists to work together to rebuild the scenes, or the digital maritime cultural landscapes, allow for the further experiential or digital phenomenological approach to viewing and understanding past environments. Current technology stops short of a completely 'life-like' experience, however the direction and integration of these disciplines is heading toward a much more immersive experience. The creation of an interpretation of the past through virtual reality.

Watterson (2015, 19) has outlined the need for archaeologists to develop a more practical approach to addressing the issues through the development of method which consider the 'multi-layered, interpretive and ambiguous processes involved in archaeological interpretation.' Equally, the way someone experiences archaeology, whether in person or virtually, matters. A virtual experience of archaeology is not the opposite of a 'real' experience, though it may be described distinctly from a physical or experience (e.g. Falconer and Scott 2018). For the moment, people can still tell the difference between a virtual and a physically real experience, though this gap can only decrease over time. The technologies and methods for archaeological survey, recording and interpretation which have been discussed in this chapter continue to develop to the advantage of archaeologists worldwide. Conceptual models for understanding landscapes, their cultural modification and change over time, and need to apply and consider digital models of ever-expanding scale and quality. This has both an aesthetic appeal, but importantly, will be increasingly functional for the integrated interpretation of material culture and physical remains within physical and cultural landscapes.

Game engines are now commonly used to bring the past to life, facilitating both public engagement with past environments as well as providing a tool to allow archaeologists to immerse themselves in their study areas, provoking new questions and inspiring new directions in research. There are a variety of game engines available: Amazon's Lumberyard, Cry Engine, Epic's Unreal Engine 4 and Unity are used by small and large-scale game development teams. The flexible licensing costs of all listed game engines have made cutting edge real time interactive 3D game design technology available to teams with a variety of available budgets and skill-sets. The example in Fig. 14.9 is a Unity-based VR environment that has been created of a coastal medieval castle in Scotland, at Ardtornish in the Sound of Mull in western Scotland. This castle is a typical power centre of the *thalassocracies* that dominated western Scotland in the later middle ages, centred on maritime castles but with a power founded on the maritime strength of huge fleets of highland galleys, also known as birlinns (Rixson 1988). This castle is in a maritime context, placed to oversee and control passage through a major seaway of the Sound of Mull. The maritime connections of the site are visible on the ground with evidence of possible boat haulages (noosts) visible in the intertidal zone around the castle. As with many maritime archaeological sites, it is difficult to appreciate the way this castle dominated and controlled this landscape and how the terrestrial power of the castle must have interacted with the maritime power of the ships that provided economic and military support. VR was employed to help understand the site in a more immersive way and to reconnect the two major elements of this power base, the castle and the ships. The castle was recorded in its landscape context in 3D using a DJI UAV (Figs. 14.1 and 14.9) including the intertidal zone. The castle itself was digitally reroofed and reconstructions of highland galleys were added to the VR environment. This simple reconstruction has shown how 3D recording is being taken beyond production of standard orthographic illustra-

Fig. 14.9 Game Engines such as Unity allow for a photo-realistic 3D virtual world to be created, based on real archaeological survey data. (**a**) An optimised 3D model created in Blender using photogrammetry data. (**b–d**) The interactive 3D environment created in Unity, navigable using a game controller, keyboard, mouse and VR headset

tions in reports and can produce assets for a wide variety of other meaningful archaeological outputs with minimal additional effort.

14.9 Conclusions

By advocating for a more technological approach to study maritime landscapes, we must be careful to avoid certain pitfalls. Relative to other forms of archaeological investigation, maritime archaeology has always had a particularly strong focus on the tools and technology and this has been a distraction from the archaeology itself. Critics of 3D recording, and representations of archaeological sites and cultural landscapes might be tempted to dismiss the technique as a gimmick; limiting the usefulness to simply a 'pretty picture.' While it is valid to point out that visual gratification does not necessarily make a scientific advancement, Verhoeven (2011, 67) is right: 'Let's face it—most people like 3D visualizations.' The reasons are not simply to do with aesthetic appeal, but rather 3D recording allows archaeologists to hold a landscape in their hands and to share their experience with others. It is also a space for enhanced archaeological interpretation, which builds on existing methods for site recording. The potential to create visual, experiential environments is also promising from an enhanced interpretation perspective as well as through visitor experience and interactivity in museums and educational sectors.

Virtual worlds can create a space for generating and testing new theories and developing those already accepted by the establishment. The interpretation of landscapes, features and setting in 3D has direct relevance to the broader discipline because archaeology deals with complex surfaces and examines traces of the human past in space. Archaeological interpretive mapping remains a main aim in the collection of such data. Rather than distracting from the subject, we agree with Chrysanthi et al. (2012, 9–10) who argue that one should become immersed in the interpretive process through a mastery of tools, and that digital tools function best when they are mastered to the extent that their own character is no longer the focus, having become an unnoticed extension or prosthesis of the maritime archaeologist.

Returning to the grand challenges facing digital archaeology and the reconsideration of MCL, the material and digital representations presented herein represent the state of the art and early adoption by archaeologists. Innovation is presented through the integrated approaches used to bridge the gap between land and sea, and particularly to record large, complex sites and landscapes through a variety of original and existing data. These are measurable, physically, in that sites can be recorded, scaled and mapped very accurately and exported into a system used in planning and development, especially GIS or other practical applications for outreach, such as VR. These relatively new techniques now present a genuine opportunity to inspire a new generation of stakeholders and end-users, not only specialists. The list of authors, all of whom have made a significant intellectual and practical contribution to the development of the material presented in this chapter also illustrates the collaborative nature of these methods and the international relevance is attested by the various case studies from around the world. In this respect, we hope to have showcased some of the advances within the discussion of Huggett et al.'s (2018, 44) 'fundamentals.' We also hope to have made an incremental advance by urging the reconsideration of a widely applied, but loosely defined theoretical framework, through a digital representation of physical landscapes, with renewed interest on the archaeological material and its relationship to landscapes over time.

For maritime archaeology, 3D capture of landscapes above and below the waterline offers the clearest, most analytical and most repeatable method for analysing and interpreting sites and contextualising material within their surrounding environments. Maritime archaeology stands to benefit, perhaps more so than any other sub-discipline for exactly these reasons; the waterline need no longer be a barrier to study the integrated cultural landscape, be that physical or cognitive. The results form a scalable, digital maritime landscape—an enhanced interpretive space in which to better examine the archaeological, anthropological, historical and environmental questions.

Acknowledgements The authors acknowledge the support of Historic Environment Scotland, Heritage Victoria, Wessex Archaeology, Flinders University (who provided an EHL Establishment Grant), Airborne Research Australia and the Australian Research Council through the Funding of the *Deep History of Sea Country* project (DP170100812). We thank two anonymous peer reviewers for their comments on an earlier draft which significantly contributed to improving this chapter.

References

Agugiaro G, Remondino F, Girardi G, Von Schwerin J, Richards-Rissetto H, De Amicis R (2011a) A web-based interactive tool for multi-resolution 3D models of a Maya archaeological site. In: Proceedings of the 4th international workshop 3D-ARCH 2011: virtual reconstruction and visualization of complex architectures

Agugiaro G, Remondino F, Girardi G, Von Schwerin J, Richards-Rissetto H, De Amicis R (2011b) Queryarch3D: querying and visualising 3D models of a Maya archaeological site in a web-based interface. Geoinformatics FCE CTU 6:10–17

Ash A (2007) The maritime cultural landscape of Port Willunga, South Australia, Flinders University Maritime Archaeology Monographs Series, vol 4. Archaeology, Flinders University, Adelaide

Astrup PM, Skriver C, Benjamin J, Ward I, Stankewicz F, Ross P, McCarthy J, Baggaley P, Ulm S, Bailey G (2019) Underwater shell middens: excavation and remote sensing of a submerged Mesolithic shell midden at Hjarnø, Denmark. Journal of Island and Coastal Archaeology. (In Press)

Austin BT, Bateman J, Jeffrey S, Mitcham J, Niven K (2009) Marine remote sensing and photogrammetry: a guide to good practice. In: Niven K (ed) Archaeology data service. http://guides.archaeologydataservice.ac.uk/g2gp/VENUS_Toc. Accessed 21 Feb 2018

Bailey G, Momber G, Bell M, Tizzard L, Hardy K, Tidbury L, Benjamin J, Bicket A, Hale A (2019) Great Britain. In: Bailey G, Galanidou N, Joens H, Lueth F, Peeters H (eds) The archaeology of Europe's submerged landscapes. Springer, Cham

Ballin TB, White R, Richardson P, Neighbour T (2010) An early Mesolithic stone tool assemblage from clachan harbour, Raasay, Scottish Hebrides. Lithics: the Journal of the Lithic Studies Society 31:94–104

Bass G (1966) Archaeology under water. Praeger, New York

Benjamin J, Bicket A, Anderson D, Hale A (2014) A multi-disciplinary approach to researching the intertidal and marine archaeology in the outer Hebrides, Scotland. Journal of Island and Coastal Archaeology 9(3):400–424. https://doi.org/10.1080/15564894.2014.934490

Bewley R, Rączkowski W (2002) Aerial archaeology: developing future practice, vol 337. NATO Science Series A. NATO

Bicket A, Shaw G, Benjamin J (2017) Prospecting for Holocene palaeolandscapes in the Sound of Harris, Outer Hebrides. In: Bailey GN, Harff J, Sakellariou D (eds) Under the sea: archaeology and palaeolandscapes of the Continental Shelf, Coastal Research Library, vol 20. Springer, Heidelberg, pp 179–195. https://doi.org/10.1007/978-3-319-53160-1

Chen X, Lynch Ogan T (2017) China's emerging Silicon Valley: how and why has Shenzhen become a global innovation centre. Eur Financ Rev (December–January):55–62

Chirayath V, Earle SA (2016) Drones that see through waves—preliminary results from airborne fluid lensing for centimetre-scale aquatic conservation. Aquat Conserv Mar Freshwat Ecosyst 26(October 2015):237–250. https://doi.org/10.1002/aqc.2654

Chrysanthi A, Murrieta Flores P, Papadopoulos C (2012) Archaeological computing: towards prosthesis or amputation? In: Chrysanthi A, Murrieta Flores P, Papadopoulos C (eds) Thinking beyond the tool: archaeological computing and the interpretive process, BAR International Series 2344. Archaeopress, Oxford, pp 7–13

Colomina I, Molina P (2014) Unmanned aerial systems for photogrammetry and remote sensing: a review. ISPRS J Photogramm Remote Sens 92:79–97. https://doi.org/10.1016/j.isprsjprs.2014.02.013

Cowley D (ed) (2011) Remote sensing for archaeological heritage management. Proceedings of the 11th EAC Heritage Management Symposium, Reykjavik, Iceland, 25–27 March 2010. Occasional Publication of the Aerial Archaeology Research Group, 3. Archaeolingua, Budapest, pp 11–14

Cowley D, Benjamin J, Martin C (2012) Aerial reconnaissance of maritime landscapes in Scotland—some preliminary observations on context, methodology and results. AARGnews 45(September):64–73

Cowley DC, Moriarty C, Geddes G, Brown GL, Wade T, Nichol CJ (2018) UAVs in context: archaeological airborne recording in a national body of survey and record. Drones 2(1):16pp. https://doi.org/10.3390/drones2010002

Delgado JP, Mendizábal T, Hanselmann FH, Rissolo D (2016) The maritime landscape of the Isthmus of Panamá. University Press of Florida, Gainesville

Dell'Unto N, Landeschi G, Leander Touati AM, Dellepiane M, Callieri M, Ferdani D (2015) Experiencing ancient buildings from a 3D GIS perspective: a case drawn from the Swedish Pompeii project. J Archaeol Method Theory 23:73–94. https://doi.org/10.1007/s10816-014-9226-7

Doneus M, Verhoeven G, Fera M, Briese C, Kucera M, Neubauer W (2011) From deposit to point cloud—a study of low-cost computer vision approaches for the straightforward documentation of archaeological excavations. Geoinformatics FCE CTU 6:81–88. https://doi.org/10.14311/gi.6.11

Doneus M, Doneus N, Briese C, Pregesbauer M, Mandlburger G, Verhoeven GJ (2013) Airborne laser bathymetry—detecting and recording submerged archaeological sites from the air. J Archaeol Sci 40(4):2136–2151

Drap P, Seinturier J, Long L (2003) Archaeological 3D modelling using digital photogrammetry and Expert System. The case study of Etruscan amphorae. In: 3IA 2003—the sixth international conference on computer graphics and artificial intelligence, Limoges

Duel L (1969) Flights into yesterday: the story of aerial archaeology. St Martin's Press, New York

Ewes S (2014) Dead men's eyes: embodied GIS, mixed reality and landscape archaeology. Archaeopress, Oxford

Eve S (2017) The embodied GIS: using mixed reality to explore archaeological landscapes. Internet Archaeol 44(44). https://doi.org/10.11141/ia.44.3

Falconer L, Scott C (2018) Phenomenology and phenomenography in virtual worlds: an example from archaeology. In: Falconer L, Gil Ortega MC (eds) Virtual worlds: concepts, applications and future directions. Nova Science Publishers, New York, pp 1–38

Firth A (2011) Marine geophysics: integrated approaches to sensing the seabed. In: Cowley D (ed) Remote sensing for archaeological heritage management, EAC Occasional Paper No 5. Europae Archaeologia Consilium, Brussels, pp 129–140

Fischer A (1995) Man and sea in the Mesolithic: coastal settlements above and below the sea. Oxbow Books, Oxford

Flatman J (2011) Places of special meaning: Westerdahl's comet, "agency," and the concept of the "Maritime Cultural Landscape". In: Ford B (ed) The archaeology of maritime landscapes. When the land meets the sea, ACUA and SHA Series, vol 2. Springer, New York

Ford A (2004) The visualisation of integrated 3D petroleum datasets in ArcGIS. In: Proceedings of 24th ESRI user conference. pp 1–11

Ford B (2018a) The shore is a bridge: the maritime cultural landscape of Lake Ontario. Texas A&M University Press, College Station

Ford B (2018b) Concluding remarks about the MCL symposium. In: Wyatt B (ed) Volume 1: presentation papers. Proceedings of the maritime cultural landscape symposium, University of Wisconsin-Madison, 14–15 October 2015

Garstki K (2017) Virtual representation: the production of 3D digital artifacts. J Archaeol Method Theory 24(3):726–750. https://doi.org/10.1007/s10816-016-9285-z

Gately I, Benjamin J (2018) Archaeology hijacked: addressing the historical misappropriations of maritime and underwater archaeology. J Marit Archaeol 13(1):15–35. https://doi.org/10.1007/s11457-017-9177-8

Georgopoulos A, Agrafiotis P (2012) Documentation of a submerged monument using improved two media techniques. Proceedings of the 2012 18th international conference on virtual systems and multimedia, VSMM 2012: virtual systems in the information society. pp 173–180

Green J, Gainsford M (2003) Evaluation of underwater surveying techniques. Int J Naut Archaeol 32:252–261. https://doi.org/10.1016/j.ijna.2003.08.007

Hacker JM, Pfennigbauer M (2017) Pushing lidar to the limits—high-resolution bathymetric lidar from slow-flying aircraft. GIM International (Feb), pp 29–31

Hacker JM, Bannehr L, Junkermann W, Neininger B, Lieff W, McGrath AJ; Zubot D, Zulueta R (2018a) Mind the gap between drones and traditional airborne platforms. Proceedings of the 2018 NRM science conference, Adelaide

Hacker JM, Brooks A, Spencer J (2018b) Mapping erosion gullies 20 minutes in the air or 2 weeks in the gullies. Proceedings of the 2018 NRM science conference, Adelaide

Hardy K, Benjamin J, Bicket A, McCarthy J, Ballin T (2015) Lub Dubh Aird: a seamless Mesolithic landscape in northwest Scotland. Proc Soc Antiqu Scotl 145:27–39

Harris L (2017) Sea ports and sea power: African maritime cultural landscapes. Springer, Cham

Henderson JC, Pizarro O, Johnson-Roberson M, Mahon I (2013) Mapping submerged archaeological sites using stereo-vision photogrammetry. Int J Naut Archaeol 42(2):243–256. https://doi.org/10.1111/1095-9270.12016

Huggett J, Reilly P, Lock G (2018) Whither digital archaeological knowledge? The challenge of unstable futures. J Comput Appl Archaeol 1(1):42–54. https://doi.org/10.5334/jcaa.7

Kotilainen AT, Kaskela AM (2017) Comparison of airborne LiDAR and shipboard acoustic data in complex shallow water environments: filling in the white ribbon zone. Mar Geol 385:250–259. https://doi.org/10.1016/j.margeo.2017.02.005

Kreij A, Scriffignano J, Rosendahl D, Nagel T, Ulm S (2018) Aboriginal stone-walled intertidal fishtrap morphology, function and chronology investigated with high-resolution close-range Unmanned Aerial Vehicle photogrammetry. J Archaeol Sci 96:148–161. https://doi.org/10.1016/j.jas.2018.05.012

Liarokapis F, Kouřil P, Agrafiotis P, Demesticha S, Chmelík J, Skarlatos D (2017) 3D Modelling and mapping for virtual exploration of underwater archaeology assets. Int Arch Photogramm Remote Sens Spat Inf Sci Arch XLII-2/W3(March):425–431. https://doi.org/10.5194/isprs-archives-XLII-2-W3-425-2017

Maas HG (2015) On the accuracy potential in underwater/multimedia photogrammetry. Sensors (Switzerland) 15(8):18140–18152

McCarthy J (2014) Multi-image photogrammetry as a practical tool for cultural heritage survey and community engagement. J Archaeol Sci 43:175–185. https://doi.org/10.1016/j.jas.2014.01.010

McCarthy J, Benjamin J (2014) Multi-image photogrammetry for underwater archaeological site recording: an accessible, diver-based approach. J Marit Archaeol 9(1):95–114. https://doi.org/10.1007/s11457-014-9127-7

McCarthy J, Benjamin J (2018) Project SAMPHIRE: crowd sourcing maritime archaeology data off Scotland's West Coast. J Island Coast Archaeol, 24pp. https://doi.org/10.1080/15564894.2017.1387620

McCarthy J, Robertson P, Mackay E (2015) Discovery and survey of a 17th–18th century shipwreck near Drumbeg, NW Scotland: an initial report. Int J Naut Archaeol 44(1):202–208. https://doi.org/10.1111/1095-9270.12087

Menna F, Nocerino E, Troisi S, Remondino F (2015) Joint alignment of underwater and above-the-water photogrammetric 3D models by independent models adjustment. Int Arch Photogramm Remote Sens Spat Inf Sci XL-5/W5:143–151. https://doi.org/10.5194/isprsarchives-XL-5-W5-143-2015

Morgan C, Wright H (2018) Pencils and pixels: drawing and digital media in archaeological field recording. J Field Archaeol 43(2):136–151. https://doi.org/10.1080/00934690.2018.1428488

Mount R (2005) Acquisition of through-water aerial survey images: surface effects and the prediction of sun glitter and subsurface illumination. Photogramm Eng Remote Sens 71(12):1407–1415

Nebiker S (2002) Design and implementation of the high-performance 3D digital landscape Server DILAS. Int Arch Photogramm Remote Sens Spat Inf Sci 34(4):391–394

Neininger B, Hacker JM (2011) Manned or unmanned – does this really matter? In: Eisenbeiss H, Kunz M, Ingensand H. Proceedings of the international conference on unmanned aerial vehicles in geomatics (UAV-g), Zurich, Switzerland, 14–16 September 2011, vol XXXVIII-1/C22. pp 223–228

Olson BR, Ryan A, Placchetti JQ, Killebrew AE (2013) The Tel Akko total archaeology project (Akko, Israel): assessing the suitability of multi-scale 3D field recording in archaeology. J Field Archaeol 38(3):244–262. https://doi.org/10.1179/0093469013Z.00000000056s

Remondino F, Barazzetti L, Nex F, Scaioni M, Sarazzi D (2011) UAV photogrammetry for mapping and 3d modeling—current status and future perspectives. The international archives of the photogrammetry, remote sensing and spatial information sciences, vol XXXVIII-1/C22, pp 25–31. ISPRS Zurich 2011 Workshop, Zurich, Switzerland, 14–16 September 2011

Remondino F (2011) Heritage recording and 3D modeling with photogrammetry and 3D scanning. Remote Sens 3(6):1104–1138. https://doi.org/10.3390/rs3061104

Richards-Rissetto H (2017) What can GIS + 2D mean for landscape archaeology? J Archaeol Sci 84:10–21. https://doi.org/10.1016/j.jas.2017.05.005s

Richards-Rissetto H, Remondino F, Agugiaro G, Von Schwerin J, Robertsson J, Girardi G (2012) Kinect and 3D GIS in archaeology. Proceedings of the 18th international conference on Virtual Systems and Multimedia (VSMM), Milan, 2012. pp 331–337. https://doi.org/10.1109/VSMM.2012.6365942

Riley DN (1982) Aerial archaeology in Britain. Shire Publications, Princes Risborough

Rixson D (1988) The West Highland galley. Birlinn, Edinburgh

Roberts A, Colwell-Pasch C, Davison L, Benjamin J (2015) *Leven Lass* historic shipwreck, Phillip Island Victoria: 2015 maritime archaeology field school technical report. Unpublished Technical Report Prepared for Heritage Victoria, Melbourne

Rönnby J (2007) Maritime durées: long-term structures in a coastal landscape. J Marit Archaeol 2(2):65–82. https://doi.org/10.1007/s11457-007-9021-7

Skriver C, Borrup P, Astrup PM (2017) Hjarnø Sund: an eroding Mesolithic site and the tale of two paddles. In: Bailey GN, Harff J, Sakellariou D (eds) Under the sea: archaeology and palaeolandscapes of the Continental Shelf, Coastal Research Library, vol 20. Springer, Heidelberg, pp 131–143. https://doi.org/10.1007/978-3-319-53160-1

Society for American Archaeology (2016) The SAA archaeological record (March) 16(2)

Spring AP, Peters C (2014) Developing a low cost 3D imaging solution for inscribed stone surface analysis. J Archaeol Sci 52:97–107. https://doi.org/10.1016/j.jas.2014.08.017

Stewart DJ (2011) Preface: putting the wheels on maritime cultural landscape studies. In: Ford B (ed) The archaeology of maritime landscapes, When the land meets the sea, ACUA and SHA series, vol 2. Springer, New York, pp vii–viii

Tait E, Laing R, Grinnall A, Burnett S, Isaacs J (2015) (Re)presenting heritage: laser scanning and 3D visualisations for cultural resilience and community engagement. J Inf Sci 42(3):420–433. https://doi.org/10.1177/0165551516636306

Tuddenham DB (2010) Maritime cultural landscapes: maritimity and quasi objects. J Marit Archaeol 5(1):5–16. https://doi.org/10.1007/s11457-010-9055-0

Uldum O, Benjamin J, McCarthy J, Feulner F, Lübke H (2017) The Late Mesolithic site of Falden, Denmark: results from underwater archaeological fieldwork and a strategy for capacity-building based on the SPLASHCOS mission. In: Bailey GN, Harff J, Sakellariou D (eds) Under the sea: archaeology and palaeolandscapes of the Continental Shelf, Coastal Research Library, vol 20. Springer, Heidelberg, pp 65–84. https://doi.org/10.1007/978-3-319-53160-1

Verhoeven G (2011) Taking computer vision aloft—archaeological three-dimensional reconstructions from aerial photographs with PhotoScan. Archaeol Prospect 18:76–73. https://doi.org/10.1002/arp.399

Verhoeven G (2017) Mesh is more—using all geometric dimensions for the archaeological analysis and interpretative mapping of 3D surfaces. J Archaeol Method Theory 24(4):999–1033. https://doi.org/10.1007/s10816-016-9305-z

Watterson A (2015) Beyond digital dwelling: re-thinking interpretive visualisation in archaeology. Open Archaeol 1:119–130. https://doi.org/10.1515/opar-2015-0006

Westerdahl C (1992) The maritime cultural landscape. Int J Naut Archaeol 21(1):5–14. https://doi.org/10.1111/j.1095-9270.1992.tb00336.x

Westerdahl C (2008) Boats apart. building and equipping an iron-age and early-medieval ship in northern Europe. Int J Naut Archaeol 37(1):17–31. https://doi.org/10.1111/j.1095-9270.2007.00170.x

Westerdahl C (2011) The maritime cultural landscape. In: Catsambis A, Ford B, Hamilton DL (eds) The Oxford handbook of maritime archaeology. Oxford University Press, Oxford, pp 733–762. https://doi.org/10.1093/oxfordhb/9780199336005.013.0032

Wheatley D, Gillings M (2002) Spatial technology and archaeology: the archaeological applications of GIS. Taylor & Francis, London

Wüst T, Nebiker S, Landolt R (2004) Applying the 3D GIS DILAS to archaeology and cultural heritage projects—requirements and first results. Int Arch Photogramm Remote Sens Spat Inf Sci XXXV(B5):407–412

Yamafune K, Torres R, Castro F (2016) Multi-Image photogrammetry to record and reconstruct underwater shipwreck sites. J Archaeol Method Theory 24(3):703–725. https://doi.org/10.1007/s10816-016-9283-1

Index

J. K. McCarthy et al. (eds.), *3D Recording and Interpretation for Maritime Archaeology*, Coastal Research Library 31,
https://doi.org/10.1007/978-3-030-03635-5

www.ingramcontent.com/pod-product-compliance
Lightning Source LLC
LaVergne TN
LVHW082207200926
848484LV00008B/2227